LINEAR AND COMBINATORIAL OPTIMIZATION
IN ORDERED ALGEBRAIC STRUCTURES

annals of discrete mathematics

50
1931 N·H 1981
P C

NORTH-HOLLAND PUBLISHING COMPANY – AMSTERDAM • NEW YORK • OXFORD

ANNALS OF DISCRETE MATHEMATICS 10

LINEAR AND COMBINATORIAL OPTIMIZATION IN ORDERED ALGEBRAIC STRUCTURES

U. ZIMMERMANN

Mathematisches Institut
Universität zu Köln

1931 N·H 1981
PC

NORTH-HOLLAND PUBLISHING COMPANY – AMSTERDAM • NEW YORK • OXFORD

Reprinted from the journal *Annals of Discrete Mathematics*, Volume 10

North-Holland ISBN for this Volume 0 444 86153 X

Published by:

NORTH-HOLLAND PUBLISHING COMPANY
AMSTERDAM • NEW YORK • OXFORD

Sole distributors for the U.S.A. and Canada:

ELSEVIER NORTH-HOLLAND, INC.
52 VANDERBILT AVENUE
NEW YORK, NY 10017

Library of Congress Cataloging in Publication Data

Zimmermann, Uwe, 1947-
Linear and combinatorial optimization in ordered algebraic structures.

(Annals of discrete mathematics ; 10)
Bibliography: p.
Includes indexes.
1. Algebra, Abstract. 2. Mathematical optimization. I. Title. II. Series.
QA162.Z55 512'.02 81-61
ISBN 0-444-86153-X

PRINTED IN THE NETHERLANDS

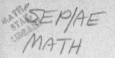

PREFACE

The object of this book is to provide an account of results and methods for linear and combinatorial optimization problems over ordered algebraic structures. In linear optimization the set of feasible solutions is described by a system of linear constraints; to a large extent such linear characterizations are known for the set of feasible solutions in combinatorial optimization, too. Minimization of a linear objective function subject to linear constraints is a classical example which belongs to the class of problems considered. In the last thirty years several optimization problems have been discussed which appear to be quite similar. The difference between these problems and classical linear or combinatorial optimization problems lies in a replacement of linear functions over real (integer) numbers by functions which are linear over certain ordered algebraic structures. This interpretation was not apparent from the beginning and many authors discussed and solved such problems without using the inherent similarity to linear and combinatorial optimization problems. Therefore results and methods which are well-known in the theory of linear and combinatorial optimization have been reinvented from time to time in different ordered algebraic structures. In this book we describe algebraic formulations of such problems which make the relationship of similar problems more transparent. Then results and methods in different algebraic settings appear to be instances of general results and methods. Further specializing of general results and methods often leads to new results and methods for a particular problem.

We do not intend to cover all optimization problems which can be treated from this point of view. We prefer to select classes of problems for which such an algebraic approach turns out to be quite useful and for which a common theory has been developed. Therefore nonlinear problems over ordered algebraic structures are not discussed; at the time being very few results on such problems are known.

We assume that the reader of this book is familiar with concepts from classical linear and combinatorial optimization. On the other hand we develop the foundations of the theory of ordered algebraic structures covering all results which are used in the algebraic treatment of linear and combinatorial optimization problems. This first part of the book is self-contained; only some representation theorems are mentioned without explicit proof. These representation theorems are not directly applied in the investigation of the optimization problems considered, but provide a general view of the scope of the underlying algebraic structure. In the second part of the book we develop theoretical concepts and solution methods for linear and combinatorial optimization problems over such ordered algebraic structures. Results from part one are explicitly and implicitly used throughout the discussion of these problems. The following is an outline of the content of the chapters in both parts.

In chapter one we introduce basic concepts from the theory of ordered sets and lattices. Further we review basic definitions from graph theory and matroid theory. Chapter two covers basic material on ordered commutative semigroups and their relationship with ordered commutative groups. Lattice-ordered commutative groups are discussed in chapter three. Without proof we expose the embedding theorems of CONRAD, HARVEY and HOLLAND [1963] and of HAHN [1907]. Chapter four contains detailed characterizations of linearly ordered commutative divisor semigroups. In particular, we prove the decomposition theorems of CLIFFORD ([1954] , [1958], [1959]) and LUGOWSKI [1964]. The discussion of the weakly cancellative case is of particular importance for the applications in part two of the book. Many examples complete this chapter. In chapter five we introduce ordered semimodules generalizing ordered rings, fields, modules and vectorspaces as known from the theory of algebra. Basic definitions of a matrix calculus lead to the formulation of linear functions and linear constraints over ordered semimodules. Chapter six contains a discussion of linearly ordered semimodules over real numbers. We provide the necessary results for the development of a duality theory for algebraic linear programs. Chapter seven is an introduction to part two. Chapter eight covers algebraic path problems. We develop results on the stability of matrices and methods for the solution of matrix equations generalizing procedures from linear algebra. Several classes of problems are considered; the shortest path problem is covered as well as the determination of all path values. Chapter nine contains algebraic eigenvalue problems which are closely related to algebraic path problems. In chapter ten extremal linear programs are considered. A weak duality theorem in ordered semimodules leads to explicit solutions in the case of extremal inequalities. The case of equality constraints can be treated using a threshold method. In chapter eleven we develop a duality theory for algebraic linear programs. Such problems are solved by generalizations of the simplex method of linear programming. We discuss primal as well as primal dual procedures. As in classical linear optimization duality theory is a basic tool for solving combinatorial optimization problems over ordered algebraic structures. Chapter twelve covers solution methods for algebraic flow problems. We develop a generalization of the primal dual solution method of FORD and FULKERSON. Clearly primal methods can be derived from the simplex method in chapter twelve. We discuss several methods for the solution of algebraic transportation and assignment problems. Chapter thirteen contains solution methods for algebraic optimization problems in independence systems; in particular, we investigate algebraic matroid and 2-matroid intersection problems. Algebraic matching problems can be solved similarly; both classes of problems are mainly differing by their combinatorial structure whereas the generalization of the algebraic structure leads to the same difficulties.

In writing this book I have benefited from the help of many people. In particular, I want to express my gratitude to Professor Dr. R.E. Burkard, who encouraged my work throughout many years. A great deal of the material on algebraic combinatorial optimization problems originated from collaborative work with him in the Mathematical Institute at the University of Cologne. He made it possible for me to visit the Department of Operations Research at Stanford University. Valuable discussions in Stanford initiated investigations of related lattice-ordered structures.

I am much indebted to the Deutsche Forschungsgemeinschaft, Federal Republic of Germany, for the sponsorship of my stay in Stanford.

The manuscript was beautifully typed by Mrs. E. Lorenz.

U.Z.

Cologne

TABLE OF CONTENTS

PART I ORDERED ALGEBRAIC STRUCTURES

1. Ordered Sets, Lattices and Matroids

In this chapter we introduce basic concepts from the theory of ordered sets and we consider certain types of lattices. Further we review basic definitions from graph theory and matroid theory as given in BERGE [1973] and WELSH [1976]. This chapter covers the necessary order-theoretic background for the discussion of the optimization problems in part II. In particular, some representation theorems from lattice theory give a general view of such structures. For further results and for some proofs in lattice theory we refer to GRÄTZER [1978].

The basic properties of the usual order relation $\leq$ of the real numbers $\mathbb{R}$ are described by the following four axioms:

(1.1)

	$a \leq a$	(*Reflexivity*)
(1.1)	$a \leq b,\quad b \leq a \;\Rightarrow\; a = b$	(*Antisymmetry*)
	$a \leq b,\quad b \leq c \;\Rightarrow\; a \leq c$	(*Transitivity*)
	$a \leq b$ or $b \leq a$	(*Linearity*)

for all $a,b,c \in \mathbb{R}$.

In general we consider a binary relation $\leq$ on a nonempty set H satisfying some of these axioms. We always assume that the binary relation is transitive. Such a binary relation is called an *ordering*. If an ordering is reflexive then it is called a *preordering* or *quasiordering*; if an ordering is antisymmetric then it is called a *pseudoordering*. A reflexive and antisymmetric ordering is called a *partial ordering*; if a partial ordering is linear then it is called a *linear* (or *total*) ordering. A system $(H,\leq)$ with such an ordering on H is called a (pre-, quasi-, pseudo-, partially, linearly) ordered set.

A preorder $\sim$ on H satisfying

$$a \sim b \iff b \sim a \qquad \text{(Symmetry)}$$

is called an *equivalence relation*. Then $R_a := \{b \in H \mid b \sim a\}$ is called the *equivalence class* of a. The set $\{R_a \mid a \in H\}$ is a *partition* of H, i.e.

$$R_a \cap R_b \neq \emptyset \iff R_a = R_b$$

and H is the union of all R_a:

$$H = \cup \{R_a \mid a \in H\}.$$

Vice versa a partition of H defines an equivalence relation on H with $a \sim b$ if and only if a and b are elements of the same element (*block*) of the partition.

Two elements a,b in an ordered set H are called *incomparable* if neither $a \leq b$ nor $b \leq a$. $a \leq b$ but $a \neq b$ is denoted by $a < b$. The *inverse* or *dual ordering* of $\leq$ is the ordering $\geq$ defined by

$$a \geq b \;:\iff\; b \leq a$$

for all $a,b \in H$. A linearly ordered subset A of H is called *maximal* if there is no linearly ordered subset B of H such that $A \subsetneqq B$.

In order to obtain a general view of ordered structures it is very useful to introduce the concept of isomorphism.

Let $(H,\leq)$ and $(H',\leq')$ denote two ordered sets. A mapping $\varphi: H \to H'$ is called an *order isomorphism* if φ is bijective and

$$a \leq b \iff \varphi(a) \leq \varphi(b)$$

for all $a,b \in H$. Then H and H' are called *order-isomorphic*. If φ is an order isomorphism between H and $\varphi(H)$ but not necessarily surjective on H', then φ is called an *order embedding* of H into H'. We say that H can be *order-embedded* into H'. For example, each finite linearly ordered set with n elements is order-isomorphic to the set $\{1,2,\ldots,n\}$ linearly ordered by the usual order relation of the natural numbers and each finite linearly ordered set with less than n elements can be order-embedded in every finite linearly ordered set with at least n elements.

Ordered sets are often visualized by graphic representations. We assume that the reader is familiar with the concepts of a *graph* and a directed graph (*digraph*). We denote a graph G by a pair (V,E) where $V = V(G)$ is the nonempty vertex set and $E = E(G)$ is the set of *edges*; a digraph G is denoted similarly by (V,A) where $V = V(G)$ is the nonempty vertex set and $A = A(G)$ the set of *arcs*. Arcs and edges are denoted as pairs (a,b) of vertices; then a,b are called *endpoints* of (a,b); in particular, for an arc (a,b) the vertices a and b are called *initial* and *terminal endpoints* of (a,b). A finite sequence of arcs (edges) $(e_1,e_2,\ldots,e_n)$ such that e_i has one common endpoint with e_{i-1} and the other endpoint in common with e_{i+1} for all $i = 2,3,\ldots,n-1$ is called a *chain*. The length of a chain is the number of its edges (arcs). *Vertices* on a chain are called *connected* (by the chain). A chain with arcs $e_i = (a_i, a_{i+1})$ for $i = 1,2,\ldots,n$ is called a *path* from a_1 to a_{n+1}. a_1 and a_{n+1} are called *initial* and *terminal endpoint* of the path. The *endpoints* of a chain are defined similarly. A chain (path) is called *simple* if its edges (arcs) are mutually distinct. A simple chain (path) with coinciding endpoints is called a *cycle* (*circuit*). A chain (path) not using any vertex twice is called *elementary*.

A *subgraph* of G *generated by* $V' \subseteq V$ is the graph (V',E') containing all edges (arcs) of G which have both endpoints in V'. A *graph* G is called *connected* if all vertices in G are connected. The binary relation $\sim$, defined on V by $a \sim b$ iff a and b are connected, is an equivalence relation on V. The elements of the induced partition $V_1, V_2, \ldots, V_k$ of V generate the *connected components* of G. A graph is called *strongly connected* if for all $a,b \in V$ there exist paths from a to b and from b to a. The strongly connected subgraphs of G are called *strongly connected components* and define another partition of the vertex set V.

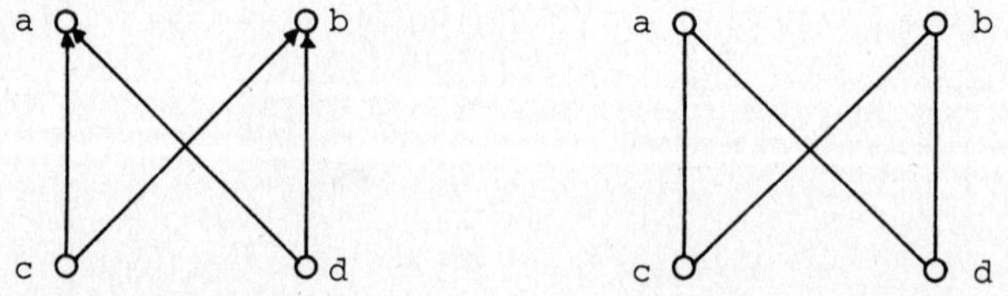

Figure 1. Directed and undirected graph

An ordered set $(H,\leq)$ is represented by the (directed) graph with vertex set H and (arc set A) edge set $E = \{(a,b) \mid a \leq b\}$.

In the case of a graph a < b is expressed by placing a below b in a drawing of the graph. Clearly, the above examples of graphs represent the same ordered set.

Let A denote a subset of an ordered set H and let b ∈ H. Then b is called an *upper (lower) bound* of A if $a \leq b$ ($b \leq a$) for all a ∈ A. An upper (lower) bound b of A is called a *least upper (greatest lower) bound* or *supremum (infimum)* of A if $b \leq c$ ($c \leq b$) for all upper (lower) bounds c of A.

An element a ∈ A is called a *maximum (minimum)* of A if $a' \leq a$ ($a \leq a'$) for all a' ∈ A. An element a ∈ A is called *maximal (minimal)* if $a \leq a'$ ($a' \leq a$) implies $a' \leq a$ ($a \leq a'$) for all a' ∈ A. The corresponding sets are denoted by

U(A), L(A)	(upper, lower bounds),
sup A, inf A	(suprema, infima),
max A, min A	(maxima, minima),
Max A, Min A	(maximal, minimal elements).

The following example illustrates these denotations. On the set of the vertices of the graph G = (V,A) in figure 2

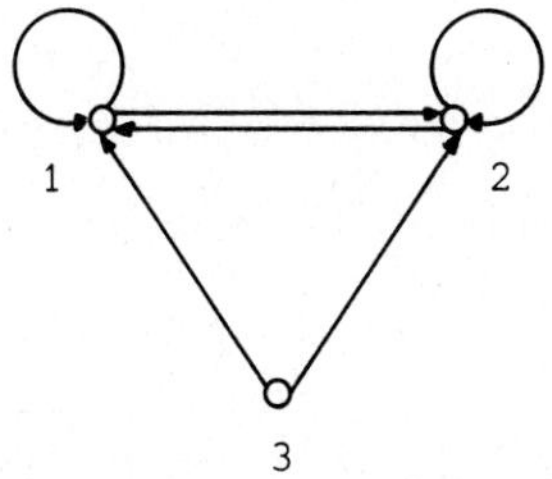

Figure 2. Digraph of an ordered set

an ordering is defined by $a \leq b$ iff (a,b) ∈ A. This ordering is only transitive. For A = {3} we find sup A = {1,2}, max A = ∅ and Max A = A. For B = {1,2} we find sup B = max B = Max B = B.

Suprema (infima) and maxima (minima) do not always exist. But if A is a finite subset of an ordered set H, then $\mathrm{Max}\, A = \emptyset$.

An ordered set is called *bounded* if it has a maximum (denotation: 1 or ∞) and a minimum (denotation: 0 or $-\infty$). Clearly, we find $1 \in \sup H = \max H = \mathrm{Max}\, H$ and $0 \in \inf H = \min H = \mathrm{Min}\, H$. If H is pseudoordered then these sets contain exactly one element.

A partially ordered set $(H,\leq)$ is called a *root system* if the set $\{b \in H \mid a \leq b\}$ is linearly ordered for all $a \in H$.

(1.2) Proposition

Let $(H,\leq)$ be a partially ordered set. Then H is a root system if and only if no pair of incomparable elements in H has a common lower bound.

Proof. The set $\{d \in H \mid c \leq d\}$ for $c \in L(\{a,b\})$ is not linearly ordered if a and b are incomparable. This immediately shows the proposed characterization of root systems. ∎

A partially ordered set $(H,\leq)$ is called a *lattice* if $\sup\{a,b\}$ and $\inf\{a,b\}$ exist for all $a,b \in H$. For convenience we use the denotations

$$a \wedge b := \inf\{a,b\}, \qquad a \vee b := \sup\{a,b\}$$

in lattices. A *sublattice* $(H',\leq)$ of a lattice $(H,\leq)$ is a subset H' of H which is closed with respect to *meet* $\wedge$ and *join* $\vee$ taken in H, i.e. $a,b \in H'$ imply $a \vee b, a \wedge b \in H'$. It is not sufficient that $(H',\leq)$ is a lattice. A lattice H is called *complete* if sup A and inf A exist for all $A \subseteq H$. We remark that sup A exists for all $A \subseteq H$ if and only if inf B exists for all $B \subseteq H$.

Two lattices H and H' are called *lattice-isomorphic* iff there exists a bijection $\varphi: H \to H'$ such that $a \leq b$ iff $\varphi(a) \leq \varphi(b)$ for all $a,b \in H$. This condition is equivalent to $\varphi(a \wedge b) = \varphi(a) \wedge \varphi(b)$ and $\varphi(a \vee b) = \varphi(a) \vee \varphi(b)$ for all $a,b \in H$. Then φ is called a lattice isomorphism. If $\varphi: H \to \varphi(H)$ is a lattice isomorphism then φ is called a *lattice embedding* of H in H'. We say that H can be *lattice-embedded* into H'. In particular, $\varphi(H)$ is a sublattice of H'.

Now let Λ be a root system and let H_λ be the linearly ordered set $\mathbb{R}$ of real numbers for all $\lambda \in \Lambda$. Let $V = V(\Lambda)$ be the following subset of the Cartesian product $\times_{\lambda\in\Lambda} H_\lambda$: $x = (\ldots x_\lambda \ldots)$ belongs to V if and only if the set $S_x = \{\lambda \in \Lambda \mid x_\lambda \neq 0\}$ contains no infinite ascending sequences. Further let

$$M_{x,y} := \{\lambda \in \Lambda \mid x_\lambda \neq y_\lambda \text{ and } x_\mu = y_\mu \text{ for all } \lambda < \mu\}$$

for $x,y \in V$. Then we define $x \leq y$ if and only if $x_\lambda \leq y_\lambda$ for all $\lambda \in M_{x,y}$. As in this definition of the binary relation $\leq$ on V we will often use the same symbol for different order relations; from the context it will always be clear which of the relations considered is meant specifically.

(1.3) Proposition

Let Λ be a root system. Then $V = V(\Lambda)$ is a lattice.

Proof. At first we show that V is a partially ordered set with respect to the above defined binary relation. Reflexivity follows from $M_{x,x} = \emptyset$. $M_{x,y} = M_{y,x}$ implies antisymmetry. Now let $x \leq y$, $y \leq z$ and let $\lambda \in M_{x,z}$. Then $x_\lambda \neq z_\lambda$ and $x_\mu = z_\mu$ for all $\mu > \lambda$. If $x_\mu = z_\mu = y_\mu$ for all $\mu > \lambda$ then $x \leq y$ and $y \leq z$ implies $x_\lambda \leq y_\lambda \leq z_\lambda$. Otherwise let $\lambda < \gamma$ with $x_\gamma = z_\gamma \neq y_\gamma$ and $x_\delta = z_\delta = y_\delta$ for all $\delta > \gamma$. Then $x \leq y$ and $y \leq z$ leads to the contradiction $y_\gamma < z_\gamma = x_\gamma < y_\gamma$. Secondly, for $a, b \in V$ we define $a \wedge b$, $a \vee b$ in the following way.

Let M be a maximal linearly ordered subset of Λ. Then the *projection* V^M of V onto M, defined by

$$V^M := \{v \in V \mid v_\lambda = 0 \quad \text{for all } \lambda \in \Lambda \setminus M\}$$

is linearly ordered. For $v \in M$ we call v^M with $v^M_\lambda := v_\lambda$ for $\lambda \in M$ and $v^M_\lambda = 0$ otherwise the *projection of* v. For $\lambda \in M$ we define

$$(1.4) \qquad (a \wedge b)_\lambda := \begin{cases} a_\lambda & \text{if } a^M \leq b^M, \\ b_\lambda & \text{if } a^M > b^M, \end{cases} \qquad (a \vee b)_\lambda := \begin{cases} a_\lambda & \text{if } a^M \geq b^M, \\ b_\lambda & \text{if } a^M < b^M. \end{cases}$$

In this way $a \wedge b$ and $a \vee b$ are well-defined on *all* maximal linear ordered sets M. We show this only for $a \wedge b$. Let $\lambda \in \Lambda$. Then the set $\{\mu \in \Lambda \mid \lambda \leq \mu\}$ is contained in all maximal linearly ordered sets M with $\lambda \in M$. If $a_\mu = b_\mu$ for all $\mu \geq \lambda$ then $(a \wedge b)_\lambda = a_\lambda = b_\lambda$ independently of the choice of the particular M considered ($\lambda \in M$). Otherwise let $\mu \geq \lambda$ with $a_\mu \neq b_\mu$ and $a_\gamma = b_\gamma$ for all $\gamma \geq \mu$. We assume w.l.o.g. $a_\mu < b_\mu$. Then $(a \wedge b)_\lambda = a_\lambda$ independently of the choice of the particular M considered ($\lambda \in M$). Next we show that $a \wedge b$ is the infimum of a and b. Since $a \leq b$ if and only if

$a^M \leq b^M$ for all maximal linearly ordered subsets M of V, we find $a \wedge b \in L(\{a,b\})$. Let $c \in L(\{a,b\})$ and let M be a maximal linearly ordered subset of V. Then $c^M \leq a^M$ and $c^M \leq b^M$. As $(a \wedge b)^M \in \{a^M, b^M\}$ we find $c^M \leq (a \wedge b)^M$. Hence $c \leq a \wedge b$. Similarly we can show that $a \vee b$ is the supremum of a and b.

■

The lattice $V(\Lambda)$ will be called the *root lattice* of the root system Λ.

In the following we consider lattices from another point of view. Since $a \wedge b, a \vee b$ are well-defined on a lattice H, the functions $\wedge, \vee: H \times H \to H$ are *internal compositions* on H. They satisfy the following system of axioms:

	$a \wedge b = b \wedge a$,	$a \vee b = b \vee a$	(*Commutativity*)
(1.5)	$a \wedge (b \wedge c) = (a \wedge b) \wedge c$,	$a \vee (b \vee c) = (a \vee b) \vee c$	(*Associativity*)
	$a \wedge (a \vee b) = a$,	$a \vee (a \wedge b) = a$	(*Absorption*)

for all $a,b,c \in H$.

As an example, we consider the left absorption identity. We know $a \leq a \wedge b$. Thus $a = a \wedge (a \vee b)$.

On the other hand we may consider a nonempty set H with two internal compositions $\wedge, \vee$ satisfying (1.5). Then we define

(1.6) $\quad a \leq b \quad :\Longleftrightarrow \quad a \wedge b = a$

for all $a,b \in H$. This binary relation is antisymmetric as $\wedge$ is commutative. Transitivity follows from associativity. (1.5) implies

(1.7) $\quad a \wedge a = a$ (*Idempotency*)

for all $a \in H$. This can be seen using the absorption identities:

$$a = a \wedge (a \vee (a \wedge a)) = a \wedge a$$

for $a \in H$. Idempotency implies reflexivity of the binary relation. Hence $(H,\leq)$ is a partially ordered set. We claim that $a \wedge B = \inf\{a,b\}$. Clearly

$$(a \wedge b) \wedge a = (a \wedge a) \wedge b = a \wedge b$$

shows $a \wedge b \leq a$. Similarly, $a \wedge b \leq b$. Now let $c \leq a$ and $c \leq b$. Then $c \wedge a = c = c \wedge b$. Therefore $c \wedge (a \wedge b) = (c \wedge a) \wedge b = c \wedge b = c$, i.e. $c \leq a \wedge b$. Hence $a \wedge b = \inf\{a,b\}$. Now observe that

(1.8) $$a \wedge b = a \iff a \vee b = b$$

for all $a,b \in H$. Assume $a \wedge b = a$. Then the right absorption identity yields

$$a \vee b = (a \wedge b) \vee b = b \vee (b \wedge a) = b.$$

The opposite implication can be shown in the same manner. Therefore $a \vee b = \sup\{a,b\}$ can be proved similarly to the proof of $a \wedge b = \inf\{a,b\}$. Thus $(H,\leq)$ is a lattice and we call a system $(H,\wedge,\vee)$ satisfying (1.5) a lattice, too.
We have shown the following result:

(1.9) <u>Theorem</u>

(1) Let $(H,\leq)$ be a lattice. Then $(H,\wedge,\vee)$ with $a \wedge b := \inf\{a,b\}$ and $a \vee b := \sup\{a,b\}$ for all $a,b \in H$ is a lattice.

(2) Let $(H,\wedge,\vee)$ be a lattice. Then $(H,\leq)$ with $a \leq b :\iff a \wedge b = a$ is a lattice.

We remark that, if the lattice $(H,\wedge,\vee)$ in (1.9.1) is used to define a further lattice $(H, \leq')$ by $a \leq' b :\iff a \wedge b = a$, then this yields the original lattice $(H,\leq)$ in (1.9.1), i.e. $a \leq b$ if and only if $a \leq' b$. Similarly, the lattice $(H,\leq)$ in (1.9.2) leads to a lattice $(H, \wedge', \vee')$, which coincides with the original lattice $(H,\wedge,\vee)$ in (1.9.2).

In the following we consider some special classes of lattices. A lattice H is called *distributive* if

(1.10) $\quad a \wedge (b \vee c) = (a \wedge b) \vee (a \wedge c)$

for all $a,b,c \in H$. We will show that (1.10) is equivalent to

(1.10') $\quad a \vee (b \wedge c) = (a \vee b) \wedge (a \vee c)$

for all $a,b,c \in H$.

(1.11) Proposition

Let $(H,\leq)$ be a lattice. Then

(1) $\quad a \wedge (b \vee c) \geq (a \wedge b) \vee (a \wedge c)$ $\quad$ for all $a,b,c \in H$;

(2) $\quad a \vee (b \wedge c) \leq (a \vee b) \wedge (a \vee c)$ $\quad$ for all $a,b,c \in H$;

(3) $\quad$ (1.10) is equivalent to (1.10').

Proof. (1) This inequality follows from the inequalities $a \wedge b \leq a$, $a \wedge b \leq b \vee c$ and $a \wedge c \leq a$, $a \wedge c \leq b \vee c$. (2) Follows similarly to (1). (3) We only show that (1.10) implies (1.10'). The opposite implication is proved in the same manner. Let $x,y,z \in H$. As $x \vee (x \wedge z) = x$ we find

$$x \vee (y \wedge z) = x \vee [(x \wedge z) \vee (y \wedge z)] .$$

Using (1.10) and $x = (x \vee y) \wedge x$ we transform the right hand side:

$$= x \vee [(x \vee y) \wedge z] = [(x \vee y) \wedge x] \vee [(x \vee y) \wedge z].$$

Using again (1.10) we find

$$= (x \vee y) \wedge (x \vee z). \qquad \blacksquare$$

It should be noted that (1.10) and (1.10') are not equivalent for the same three elements. Equivalence holds only with respect to *all* elements.

Let S denote a subset of the set P(A) of all subsets of a given set A. If $X \cap Y \in S$ and $X \cup Y \in S$ for all $X, Y \in S$ then the system $(S, \cap, \cup)$ is called a *ring of sets*. A ring of sets is obviously a distributive lattice with partial order $\subseteq$ (set inclusion).

(1.12) Theorem

A lattice is distributive if and only if it is lattice-isomorphic to a ring of sets.

Proof. A ring of sets is obviously a distributive lattice. We show the opposite implication only in the finite case. For a proof in the infinite case we refer to GRÄTZER [1978].

Now let H be a distributive lattice with finite set H. An element a of H is called *join-irreducible* if $a = b \vee c$ implies $a = b$ or $a = c$ for all $b, c \in H$. Let I denote the set of all join-irreducible elements of H. Let $P^*(I)$ denote the set of all subsets A of I with $a \in A$ implies $\{x \in H \mid x \leq a\} \subseteq A$. Then we claim that $r: H \to P^*(I)$ defined by

$$r(a) := \{x \in I \mid x \leq a\}$$

is a lattice isomorphism. As $P^*(I)$ is a ring of sets we can prove the theorem in the finite case in this way. Now r is an injective mapping as a is the join of all elements in r(a).

The same argument yields $r(a) \cap r(b) = r(a \wedge b)$. In order to show surjectivity let $J \in P^*(I)$ and let a be the join of all elements in J. Clearly $J \subseteq r(a)$. For $x \in r(a)$ we have $x = x \wedge a$. Distributivity shows $x = \sup\{x \wedge b \mid b \in J\}$. As $x \in I$ we find $x = x \wedge b$ for some $b \in J$, i.e. $x \leq b$. Due to the definition of $P^*(I)$ this implies $x \in J$. Thus $r(a) \subseteq J$. Finally we show $r(a) \vee r(b) = r(a \vee b)$. Clearly, $r(a) \cup r(b) \subseteq r(a \vee b)$. Let $x \in r(a \vee b)$. Distributivity leads to $x = x \wedge (a \vee b) = (x \wedge a) \vee (x \wedge b)$. Then $x \in I$ implies $x = x \wedge a$ or $x = x \wedge b$, i.e. $x \in r(a)$ or $x \in r(b)$. Thus $r(a \vee b) \subseteq r(a) \cup r(b)$.

Theorem (1.12) characterizes all distributive lattices up to isomorphism. Every distributive lattice can be visualized as a ring of sets. A simple finite example is given in figure 3.

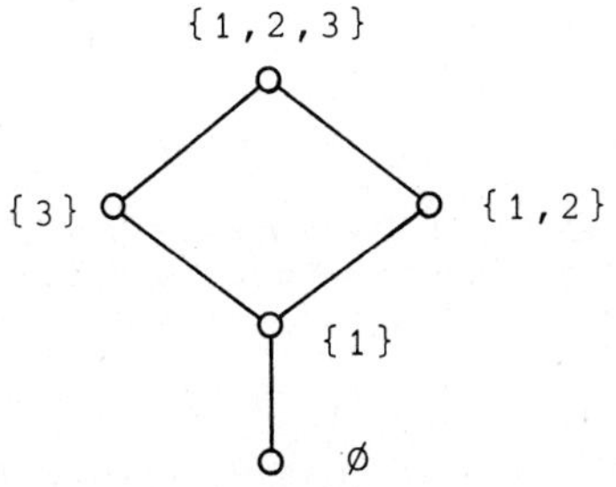

Figure 3. Ring of sets

A finite lattice is always bounded. In a bounded lattice H an element $b \in H$ is called a *complement* of $a \in H$ iff $a \wedge b = 0$ and $a \vee b = 1$. In the lattice in figure 3 the elements $\{1,2,3\}$ and $\emptyset$ are complements of each other; all other elements have no complements. A lattice is called *complemented* if all elements have complements.

(1.13) Proposition

Let $(H,\leq)$ be a bounded, distributive lattice. Then $a \in H$ has at most one complement.

Proof. Suppose that $b,b' \in H$ are complements of $a \in H$. Then $b = b \wedge 1 = b \wedge (a \vee b') = (b \wedge a) \vee (b \wedge b') = 0 \vee (b \wedge b') = b \wedge b'$. Similarly $b' = b \wedge b'$. Thus $b = b'$.

■

A bounded, complemented and distributive lattice is called a *Boolean* lattice. In a Boolean lattice the unique complement of a is denoted by a^*. Boolean lattices are characterized in a similar manner as distributive lattices. Let $P(A)$ denote the set of all subsets of a nonempty set A. Then for $X \in P(A)$ the set $\bar{X} := A \setminus X$ is the complement of X. A ring of sets $S \subseteq P(A)$ is called a *field of sets* iff $\bar{X} \in S$ for all $X \in S$. A field of sets is obviously a Boolean lattice with partial order $\subseteq$.

(1.14) Theorem

A lattice is Boolean if and only if it is lattice-isomorphic to a field of sets.

Proof. Again we consider only the finite case and refer to GRÄTZER [1978] for the infinite case. Let H be a finite Boolean lattice. We use the same denotations as in the proof of theorem (1.12). Let $a \in I$. Then a *covers* 0, i.e. $0 < a$ and $0 \leq x \leq a$ implies $x \in \{0,a\}$ for all $x \in H$. Suppose $0 < x < a$. Then $a = a \wedge (x \vee x^*) = (a \wedge x) \vee (a \wedge x^*) = x \vee (a \wedge x^*)$ shows that a is not join-irreducible contrary to $a \in I$. Thus $a,b \in I$ with $a \neq b$ are incomparable. Therefore $P^*(I) = P(I)$. This means that the lattice isomorphism r maps H onto the fields of sets $P(I)$.

■

From the proof of (1.14) we see that in the finite case a lattice is Boolean iff it is isomorphic to the Boolean lattice of all subsets of a finite set.

Next we consider a class of lattices which are distributive but not Boolean, in general. Let H be a bounded lattice. Then H is called *pseudo-Boolean* if for all $a, b \in H$ there exists an element $c \in H$ such that

(1.15) $$a \wedge x \leq b \iff x \leq c$$

for all $x \in H$. If such an element exists then it is unique and will be denoted by $b:a$.

(1.16) Proposition

Let H be a Boolean lattice. Then H is pseudo-Boolean and $b:a = b \vee a^*$ for all $a, b \in H$.

Proof. Let $a \wedge x \leq b$ for $x \in H$. Then $a^* \vee (a \wedge x) \leq a^* \vee b$. Now $a^* \vee (a \wedge x) = (a^* \vee a) \wedge (a^* \vee x) = 1 \wedge (a^* \vee x) \geq x$. The opposite inequality follows from $x \leq b \vee a^*$ as then $a \wedge x \leq a \wedge (b \vee a^*)$ $= (a \wedge b) \vee (a^* \wedge a) = a \wedge b \leq b$.

■

The relationship between distributive and pseudo-Boolean lattices is given in the following proposition.

(1.17) Proposition

A pseudo-Boolean lattice H is distributive.

Proof. Proposition (1.11) shows that it suffices to prove

$$a \wedge (b \vee c) \leq (a \wedge b) \vee (a \wedge c)$$

for all $a,b,c \in H$. Let $x := (a \wedge b) \vee (a \wedge c)$. Then $a \wedge b \leq x$, $a \wedge c \leq x$. Therefore $b \leq x{:}a$, $c \leq x{:}a$. Then $b \vee c \leq x{:}a$ which is equivalent to $a \wedge (b \vee c) \leq x$.

■

In particular we consider the element $0{:}a$ for $a \in H$ in pseudo-Boolean lattices H. Then $a \wedge x = 0$ is equivalent to $x \leq 0{:}a$. In a Boolean lattice $a^* = 0{:}a$ and therefore

(1.18) $$a \vee (0{:}a) = 1$$

for all $a \in H$.

(1.19) Proposition

Let H be a pseudo-Boolean lattice. Then H is a Boolean lattice if and only if it satisfies (1.18).

Proof. Due to (1.17) we know that H is distributive. If (1.18) is satisfied then $0{:}a$ is a complement of a. Thus H is Boolean.

■

A simple example of a pseudo-Boolean lattice is a bounded, linearly ordered set $(B,\leq)$. Then $a \wedge b = \min(a,b)$ and $a \vee b = \max(a,b)$. Further

$$b{:}a = \begin{cases} 1 & \text{if } b \geq a, \\ b & \text{if } b < a, \end{cases}$$

for all $a,b \in B$. In general, B is not Boolean as for $0 < a < 1$ we find $a \vee (0{:}a) = a \vee 0 = a < 1$.

The following example shows that a distributive lattice is not necessarily pseudo-Boolean. Let $N_\infty := \mathbb{N} \cup \{\infty\}$ and let $a \leq b$ iff there exists $c \in N_\infty$ with $a \cdot c = b$. Then $(N_\infty,\leq)$ is a bounded distributive lattice with minimum 1 and maximum ∞ $(a \cdot \infty = \infty)$.

We define $a \wedge x := \text{g.c.d.}(a,x)$ (*greatest common divisor of* a *and* x). Then $a \wedge x = 1$ for $a \notin \{1,\infty\}$ has the set of solutions $\{x \mid \text{g.c.d.}(a,x) = 1\}$ which does not have a maximum. Thus $1:a$ does not exist and N_∞ is not pseudo-Boolean.

On the other side we show now that every finite distributive lattice H is pseudo-Boolean. Then H is bounded and $1 = \sup H$, $0 = \inf H$. Let $a,b \in H$ and $X := \{x \mid a \wedge x \leq b\}$. $0 \in X$ implies $X \neq \emptyset$. Let $c := \sup X$. Then $a \wedge \sup X = \sup\{a \wedge x \mid x \in X\} \leq b$. Thus $x \leq c$ is equivalent to $x \in X$. The same arguments show that a complete lattice H is pseudo-Boolean provided that the *general law of distributivity*

(1.20) $\quad a \wedge \sup X = \sup\{a \wedge x \mid x \in X\}$

holds for all $a \in H$ and for all $X \in P(H)$. Obviously,(1.20) implies distributivity.

Next we discuss a class of finite lattices. A finite lattice always has a minimum 0 and a maximum 1. An element a *covers* an element b if $a > b$, and if $a \geq x \geq b$ implies $x \in \{a,b\}$. An element a is called an *atom* if a covers 0. A finite lattice H is called *semimodular* if

(1.21) $\quad$ a and b cover $a \wedge b \Rightarrow a \vee b$ covers a and b

for all $a,b \in H$ with $a \neq b$. A finite lattice H is called *geometric* if it is semimodular and every element is the join of a set of atoms.

Let H be a finite lattice and let $G^H = (H,E)$ be a digraph representing H with $E := \{(a,b) \mid b \text{ covers } a\}$. For $x \in H$ let P_x

denote the set of all elementary paths from O to x. The *length* of a path p is denoted by *l(p)*. Then the *height function* $h: H \to \mathbb{N} \cup \{O\}$ is defined by

$$h(x) := \max\{l(p) \mid p \in P_x\}$$

for all $x \in H$. H satisfies the *Jordan-Dedekind chain condition* if for all $a, b \in H$ all paths from a to b have the same length.

(1.22) Theorem

A finite lattice is semimodular if and only if

(1) it satisfies the Jordan-Dedekind chain condition,

(2) $h(x) + h(y) \geq h(x \vee y) + h(x \wedge y)$ for all $x, y \in H$.

Proof. At first we assume that H is semimodular.

(1) Let $a, b \in H$ and assume w.l.o.g. $a < b$. We claim that if there exists a path from a to b of length k then all paths from a to b have length k. This is obvious for $k = 1$. We assume that the assertion holds for all $k \leq m$ for some $m \geq 1$. Let $p = (a, x, \ldots, b)$ and $q = (a, y, \ldots, b)$ be paths from a to b in G^H (both denoted by the sequence of their vertices). Let $l(p) = m+1$. If $x = y$ then the inductive hypothesis shows that every path from x to b has length m and therefore $l(p) = l(q) = m + 1$. Otherwise $x \neq y$ and semimodularity shows that $x \vee y$ covers x and y. Using the inductive hypothesis we find that every path of the form $(x, x \vee y, \ldots, b)$ has length m; every path of the form $(x \vee y, \ldots, b)$ has length $m - 1$; every path of the form $(y, x \vee y, \ldots, b)$ has length m. Thus $l(q) = m + 1$.

(2) Due to (1) h is well-defined. Let p and q be paths from $x \wedge y$ to x and from y to $x \vee y$. Then $l(p) = h(x) - h(x \wedge y)$ and

$l(q) = h(x \vee y) - h(y)$. Let $p = (a_1, a_2, \ldots, a_m)$. As a_{i+1} covers a_i for $i = 1,2,\ldots,m-1$ we find that either $a_{i+1} \vee y$ covers $a_i \vee y$ or $a_{i+1} \vee y = a_i \vee y$ for $i = 1,2,\ldots,m-1$. Therefore $l(q) \leq l(p)$.

Secondly, we assume that (1) and (2) are fulfilled in H. Let $a,b \in H$ with $a \neq b$ and let a and b cover $a \wedge b$. Then $h(a) = h(b) = h(a \wedge b) + 1$. Then (2) implies $h(a \vee b) \leq h(a) + 1$. Hence $a \vee b$ covers a. Similarly we find that $a \vee b$ covers b.

■

Let a,b be elements of a lattice H and let $a \leq b$. Then $[a,b] := \{x \in H \mid a \leq x \leq b\}$ is called an *interval* of H. [a,b] is a sublattice of H. If H is a finite, semimodular lattice then theorem (1.22) leads to the fact that [a,b] is semimodular. If H is a finite, geometric lattice with set of atoms D then

$$\{a \vee d \mid d \in D,\ d \nleq a,\ d \leq b\}$$

is the set of the atoms of an interval [a,b]. This implies that [a,b] is geometric.

Next we review basic results of matroid theory. We assume that the reader is familiar with the concept of a matroid and for proofs we refer to WELSH [1976].

Let F denote a nonempty subset of the set P(N) of all subsets of $N := \{1,2,\ldots,n\}$. Then F is called an *independence system* if

(1.23) $\quad I \in F$ and $J \subseteq I \Rightarrow J \in F$.

The elements of F are called *independent sets*. Obviously F is partially ordered with respect to set inclusion. An element of F is called *basis* if it is maximal in F. An element of P(N) is called *dependent* if it is not an element of F. A minimal dependent set is called a *circuit*. The function $r: P(N) \to \mathbb{Z}_+$ (nonnegative integers) defined by

$$r(I) = \max\{|J| \mid J \in F \text{ and } J \subseteq I\}$$

for all $I \in P(N)$ is called *rank function* of F. The function $\sigma: P(N) \to P(N)$ defined by

$$\sigma(I) = \{i \in N \mid r(I \cup \{i\}) = r(I)\}$$

for all $I \in P(N)$ is called *closure operator*. $\sigma(I)$ is called the *closure of* I.

An independence system F is called a *matroid* if

(1.24) for all $I, J \in F$ with $|I| < |J|$ there exists $j \in J \setminus I$ such that $I \cup \{j\} \in F$

is satisfied. Matroids can be described in terms of bases, dependent sets, circuits, rank function or closure operator. We can give axiomatic definitions of a matroid in terms of each of these concepts.

(1.25) Theorem (Base axioms)

A nonempty collection $B \subseteq P(N)$ is the set of bases of a matroid on N if and only if it satisfies the following condition:

(B1) If $I, J \in B$ and $i \in I \setminus J$ then there exists $j \in J \setminus I$ such that $(I \cup \{j\}) \setminus \{i\} \in B$.

(1.26) Theorem (Rank axioms)

A function $r\colon P(N) \to \mathbb{Z}_+$ is the rank function of a matroid on N if and only if for all subsets I,J of N:

(R1) $r(I) \leq |I|$,

(R2) $I \subseteq J \Rightarrow r(I) \leq r(J)$,

(R3) $r(I \cup J) + r(I \cap J) \leq r(I) + r(J)$.

(1.27) Theorem (Closure axioms)

A function $\sigma\colon P(N) \to P(N)$ is the closure operator of a matroid on N if and only if for all $I,J \in P(N)$ and for all $i,j \in N$:

(S1) $I \subseteq \sigma(I)$,

(S2) $I \subseteq J \Rightarrow \sigma(I) \subseteq \sigma(J)$,

(S3) $\sigma(I) = \sigma(\sigma(I))$,

(S4) if $j \notin \sigma(I)$ but $j \in \sigma(I \cup \{i\})$ then $i \in \sigma(I \cup \{j\})$.

(1.28) Theorem (Circuit axioms)

A collection $C \subseteq P(N) \setminus \{\emptyset\}$ is the set of circuits of a matroid on N if and only if for all distinct $I,J \in C$ and for all $i \in N$

(C1) $I \nsubseteq J$,

(C2) $i \in I \cap J$ implies the existence of $K \in C$ such that $K \subseteq (I \cup J) \setminus \{i\}$.

A matroid will often be denoted by M independently of the choice of the particular description considered.
The bases of a matroid have all the same cardinality $r(N)$; furthermore for a set $J \in P(N)$ all maximal independent sets $I \subseteq J$ have the same cardinality $r(J)$ and $r(\sigma(J)) = r(J)$. If J is independent then

$$\sigma(J) = \{j \in N \setminus J \mid J \cup \{j\} \notin F\}$$

and for all $j \in \sigma(J) \setminus J$ there exists a unique circuit in $J \cup \{j\}$ denoted by $C(j,J)$.

If $\{i,j\}$ is independent for all $i,j \in N$ then the matroid is called *simple*.

Let B be the set of all bases of a matroid M on N. Then $\{N \setminus I \mid I \in B\}$ again is the set of the bases of a matroid, which is called the *dual matroid* M* of M. Bases, circuits and rank function of M* are called *cobases*, *cocircuits* and *corank function* of M.

Let SC and SB denote the set of all circuits and cocircuits of a matroid M. Then M is called *regular* or *orientable* if there exist partitions C^+, C^- of each $C \in SC$ and B^+, B^- of each $B \in SB$ such that

$$|C^+ \cap B^+| + |C^- \cap B^-| = |C^+ \cap B^-| + |C^- \cap B^+|.$$

Now we consider a certain partially ordered subset of P(N) induced by a matroid M on N. $I \in P(N)$ is called a *flat* or *closed set* of M if $\sigma(I) = I$.

(1.29) Theorem

The set H of all flats of a matroid M is a finite geometric lattice.

Proof. H is partially ordered by set inclusion. The intersection of two flats is a flat; therefore join and meet exist and are given by

$$I \wedge J = I \cap J,$$

$$I \vee J = \cap \{K \in H \mid I \cup J \subseteq K\} = \sigma(I \cup J),$$

for all $I, J \in H$. H has minimum $\emptyset$ and maximum N. The rank function r of M is the height function of H. This follows from the fact that a flat I covers a flat J if and only if $J \subseteq I$ and $r(I) = r(J) + 1$. Thus the length of a path from $\emptyset$ to I is bounded by $r(I)$. In fact, each such path has length $r(I)$. Therefore H satisfies the Jordan-Dedekind chain condition. From (R3) we know $r(I) + r(J) \geq r(I \cup J) + r(I \cap J)$. If $I, J \in H$ then $r(I \wedge J) = r(I \cap J)$ and $r(I \cup J) = r(\sigma(I \cup J)) = r(I \cup J)$. Then theorem (1.22) shows that H is semimodular. Finally let $I \in H$ with $r(I) = k$. Then there exists an independent set $J := \{j_1, j_2, \ldots, j_k\} \subseteq I$. Clearly, $\sigma(j_\mu) \cap J = \{j_\mu\}$ for all $\mu = 1, 2, \ldots, k$. Thus $\sigma(j_\mu)$, $\mu = 1, 2, \ldots, k$ are distinct atoms and $\sup\{\sigma(j) \mid j \in J\} = I$. ■

Now we prove a representation theorem for finite geometric lattices.

(1.30) <u>Theorem</u>

A finite lattice H is geometric if and only if it is lattice-isomorphic to the lattice of flats of a matroid.

<u>Proof</u>. Due to theorem (1.29) it suffices to find a suitable representation of a geometric lattice H. Let A denote the set of all atoms of H. Let $F \subseteq P(A)$ consist in all subsets I of A with $h(\sup I) = |I|$ with respect to the height function h of H. Theorem (1.22) shows that for $I = \{i_1, i_2, \ldots, i_k\} \subseteq A$ the inequality $h(\sup I) \leq \Sigma_\mu h(i_\mu) = |I|$ is valid. Let $I, J \subseteq A$. Then

$I \subseteq J$ implies $h(\sup I) \leq h(\sup J)$. Further $h(\sup I) + h(\sup J) \geq h((\sup I) \vee (\sup J)) + h((\sup I) \wedge (\sup J)) \geq h(\sup(I \cup J)) + h(\sup(I \cap J))$. The last inequality follows from $(\sup I) \wedge (\sup J) \geq \sup(I \cap J)$. Therefore the function $r(I) := h(\sup I)$ satisfies the rank axiomatics of a matroid. Hence r is the rank function of a matroid $M(H)$ on A.

Now let σ be the closure operator of $M(H)$ and let H' be the geometric lattice of the flats of $M(H)$. Let $\varphi: H \to H'$ defined for any $I \in H$ by $\varphi(I) := \sigma\{i_1, i_2, \ldots, i_k\}$ where $\{i_1, i_2, \ldots, i_k\}$ is the set of all atoms below I. Then φ is a lattice-isomorphism. ■

Two matroids M, M' on N and on N' are called *isomorphic* if there exists a bijective mapping $\varphi: N \to N'$ preserving independence. Equivalently, we may assume that φ preserves the rank functions, circuits and so on. Now two matroids with lattice-isomorphic geometric lattices of flats are not necessarily isomorphic as matroids. It should be noted that the rank function of the matroid $M(H)$ for a geometric lattice H satisfies

$$r\{i\} = 1, \quad r\{i,j\} = 2$$

for all distinct atoms $i, j \in H$. Hence the lattice H' of the flats of $M(H)$ is a simple matroid. Conversely, two simple matroids are isomorphic if and only if their geometric lattices of flats are lattice-isomorphic. Thus the study of simple matroids is just the study of finite geometric lattices.

A lattice is called *relatively complemented* if any interval $[a,b]$ is complemented.

(1.31) Proposition

A finite geometric lattice is relatively complemented.

Proof. It suffices to consider the lattice H of the flats of a simple matroid M on N. Let A be a flat and let I be a maximal independent set with $I \subseteq A$. Let B be a base containing I. Then we define $A^* := \sigma(B \setminus I)$. We find $\sigma(A \cup A^*) = N$ and $A \cap A^* = \emptyset$. In the language of lattices this means that H is complemented (cf. definitions of $\wedge, \vee$ in the proof of (1.29)). Now every interval of H is a finite, geometric lattice. Therefore, H is relatively complemented. ■

A special class of finite geometric lattices is derived from the partitions of a finite set N. Let $\pi(N)$ denote the set of all partitions of N. Let $R = \{R_i \mid i \in I\}$ and $S = \{S_j \mid j \in J\}$ denote two partitions of N. We define $R \leq S$ iff there exists a function $\varphi: I \to J$ such that $R_i \subseteq S_{\varphi(i)}$ for all $i \in I$. Meet and join of this partial order are given in the following. At first

$$R \wedge S = \{R_i \cap S_j \mid i \in I, j \in J\}.$$

Now consider a graph with vertex set $R \cup S$. Two vertices R_i and S_j are joined by an edge if $R_i \cap S_j \neq \emptyset$. Let $(C_k, k \in K)$ denote the family of the vertices of the connected components of G and define $T_k := \cup\{R_i \mid R_i \in C_k\} = \cup\{S_j \mid S_j \in C_k\}$ for $k \in K$. Then

$$R \vee S = \{T_k \mid k \in K\}.$$

$\pi(N)$ is called the *partition lattice* of N.

A recent result of PUDLÁK and TŮMA ([1977], [1980]) answers a question raised by BIRKHOFF [1967]:

(1.32) Theorem

Every finite lattice can be lattice-embedded in a partition lattice over a finite set.

For a proof we refer to PUDLÁK and TŮMA [1980]. Finite partition lattices are special finite geometric lattices. We consider the *complete graph* K_n, i.e. the graph with vertex set $N = \{1,2,\dots,n\}$ and set of edges E containing all pairs (i,j) with $i \neq j$. Here (i,j) and (j,i) denote the same edge for $i \neq j$. A subset I of E is called independent if it contains no cycle. These sets are the independent sets of the *cycle matroid* of K_n.

(1.33) Theorem

The partition lattice of the finite set $N = \{1,2,\dots,n\}$ is a geometric lattice which is lattice-isomorphic to the lattice of the flats of the cycle matroid of the complete graph K_n.

Proof. Let $(V_\mu,\ \mu = 1,2,\dots,k)$ be a family of mutually disjoint subsets of N. Then the subgraphs G_μ generated by V_μ are complete subgraphs of K_n. Let E_μ denote the corresponding set of edges. Then

$$E' = E_1 \cup E_2 \cup \dots \cup E_k$$

is a flat of the cycle matroid M. All flats of M have this form. Now $\{E_\mu \mid \mu = 1,2,\dots,k\} \cup (N \setminus E')$ is a partition of N. This 1-1 correspondence of flats and partitions leads to a lattice isomorphism between $\pi(N)$ and the finite geometric lattice of all flats of M.

■

Together with the theorem of PUDLÁK and TŮMA theorem (1.33) shows that every finite lattice can be lattice-embedded in the finite, geometric lattice of the flats of the cycle matroid over a suitable complete graph K_n.

As an example we consider a finite, distributive lattice H of n elements. ORE [1942] shows that $\varphi: H \to \pi(H)$ defined by $\varphi(a) := \{\{b \in H \mid a \vee b = c\} \mid c \in H\}$ is a lattice-embedding of H in $\pi(H)$. For a proof let

$$R_c^a := \{b \in H \mid a \vee b = c\} .$$

For $c \neq c'$ we find immediately $R_c^a \cap R_{c'}^a = \emptyset$. Further for $b \in H$ we know $b \in R_{a \vee b}^a$. Therefore $\varphi(a) \in \pi(H)$. Now let $a \leq a'$ and let $b, c \in H$ with $a \vee b = c$. Then $a' \vee b = a' \vee a \vee b = a' \vee c$. Therefore $R_c^a \subseteq R_{a' \vee c}^{a'}$ which implies $\varphi(a) \leq \varphi(a')$. Finally a finite distributive lattice has always a minimum 0. Therefore $0 \in R_a^a$, i.e. $R_a^a \neq \emptyset$. Thus $a = \max R_a^a$. Let now $a, a' \in H$ with $\varphi(a) \leq \varphi(a')$. Then $R_a^a \subseteq R_{a'}^{a'}$ and thus $a \leq a'$.

Finally in this section we discuss the role of certain duality principles in ordered structures $(H, \leq)$. Its dual is $(H, \geq)$ with respect to the inverse or dual order of $\leq$. Reflexivity, antisymmetry, transitivity and linearity hold in $(H, \leq)$ if and only if the same property holds in $(H, \geq)$. The *dual of a lattice* $(H, \wedge, \vee)$ again is a lattice denoted by $(H, \vee, \wedge)$, i.e. join and meet are exchanged. If a lattice $(H, \wedge, \vee)$ has a minimum (maximum) then its dual $(H, \vee, \wedge)$ has a maximum (minimum). The dual of a distributive lattice is distributive. This follows from (1.11.3). The law of distributivity (1.10) implies (1.10') as (1.10') can be derived as law of distributivity in $(H,\vee,\wedge)$, i.e. by exchanging meet and join.

Now the *duality principle* of ordered sets is the following. It can be applied to ordered sets $(H,\leq)$ of special type iff the dual $(H,\geq)$ is of the same type. Let A be a statement involving $\leq,\wedge,\vee,0,1$ which holds in all ordered sets of the special type considered. Then the dual A' of this statement is found by replacing $\leq$ by $\geq$, $\wedge$ by $\vee$, $\vee$ by $\wedge$, 0 by 1 and 1 by 0 in A. Then A' holds in all ordered sets of the special type considered. This follows from the fact that A' is the same as A formulated for the dual $(H,\geq)$.

We can apply the duality principle for example to preordered or quasiordered or partially ordered or linearly ordered sets, to lattices, and distributive lattices. In particular, (1.10') is the dual of (1.10).

The dual of a Boolean lattice is distributive and bounded. Let a^* be the complement of a. Then $a \wedge b = 0$ and $a \vee b = 1$. These equations show that with respect to the dual lattice a^* is a complement of a, too. Thus the dual of a Boolean lattice is Boolean.

On the other hand the dual of a pseudo-Boolean lattice is not necessarily a pseudo-Boolean lattice. This is due to an asymmetry in (1.15). The dual of (1.15) is

$$(1.15') \qquad a \vee b \geq b \iff x \geq c \qquad \forall\, a,b \in H,\ \exists\, c \in H.$$

If such an element c exists then it is uniquely determined and is denoted by b/a. A bounded, linearly set $(B,\leq)$ is pseudo-Boolean and (1.15') is satisfied with

$$b/a := \begin{cases} b & \text{if } a < b, \\ 0 & \text{if } a \geq b, \end{cases}$$

for all $a,b \in B$. (1.15') is satisfied for any finite distributive lattice. Therefore the dual of a finite, pseudo-Boolean lattice is pseudo-Boolean.

(1.22) shows that the dual of a finite, semimodular lattice H satisfies the Jordan-Dedekind chain condition but the dual height function h^* given by $h^*(x) = h(1) - h(x)$ for $x \in H$ satisfies

$$h^*(x) + h^*(y) \leq h^*(x \vee y) + h^*(x \wedge y)$$

and thus the dual lattice is not necessarily semimodular.

2. Ordered Commutative Semigroups

In this chapter we consider ordered sets H in which a *binary operation (internal composition)* $*: H \times H \to H$ is defined. Basic concepts and definitions are introduced and discussed with respect to possible specializations. Commutativity is assumed throughout this chapter although commutativity is not necessary for some results. This assumption is motivated by the optimization problems considered in the second part. Further commutativity often leads to simpler proofs and is necessary for some important results. For a more detailed discussion of ordered algebraic structures we refer to FUCHS [1966], KOKORIN and KOPYTOV [1974], MURA and RHEMTULLA [1977] and SATYANARAYANA [1979].

A nonempty set H with internal composition $*: H \times H \to H$ is called a *semigroup*, if

(2.1) $\quad a * (b * c) = (a * b) * c$ (*Associativity*)

for all $a, b, c \in H$. A semigroup is called a *monoid* if it contains an element e with $e * a = a * e = a$ for all $a \in H$. Such an element is always uniquely determined and is called the *neutral element (identity)* of H. A semigroup containing a neutral element is called a *monoid*. A semigroup is called *commutative* if

(2.2) $\quad a * b = b * a$ (*Commutativity*)

for all $a, b \in H$.

An ordered set H, which is a commutative semigroup, is called an *ordered commutative semigroup* if

(2.3) $a \leq b \Rightarrow a * c \leq b * c$ (*Monotonicity*)

for all $a,b,c \in H$. Axiom (2.3) describes the interaction of internal composition and order relation. Thus we consider only binary operations which are *isotone* in both arguments.

Let $a_i \in H$ for $i \in N := \{1,2,\ldots,n\}$. Then we define

$$\mathop{*}_{i\in N} a_i := a_1 * a_2 * \ldots * a_n .$$

For $a_i \leq b_i$, $i \in N$, monotonicity leads to

$$\mathop{*}_{i\in N} a_i \leq \mathop{*}_{i\in N} b_i .$$

In particular, let $a^n := * a_i$ if $a_i = a$ for all $i \in N$ ($a^o := e$). Then monotonicity implies $a^n \leq b^n$ if $a \leq b$.

In chapter 1 we have seen that in a lattice $(H,\leq)$ meet $\wedge$ and join $\vee$ are binary operations. Due to (1.5) we know that $(H,\wedge)$ and $(H,\vee)$ are commutative semigroups. Furthermore $(H,\vee,\leq)$ and $(H,\wedge,\leq)$ are lattice-ordered commutative semigroups, since $a \leq b$ implies $a \wedge c \leq b \wedge c$ and $a \vee c \leq b \vee c$ for all $a,b,c \in H$.

Two further examples are derived from real numbers. Besides $(\mathbb{R},+,\leq)$ with usual addition + and usual order relation $\leq$ we find the totally ordered commutative semigroups $(\mathbb{R},\max,\leq)$ and $(\mathbb{R},\min,\leq)$ with internal composition $\max(a,b)$ and $\min(a,b)$.

We call a commutative monoid G a *commutative (abelian) group* if for all $a \in G$ there exists an element $b \in G$ with $a * b = e$. Such an element is always uniquely determined and denoted by a^{-1}. An ordered commutative monoid G is called an *ordered commutative group* if G is a group.

Ordered commutative groups have many properties which do not hold in ordered commutative semigroups. In particular the *cancellation law*

(2.4) $a * c = b * c \Rightarrow a = b$

for all $a,b,c \in G$ is valid in a group G and implies a stronger version of (2.3), i.e.

(2.5) $a \leq b \iff a * c \leq b * c$

for all $a,b,c \in G$. Therefore $a < b$ implies $a * c < b * c$ in ordered commutative groups. A semigroup H satisfying (2.4) is called *cancellative*.

Let H be an ordered commutative semigroup. $a \in H$ is called *positive* if

(2.6) $b \leq a * b$

for all $b \in H$. If all elements in H are positive then H is called *positively ordered*. We remark that "positively ordered" often includes "partially ordered" which is not assumed here.

Strictly positive, *negative* and *strictly negative* elements are defined similarly by inequality (2.6) with $\leq$ replaced by $<$, $\geq$, and $>$. The set of all positive (negative) elements of H is denoted by H_+ (H_-). $a \in H_+$ ($a \in H_-$) and $a \leq b \in H$ ($a \geq b \in H$) imply $b \in H_+$ ($b \in H_-$). Therefore H_+ (H_-) is called the *positive (negative cone* of H. A positively ordered commutative semigroup H is called *naturally ordered* if

(2.7) $a < b \Rightarrow \exists\, c \in H: a * c = b$

for all $a,b \in H$. We remark that "naturally ordered" often includes "partially ordered" which is not assumed here.

In an ordered group G an element a is positive if and only if $a \geq e$. The ordering on G is completely determined by the positive cone G_+ as $a \leq b$ if and only if $b * a^{-1} \in G_+$ for all $a,b \in G$. Reflexivity is equivalent to $e \in G_+$, antisymmetry to $G_+ \cap G_+^{-1} = e$ $(G_+^{-1} := \{a^{-1} \mid a \in G\})$ and linearity to $G_+ \cup G_+^{-1} = G$. For subsets P,Q of an ordered algebraic structure we define $P * Q := \{a * b \mid a \in P,\ b \in Q\}$ and $P \leq Q$ by $a \leq b$ for all $a \in P$, $b \in Q$. For $P = \{a\}$ we use $a * Q$ and $a \leq Q$ meaning $\{a\} * Q$ and $\{a\} \leq Q$. Then transitivity in G is equivalent to $G_+ * G_+ \subseteq G_+$. Vice versa a subset P of G defines a suitable ordering on a group G if at least $P * P \subseteq P$. It should be remarked that commutativity implies monotonicity if P defines an ordering by

$$a \leq b \iff b * a^{-1} \in P.$$

If G is an ordered commutative group then G_+ is naturally ordered.

A bijective mapping $\varphi: H \to H'$ for two semigroups $(H,*)$, $(H,*')$ is called a *semigroup isomorphism* if

$$\varphi(a * b) = \varphi(a) *' \varphi(b)$$

for all $a,b \in H$. If $\varphi: H \to \varphi(H)$ is a semigroup isomorphism but not necessarily surjective then φ is called a *semigroup embedding* of H into H'. Similar denotations are used for special classes of semigroups (for example, *group isomorphism*). If H and H' are ordered semigroups and $\varphi: H \to H'$ is an order isomorphism then φ is called an *isomorphism* between $(H,*,\leq)$ and $(H',*',\leq')$. An embedding is defined accordingly. In general, an isomorphism between ordered algebraic structures is an isomorphism with respect to all relevant compositions and orderings.

(2.8) Theorem

A naturally ordered commutative monoid H fulfilling the cancellation rule (2.4) is isomorphic to the positive cone of an ordered commutative group.

Proof. Define a binary relation $\sim$ on $H \times H$ by $(a,b) \sim (c,d)$ iff $a * d = b * c$. Then $\sim$ is an equivalence relation. Reflexivity and symmetry are obvious. Let $(a,b) \sim (c,d)$ and $(c,d) \sim (e,f)$. Then $a * d = b * c$ and $c * f = d * e$. Composition with f resp. b shows $a * d * f = d * e * b$. Cancellation yields $(a,b) \sim (e,f)$. Thus $\sim$ is transitive. Let G denote the induced partition of $H \times H$. Blocks of G are represented by elements as usual.

Now $(a,b) \oplus (c,d) = (a*c, b*d)$ defines an internal composition $\oplus : G \times G \to G$ which is commutative and associative. Further (e,e) is the neutral element and (b,a) the inverse element of (a,b). Thus $(G, \oplus)$ is a group and $\varphi: H \to G$ defined by $\varphi(a) = (a,e)$ fulfills $\varphi(a * b) = \varphi(a) \oplus \varphi(b)$.

Let $P := \varphi(H)$. Then $P \oplus P \subseteq P$. Therefore we can define an order relation with positive cone P. If $a \leq b$ then there exists $c \in H$ such that $a * c = b$ (if $a = b$ then $c := e$ for example). Therefore $\varphi(b) \oplus \varphi(a)^{-1} = (b,a) \sim (c,e) \in P$. Finally let $(a,e) \leq (b,e)$. Then $(b,a) \in P$ implies the existence of $c \in H$ with $(b,a) \sim (c,e)$. This means $a * c = b$. As H is positively ordered $a \leq a * c$ holds. Then $a \leq b$. Thus we have shown

$a \leq b \iff \varphi(a) \leq \varphi(b)$.

■

The group $(G, \oplus, \leq)$ in the proof of theorem (2.8) is obviously

reflexively, antisymmetrically or linearly ordered iff the same holds for $(H,*,\leq)$. In the *linearly ordered case* $G = (G_- \setminus \{e\}) \dot{\cup} \{e\} \dot{\cup} (G_+ \setminus \{e\})$ and elements of the negative cone of G are just the inverse elements of elements of the positive cone, i.e. $G_- \setminus \{e\} = \{a^{-1} \mid a \in G_+ \setminus \{e\}\}$. In view of theorem (2.8) we may identify H and G_+. Therefore
$G = \{a^{-1} \mid a \in H,\ a \neq e\} \dot{\cup} \{e\} \dot{\cup} (H \setminus \{e\})$.
The assumption of the existence of a neutral element in theorem (2.8) can often be fulfilled by adjoining an extra element.

(2.9) Proposition

Let $(H,*,\leq)$ be an ordered commutative semigroup without a neutral element. Let $e \notin H$ and $H' := H \cup \{e\}$, define $*' : H' \times H' \to H'$ by $a *' b = a * b$ for $a,b \in H$ and

$$a *' e = e *' a = a$$

for all $a \in H$. Let $\leq'$ denote the binary relation on H' defined by $a \leq' b$ iff $a \leq b$ for all $a,b \in H$,

$$a <' e <' b$$

for all $a \in H_-$, $b \in H_+$ and, if $\leq$ is reflexive or linear, $e \leq' e$.
Then

(1) H' is an ordered commutative monoid,

(2) if $\leq$ is reflexive, antisymmetric or linear then the same holds for $\leq'$,

(3) if H is positively or naturally ordered then the same holds for H',

(4) if H is cancellative then the same holds for H'.

Proof. We only remark that $H_- < H_+$. Then all properties can easily be verified. ■

In particular (2.9.3) and (2.9.4) show that it is sufficient to assume that H is a naturally ordered commutative semigroup in (2.8). If H has no neutral element then it is isomorphic to the strict positive cone $G_+ \setminus \{e\}$ of an ordered commutative group $(G, \oplus, \leq)$.

Some examples are considered in the following. We begin with $(\mathbb{N}, \cdot, \leq)$ with $a \leq b$ iff $b/a \in \mathbb{N}$ (*divisibility*). This is a naturally and lattice-ordered, commutative monoid. Clearly it is isomorphic to the positive cone of the lattice ordered group of positive rational numbers $\{a \in \mathbb{Q} \mid a > 0\}$ endowed with multiplication and ordered by divisibility.

Secondly let $(G,+)$ denote the additive group of all real functions $f\colon [0,1] \to \mathbb{R}$ with $(f+g)(x) := f(x) + g(x)$ and $f \leq g$ iff $f(x) \leq g(x)$ for all $x \in [0,1]$. Then $(G,+,\leq)$ is a lattice ordered commutative group.

The third example does not look so familiar. For $k \in \mathbb{N}$ let V denote the set of all subsets A of $\mathbb{N} \cup \{0\}$ with $|A| \leq k$. For arbitrary $B \subseteq \mathbb{N} \cup \{0\}$ let $k - \min(B)$ denote the set of the k smallest elements of B (in particular, for $|B| < k$ let $k - \min(B) = B$). Now for $A, B \in V$ let

$$A \leq B \quad :\Longleftrightarrow \quad k - \min(A \cup B) = A \; .$$

This binary relation defines a lattice on V. Meet is given by

$$A \wedge B = k - \min(A \cup B) \; ,$$

and join is (under consideration of its commutativity) given by

$$(2.10)\qquad A \vee B = (A \cap B) \cup \begin{cases} \emptyset & \text{if } |A|,|B| < k, \\ \{b \in B \mid b > \alpha\} & \text{if } |A| = k,\ |B| < k, \\ \{b \in B \mid b > \alpha\} \cup \{\beta+1,\dots,\beta+r\} & \text{if } |A|,|B| = k;\ \alpha \le \beta \end{cases}$$

with $\alpha = \max A$, $\beta = \max B$ and $r = k - |(A \cap B) \cup \{b \in B \mid b > \alpha\}|$. The meet formula is obvious and the join formula is proved in the following proposition. Beforehand we define

$$A * B := k - \min(A + B)$$

with $A + B := \{a + b \mid a \in A,\ b \in B\}$ for all $A, B \in V$.

(2.11) Lemma

$(V, *, \le)$ is a lattice-ordered commutative monoid with neutral element $\{0\}$, least element $Z = \{0, 1, \dots, k-1\}$ and greatest element $\emptyset$.

Proof. The properties of the neutral element, the least element and the greatest element are easily verified. The same holds for the properties of the partial order, the meet and for commutativity, associativity and monotonicity of the internal composition. The remaining part is the verification of the join-formula in (2.10).

At first we remark that $A \le B$ is equivalent with

$$B \subseteq A \quad \text{or} \quad (|A| = k \text{ and } b \in B \setminus A \text{ implies } b > \max A).$$

Further let $C := A \vee B$, $\alpha = \max A$, $\beta = \max B$, $\gamma = \max C$ ($\max \emptyset = \infty$). Secondly we prove $A \cap B \subseteq C$. As $A \cap B$ is an upper bound of A and B we find $C \le A \cap B$. Assume $A \cap B \not\subseteq C$. Then $|C| = k$ and for each $d \in (A \cap B) \setminus C$ we find $\alpha, \beta \ge d > \gamma$. The assumption $C \subseteq A \cap B$ leads to $A \cap B \le C$ and thus to the contradiction $A \cap B = C$. Therefore there exists $c \in C \setminus (A \cap B)$. W.l.o.g. let $c \notin A$. Together with $A \leq C$ this implies $c > \alpha$ contrary to $\alpha > \gamma$.

Thirdly consider the different cases in formula (2.10). If $|A|, |B| < k$ then $C \subseteq A$ and $C \subseteq B$ shows $C \subseteq A \cap B$. As we know $A \cap B \subseteq C$ this proves $C = A \cap B$. In the second case $|A| = k$ and $|B| < k$ we know that $c \in C \setminus A$ implies $c > \alpha$ and that $C \subseteq B$. Therefore

$$C \subseteq (A \cap B) \cup \{b \in B \mid c > \alpha\} =: D.$$

Thus $D \leq C$. As D is an upper bound of A and B we find $D = C$. In the third case $|A| = |B| = k$ we know that $c \in C \setminus A$ implies $c > \alpha$ and that $c \in C \setminus B$ implies $c > \beta$. Therefore $(\alpha \leq \beta)$

$$C \subseteq (A \cap B) \cup \{b \in B \mid c > \alpha\} \cup \{d \in \mathbb{N} \cup \{0\} \mid d > \beta\} =: D.$$

Then $\tilde{D} = k\text{-}\min(D)$ fulfills $\tilde{D} \leq C$. Further $\tilde{D}$ is an upper bound of A and B. Thus $\tilde{D} = C$. ∎

The order relation in this example is neither positively ordered nor is (2.7) fulfilled. Therefore $(V, *, \leq)$ may not be embedded in the positive cone of an ordered commutative group.

An ordered commutative semigroup H is called *residuated* if for all $a, b \in H$ there exists $c \in H$ such that

$$(2.12) \qquad a * x \leq b \iff x \leq c\,.$$

If such an element c exists for $a, b \in H$ then it is called a *residual* of b with respect to a. If the order relation in H is antisymmetric then a residual is uniquely determined and denoted by $b{:}a$. If for all $a, b \in H$ there exists $c \in H$ such that

(2.12') $a * x \geq b \iff x \geq c$

then H is called *dually residuated*. Such an element c is called a *dual residual* and is denoted by b/a. (2.12') is the dual of (2.12) in the lattice-theoretic sense. A (dual) residual b:a (b/a) is the greatest (smallest) solution of the inequality $a * x \leq b$ $(a * x \geq b)$. Therefore the (dual) residual exists if and only if the set $\{x \in H \mid a * x \leq b\}$ $(\{x \in H \mid a * x \geq b\})$ has a greatest (smallest) element.

An important implication of the existence of residuals is the distributivity of the least upper bound.

(2.13) <u>Proposition</u>

Let H be a residuated, lattice-ordered commutative semigroup. Then $a * (b \vee c) = (a * b) \vee (a * c)$ for all $a,b,c \in H$.

<u>Proof</u>. Let $a,b,c \in H$. Monotonicity implies $a * b, a * c \leq a * (b \vee c)$. Therefore $(a * b) \vee (a * c) \leq a * (b \vee c)$. Let $x := (a * b) \vee (a * c)$. Then $a * b \leq x$ and $a * c \leq x$. Then $b \leq x{:}a$ and $c \leq x{:}a$. Thus $b \vee c \leq x{:}a$ which implies $a * (b \vee c) \leq x$.

∎

On the other hand the greatest lower bound is not necessarily distributive.

(2.14) <u>Proposition</u>

Let G be a lattice-ordered group. Then G is residuated and dually residuated and greatest lower and least upper bound are distributive.

Proof. In a group $b:a = b/a = b * a^{-1}$. Therefore distributivity of $\vee$ follows from (2.13) and distributivity of $\wedge$ follows from (2.13) applied to $(G,*)$ ordered with respect to the dual of the lattice $(G,\leq)$. ∎

(2.14) shows that lattice-ordered commutative groups are examples for residuated lattice-ordered commutative semigroups. A more detailed discussion is given in chapter 3.

Let $(H,\wedge,\leq)$ denote the lattice-ordered commutative semigroup derived from a bounded lattice $(H,\leq)$. If $(H,\wedge,\leq)$ is residuated then from (2.13) we know that for all $a,b,c \in H$

$$a \wedge (b \vee c) = (a \wedge b) \vee (a \wedge c).$$

We have found a stronger result in chapter 1. (2.12) means in this case that $(H,\leq)$ is a pseudo-Boolean lattice. Therefore $(H,\leq)$ fulfills

$$a \vee (b \wedge c) = (a \vee b) \wedge (a \vee c)$$

for all $a,b,c \in H$, too. If the lattice $(H,\leq)$ is Boolean then the residuals of $(H,\wedge,\leq)$ are given by

$$b:a = b \vee a^*$$

(see 1.15). Clearly, for pseudo-Boolean lattices the ordered commutative semigroup $(H,\vee,\leq)$ fulfills (2.12'), i.e. for all $a,b \in H$ there exists $c \in H$ such that

$$a \vee x \geq b \iff x \geq c.$$

3. Lattice-Ordered Commutative Groups

In this chapter we consider lattice-ordered commutative groups $(G, *, \leq)$. A characterization of such groups can be found in CONRAD, HARVEY and HOLLAND [1963]. For a detailed discussion we refer to CONRAD [1970]. We introduce basic concepts and properties which allow a precise formulation of this characterization.

Due to proposition (2.14) we know that a lattice-ordered commutative group is *distributive*, i.e. least upper and greatest lower bound are distributive over the internal composition.

(3.1) Proposition

(1) A partially ordered commutative group G is lattice-ordered if and only if $a \vee e$ exists for all $a \in G$.

Let G be a lattice-ordered commutative group. Let $a, b \in G$ and $n \in \mathbb{N}$. Then

(2) $a^n \geq b^n \iff a \geq b$ (G is *isolated*),

(3) $(a \vee e)^n = a^n \vee e$.

Proof. (1) Necessity is obvious. Otherwise let $\alpha := [(a * b^{-1}) \vee e] * b$. Then $\alpha = a \vee b$. Further let $\beta := (a^{-1} \vee b^{-1})^{-1}$. Then $\beta = a \wedge b$.

(2) $a^n \geq b^n$ is equivalent to $(a * b^{-1})^n \geq e$. Thus it suffices to show

$$a^n \geq e \iff a \geq e.$$

Let $a \geq e$. Then monotonicity leads to $a^n \geq e$. Let now $a^n \geq e$. $n = 1$ is trivial. Let $n > 1$. Then distributivity leads to

$$(a \wedge e)^n = a^n \wedge a^{n-1} \wedge \ldots \wedge e = a^{n-1} \wedge a^{n-2} \wedge \ldots \wedge e = (a \wedge e)^{n-1}.$$

Cancellation implies $a \wedge e = e$; i.e. $a \geq e$.

(3) For $0 < k \leq n$ we find

$$(a^{n-k} \vee a^{-k})^n = a^{(n-k)n} \vee \ldots \vee a^{(n-k)k-k(n-k)} \vee \ldots \vee a^{-kn} \geq e.$$

Therefore (2) implies $a^{n-k} \vee a^{-k} \geq e$, i.e. $a^n \vee e \geq a^k$. Then

$$(a \vee e)^n = a^n \vee a^{n-1} \vee \ldots \vee e = a^n \vee e.$$

■

A semigroup (group) $(H,*)$ is called *divisible* or *radicable* if for all $a \in H$ and all $n \in \mathbb{N}$ there exists $b \in H$ such that $a = b^n$.

(3.2) Proposition

A lattice-ordered commutative group G can be embedded into a divisible, lattice-ordered commutative group $\bar{G}$.

Proof. Let $\bar{G}$ denote the set of all equivalence classes of the set $G \times \{\frac{1}{n} \mid n \in \mathbb{N}\}$ with respect to the equivalence relation $\sim$ defined by

$$(a,\tfrac{1}{n}) \sim (b,\tfrac{1}{m}) :\Longleftrightarrow \quad a^m = b^n .$$

Then $(\bar{G}, \otimes)$ is a commutative group with internal composition

$$(a,\tfrac{1}{n}) \otimes (b,\tfrac{1}{m}) := (a^m b^n , \tfrac{1}{mn})$$

where elements in $\bar{G}$ are represented in the usual form. This definition is independent of the choice of the representing elements as G is cancellative.

Let φ: $G \to \bar{G}$ defined by $\varphi(a) := (a,1)$. Then

$$\varphi(a * b) = \varphi(a) \otimes \varphi(b).$$

The lattice ordering on G is given by the positive cone G_+ of G. Now $\bar{G}_+ := \{(a,\frac{1}{n}) \mid a \in G_+\}$ defines a partial order on $\bar{G}$. This definition is independent of the choice of the representing elements as G is isolated. Clearly, $\varphi(G_+) = \varphi(G) \cap \bar{G}_+$. Therefore

$$a \leq b \iff \varphi(a) \leq \varphi(b)$$

for all $a, b \in G$. Monotonicity follows from (3.1.2) and cancellation arguments. Thus $(\bar{G}, \otimes, \leq)$ is a partially ordered commutative group. From (3.1.1) we know that it suffices now to show the existence of $(e,1) \vee (a,\frac{1}{n})$ in $\bar{G}$. Let $(x,\frac{1}{m})$ denote an upper bound of $(e,1)$ and $(a,\frac{1}{n})$. Then $x \geq e$ and $x^n \geq a^m$. Therefore using (3.1.3) $x^n \geq e \vee a^m = (e \vee a)^m$, i.e. $(x,\frac{1}{m}) \geq (e \vee a,\frac{1}{n})$. Similarly we find that $(e \vee a,\frac{1}{n})$ is such an upper bound. Thus

$$(e,1) \vee (a,\tfrac{1}{n}) = (e \vee a,\tfrac{1}{n}) .$$

■

A special divisible, lattice-ordered commutative group can be constructed from the root lattice $V = V(\Lambda)$ in chapter 1. The components x_λ for $\lambda \in \Lambda$ of an element $x \in V$ are real numbers. Therefore an internal composition + on V can be defined by componentwise addition of real numbers, i.e.

$$(a+b)_\lambda := a_\lambda + b_\lambda$$

for all $\lambda \in \Lambda$ and $a, b \in V$. Then $(V,+)$ is a divisible, commutative group. $V(\Lambda)$ is a subset of the cartesian product $\times_{\lambda\in\Lambda} H_\lambda$ with $H_\lambda = \mathbb{R}$ for all $\lambda \in \Lambda$. If $(H_\lambda,+)$ is a subgroup of $(\mathbb{R}, +)$ then the same construction leads to the *root lattice* $\bar{V} = \bar{V}(H_\lambda,\Lambda)$ *with respect to the family* (H_λ,Λ) and to a commutative group $(\bar{V},+)$ which is not necessarily divisible.

(3.3) Proposition

$\overline{V}$ (V) is a (divisible) lattice ordered commutative group.

Proof. It suffices to show monotonicity. Let $a \leq b$. From the proof of (1.3) we know that $a^M \leq b^M$ for each maximal linearly ordered subset of Λ. This means that either $L := \{\mu \in M \mid a_\mu \neq b_\mu\}$ is empty or $a_\lambda \leq b_\lambda$ for $\lambda = \max L$. Now let $c \in V$. Then

$$L = \{\mu \in M \mid a_\mu + c_\mu \neq b_\mu + c_\mu\}$$

and therefore $a^M + c^M \leq b^M + c^M$. As this is satisfied for each maximal linearly ordered subset of Λ we find $a + c \leq b + c$. ■

V is called a *Hahn-type* group and is considered in CONRAD, HARVEY and HOLLAND [1963]. They prove the much deeper result that each lattice-ordered commutative group can be embedded into a Hahn-type group. A proof of this important theorem is beyond the scope of our book; but in order to understand a more detailed formulation of this result we introduce some basic concepts.

In the following G is a divisible, lattice-ordered commutative group. Divisibility can be assumed w.l.o.g. due to proposition (3.2).

A subset G' of G is called *convex* if $a, b \in G'$ implies that $[a,b] \subseteq G'$. A convex subgroup G' which is a sublattice of G is called a *lattice ideal* of G. The set I of all lattice ideals is partially ordered by set inclusion. A lattice ideal G' is called *regular* if there exists an element $g \in G$ such that G' is maximal in I with respect to the property $g \notin G'$. Let (G_λ, Λ) denote the family of all regular lattice ideals of G.

CONRAD, HARVEY and HOLLAND [1963] show the existence of a unique lattice ideal G^{λ} covering G_{λ} for $\lambda \in \Lambda$. G^{λ} is called the *cover* of G_{λ} for $\lambda \in \Lambda$.

As a simple example we consider the Hahn-type group $V(\Gamma)$ with $\Gamma = \{1,2,3\}$. The root system Γ is ordered as described in figure 4.

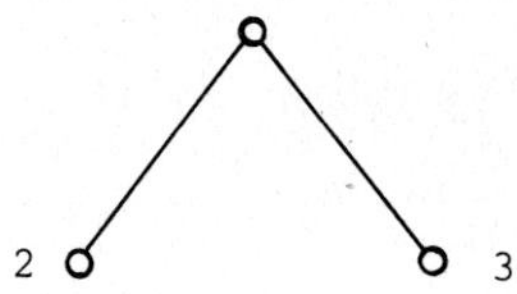

Figure 4. The root system Γ

Then $V = \mathbb{R}^3$. We find three regular lattice ideals

$$G_1 = \{(0,y,z) \mid y,z \in \mathbb{R}\} ,$$
$$G_2 = \{(0,0,z) \mid z \in \mathbb{R}\} ,$$
$$G_3 = \{(0,y,0) \mid y \in \mathbb{R}\}$$

and $G^1 = V$, $G^2 = G^3 = G_1$. Therefore we define $\Lambda := \Gamma$.

Due to commutativity $a * G_{\lambda} * a^{-1} = G_{\lambda}$ for all $a \in G^{\lambda}$ and $\lambda \in \Lambda$ (G_{λ} is called a *normal* subgroup of G^{λ}). Therefore the *factor group* G^{λ}/G_{λ} defined by

$$G^{\lambda}/G_{\lambda} := \{a * G_{\lambda} \mid \lambda \in \Lambda\}$$

is well-defined with respect to the internal composition

$$(a * G_{\lambda}) \otimes (b * G_{\lambda}) = (a * b) * G_{\lambda} .$$

Due to convexity G^{λ}/G_{λ} is totally ordered by the well-defined induced ordering given by

$$(a * G_{\lambda}) \leq (b * G_{\lambda}) \iff a \leq b .$$

Using the fact that G^{λ} covers G_{λ} it can be seen that the factor group G^{λ}/G_{λ} is isomorphic to a subgroup of the additive group of real numbers. In our above example every G^{λ}/G_{λ} is isomorphic to $\mathbb{R}$.

For $\lambda,\mu \in \Lambda$ let $\lambda \geq \mu$ if and only if $\lambda = \mu$ or $G_{\lambda} \supseteq G^{\mu}$. It can be seen that $(\Lambda,\leq)$ is a root system, which has the following two properties:

(1) $\forall\, g \in G,\ g \neq e \quad \exists\, \lambda \in \Lambda \quad : \quad g \in G^{\lambda} \setminus G_{\lambda}$,

(2) $g \notin G^{\mu} \quad \Rightarrow \quad \exists\, \lambda \in \Lambda,\ \lambda > \mu : \quad g \in G^{\lambda} \setminus G_{\lambda}$.

An index λ in (1) is called a *value* of g. A subset $\tilde{\Lambda}$ of Λ satisfying (1) and (2) is called *plenary*. It can be seen that $\tilde{\Lambda}$ is a root system. Therefore $\bar{V}(G^{\lambda}/G_{\lambda},\tilde{\Lambda})$ is a lattice-ordered commutative group (cf. proposition 3.3).

In the above example Λ is order-isomorphic to Γ. Properties (1) and (2) are easily checked. Clearly, $\lambda = 1$ is a value of (1,0,0). Further Λ is the only plenary set and $\bar{V}(G^{\lambda}/G_{\lambda},\Lambda)$ is isomorphic to $V(\Gamma)$.

In general, a group-embedding φ: $G \to \bar{V}(G^{\lambda}/G_{\lambda},\tilde{\Lambda})$ is called *value-preserving* if $\lambda \in \tilde{\Lambda}$ is a value of $g \in G$ if and only if $\varphi(g)_{\lambda}$ is a maximal nonvanishing component of $\varphi(g)$ and, in this case, $\varphi(g)_{\lambda} = g * G_{\lambda}$. Such an embedding φ is always a lattice embedding; in particular $\varphi(G)$ is a sublattice of $\bar{V}$.

Now we can give a detailed version of the embedding result of CONRAD, HARVEY and HOLLAND [1963]. For a proof we refer to CONRAD [1970].

(3.4) Theorem

Let G be a divisible, lattice-ordered commutative group and let (G_λ, Λ) be the family of all regular lattice ideals of G. Let $\widetilde{\Lambda}$ be a plenary subset of Λ. Then G can be group-embedded into $V = V(\widetilde{\Lambda})$. This embedding is value-preserving and therefore the image of G is a sublattice of V.

In general, there may exist many different plenary sets and for each of them we get an embedding. The discussion in CONRAD, HARVEY and HOLLAND [1963] shows on the other hand that if Λ is finite then Λ is the only plenary set and G is isomorphic to $\bar{V}(G^\lambda/G_\lambda, \Lambda)$. Thus, for finite Λ lattice-ordered commutative groups are represented by $\bar{V}(H_\lambda, \Lambda)$, where H_λ is a subgroup of the additive group of real numbers and Λ a root system. Other examples with interesting properties are discussed in CONRAD, HARVEY and HOLLAND [1963].

If the root system Λ is linearly ordered then $V(\Lambda)$ is called a *Hahn-group*. Then $V(\Lambda)$ is a linearly ordered commutative group. In linearly ordered groups a lattice ideal is called *ideal*. Let G be a divisible, linearly ordered commutative group and let $(G_\lambda, \lambda \in \Lambda)$ be the family of all regular ideals of G. Then Λ is linearly ordered. Theorem (3.4) reduces to the theorem of HAHN [1907].

(3.5) Theorem

Let G be a divisible, linearly ordered commutative group and let $(G_\lambda, \lambda \in \Lambda)$ be the family of all regular ideals of G. Then G can be embedded in the divisible, linearly ordered commutative group $V(\Lambda)$.

For a particular proof of this theorem we refer to FUCHS [1966]. Theorem (3.4) and (3.5) are valuable characterizations of lattice-(linearly) ordered commutative groups and provide a general view of such ordered groups.

Our simple example illustrating the various definitions and denotations above can easily be extended to finite root systems Λ. In particular, if Λ is a linearly ordered set with n elements then $V(\Lambda)$ is isomorphic to the additive group $\mathbb{R}^n$ of real vectors which is linearly ordered by the *lexicographic order relation* of vectors defined by

$$(3.6) \qquad x \preccurlyeq y \quad :\Longleftrightarrow \quad x = y \quad \text{or} \quad x_i < y_i \quad \text{for } i = \min\{j \mid x_j \neq y_j\}.$$

We remark that for $\Lambda = \{\lambda_1, \lambda_2, \ldots, \lambda_n\}$ with $\lambda_1 > \lambda_2 > \ldots > \lambda_n$ the component x_{λ_i} for $x \in V(\Lambda)$ corresponds to x_i for $x \in \mathbb{R}^n$.

An infinite example is the additive group of all real functions $f: [0,1] \to \mathbb{R}$ with $(f+g)(x) := f(x) + g(x)$ and $f \leq g$ defined by $f(x) \leq g(x)$ for all $x \in [0,1]$. This is a divisible, lattice-ordered commutative group isomorphic to $V(\Lambda)$ for the root system $\Lambda = [0,1]$ trivially ordered by $x \leq y$ iff $x = y$.

Finally we consider a subgroup of the additive group of all

real functions $f: [0,1] \to \mathbb{R}$ with $f(0) = 0$. Let $x \in [0,1]$. A finite partition P_x of the interval $[0,x]$ can be represented by the set $\{\alpha_0, \alpha_1, \ldots, \alpha_r\}$ of its interval-endpoints; in particular $0 = \alpha_0 < \alpha_1 < \ldots < \alpha_r = x$. Now define

$$V(x,P_x) := \sum_{i=1}^{r} |f(\alpha_i) - f(\alpha_{i-1})|$$

for such a real function f. If $V(b,P_b)$ is bounded by a real constant β independent of P_b then

$$V(x) := \sup\{V(x,P_x) \mid P_x\}$$

exists and is called the *total variation* of f in $[0,x]$. In this case we say that f is of *bounded variation.*

Let G denote the set of all real functions $f: [0,1] \to \mathbb{R}$ of bounded variation with $f(0) = 0$. Clearly, this is a subgroup of the additive group of all real functions $f: [0,1] \to \mathbb{R}$. Its neutral element ($f \equiv 0$) will be denoted by 0.

A function $f: [0,1] \to \mathbb{R}$ is called *monotonically non-decreasing* if $x \leq y$ implies $f(x) \leq f(y)$. Such a function with $f(0) = 0$ is of bounded variation, in particular $V(x) = f(x)$ for all $x \in [0,1]$. The set G_+ of all such functions induces a partial order $\leq$ on G as $G_+ + G_+ \subseteq G_+$, $0 \in G_+$ and $G_+ \cap -G_+ = \{0\}$. In order to show that G is a lattice-ordered group it now suffices to show that $f \vee 0$ exists for all $f \in G$. Let

$$f_\pm(x) := \frac{1}{2}(V(x) \pm f(x)) \tag{3.7}$$

for $f \in G$. Clearly $f = f_+ - f_-$. Further $f_+, f_- \in G_+$ as

$$V(x) + |f(y) - f(x)| \leq V(y)$$

for all x,y with $x \leq y$. Therefore $f_+ \geq 0$, $f_+ \geq f$. Now assume that g is an arbitrary upper bound of f and 0. We will show $g \geq f_+$. As $g - f_+ = \frac{1}{2}(g - f) + \frac{1}{2}(g - V)$ and $g - f \geq 0$ it suffices to prove $g - V \geq 0$. Let $x \leq y$. Then $g - f \geq 0$, $g \geq 0$ implies

$$|f(y) - f(x)| \leq g(y) - g(x) .$$

Now for a partition P_y we define a corresponding partition $P_x := (P_y \cap [0,x]) \cup \{x\}$ of $[0,x]$. Then the above inequality used for points in $(P_y \cap [x,y]) \cup \{x\}$ implies

$$V(y,P_y) \leq V(x,P_x) + g(y) - g(x)$$

and therefore $g - V \geq 0$. Summarizing we have proved the following proposition.

(3.8) Proposition

Let G denote the additive group of all real functions $f: [0,1] \to \mathbb{R}$ of bounded variation with $f(0) = 0$. Let G_+ denote the set of all monotonically non-decreasing functions of G. Then G is a divisible, lattice-ordered commutative group with respect to the positive cone G_+. Meet and join are given by

$$f \wedge g = (f + g + V(f - g)) / 2 ,$$

$$f \vee g = (f + g - V(f - g)) / 2 .$$

For further examples of lattice-ordered commutative groups we refer to CONRAD [1970].

4. Linearly Ordered Commutative Divisor Semigroups

In this chapter we develop a characterization of linearly ordered commutative semigroups $(H,*,\leq)$ which satisfy the additional axiom

(4.1) $\quad a < b \Rightarrow \exists\, c \in H: \quad a * c = b \qquad$ (*Divisor rule*)

for all $a,b \in H$. An element $a \in H$ is called a *divisor* of $b \in H$ if there exists $c \in H$ with $a * c = b$. Thus (4.1) means that an element a is a divisor of all strictly greater elements. Therefore we call such semigroups *divisor semigroups*. For convenience we denote a linearly ordered commutative divisor semigroup shortly as *d-semigroup*.

Positively ordered d-semigroups have been characterized by CLIFFORD ([1954], [1958] and [1959]). D-semigroups which are not positively ordered were characterized by LUGOWSKI [1964] in a similar manner extending the results of CLIFFORD. The positively ordered case is also covered by the discussion of positively totally ordered semigroups in the monograph of SATYANARAYANA [1979].

From proposition (2.9) we know that we can assume w.l.o.g. the existence of a neutral element in a positively ordered d-semigroup. Later on we will see that a d-semigroup containing elements which are not positive is always a d-monoid. Therefore the discussion of d-monoids and d-semigroups leads to the same results.

A positive element $a \in H$ satisfies $b \leq a * b$ for all $b \in H$ (cf. 2.6). Since d-semigroups are linearly ordered there exists $b \in H$ with $b > a * b$ if an element a is not positive.

(4.2) Proposition

Let H be a d-semigroup and $a \in H$. Then the following two properties of a are equivalent:

(1) a is not positive,

(2) $a > a * a$.

Further both properties imply

(3) $b \geq a * b$ for all $b \in H$.

Proof. (1) ⇒ (2). Then $b > a * b$ for some $b \in H$ and $a * b * c = b$ for some $c \in H$. Therefore $a * b * c > a * (a * b * c)$ which implies (2) by cancellation of $b * c$. (2) ⇒ (1) is obvious.
(2) ⇒ (3). The case $a = b$ is obvious. Let $b < a$. Then $a = b * c$ for some $c \in H$. Thus $b * c > a * b * c$. Cancellation of c implies $b > a * b$. Let $a < b$. Then $a * c = b$ for some $c \in H$. Suppose $b < a * b$. Then $a * c < a * a * c$ implies $a < a * a$ contrary to (2).

■

Proposition (4.2) shows that an element a is negative if it is not positive. In general, if the d-semigroup is not a group, such an element is not strictly negative. Motivated by (4.2.2) we call it *self-negative*.

Let $(\Lambda, \leq)$ be a nonempty, linearly ordered set and let $(H_\lambda, *_\lambda, \leq_\lambda)$ be a linearly ordered commutative semigroup for each $\lambda \in \Lambda$. The sets H_λ are assumed to be mutually disjoint. Let H be the union of all H_λ. Then we continue internal compositions and order relations of the H_λ on H by

(4.3) $a < b$ and $a * b = b * a = b$

for all $a \in H_\lambda$, $b \in H_\mu$ with $\lambda < \mu$. $(H, *, \leq)$ is called the *ordinal sum* of the family $(H_\lambda;\ \lambda \in \Lambda)$.

It can easily be seen that $(H,*)$ is a commutative semigroup and $(H,\leq)$ is linearly ordered.

(4.4) Proposition

Let H be the ordinal sum of the family $(H_\lambda;\ \lambda \in \Lambda)$. Then

(1) H is a linearly ordered commutative semigroup if and only if all H_μ are positively ordered for $\mu \neq \min \Lambda$.

If H is a linearly ordered commutative semigroup, then

(2) H is positively ordered iff all H_λ are positively ordered,

(3) H is a d-semigroup iff all H_λ are d-semigroups.

Proof. (1) Let H be a linearly ordered commutative semigroup and let $a \in H_\lambda$, $b,c \in H_\mu$ with $\lambda < \mu$. Then $a < b$ implies $c = a * c \leq b * c$. For the reverse direction it suffices to show monotonicity. This follows easily by considering the possible cases $a,b \in H_\lambda$ and $c \in H$; $a \in H_\mu$, $b \in H_\lambda$ with $\mu < \lambda$ and $c \in H$. We always find that $a \leq b$ implies $a * c \leq b * c$.

(2) H positively ordered immediately implies that H_λ as a subset is positively ordered. As $a \in H_\lambda$, $b \in H_\mu$ with $\lambda < \mu$ implies $a < b = a * b$, the reverse direction immediately follows.

(3) Let H be a d-semigroup and let $a,b \in H_\lambda$ with $a < b$. Then $a * c = b$ for some $c \in H$. Now $c \notin H_\lambda$ implies $a * c = a$ or $a * c = c$, i.e. a contradiction. Thus H_λ is a d-semigroup. The reverse direction follows from $a * b = b$ for $a \in H_\lambda$, $b \in H_\mu$ with $\lambda < \mu$.

■

A d-semigroup is called *ordinally irreducible* or shortly *irreducible* if it is not the ordinal sum of two or more subsemigroups. The following theorem shows that each d-semigroup can uniquely be decomposed into irreducible subsemigroups; in other words each d-semigroup has a unique representation as the ordinal sum of a family of subsemigroups. Such a theorem was given for positively ordered commutative semigroups by KLEIN-BARMEN ([1942], [1943]) and CLIFFORD [1954], for naturally ordered commutative semigroups by CLIFFORD [1954], and for d-semigroups by LUGOWSKI [1964].

(4.5) Theorem

(1) Each d-semigroup H has a unique representation as ordinal sum of a family of irreducible d-subsemigroups.

(2) If H contains a self-negative element then there exists a first irreducible semigroup H_{λ_o} containing all self-negative elements of H.

Proof. We consider certain partitions (L,U) of H, i.e. $L \cap U = \emptyset$ and $L \cup U = H$. We call (L,U) a *cut* if $L < U$, if L is a subsemigrou and if $a * b = b$ for all $a \in L$, $b \in U$. The set of all cuts is totall ordered by $(L,U) < (V,W)$ iff $L \subsetneqq V$. A pair of cuts $(L,U) < (V,W)$ such that $(L,U) \leq (R,S) \leq (V,W)$ implies $(L,U) = (R,S)$ or $(R,S) = (V,W)$ is called *essential*. Then (V,W) is called the *immediate successor* of (L,U). Let Λ denote the set of all essential pairs. For $\lambda = [(L,U),(V,W)] \in \Lambda$ define the subset $H := V \cap U$.

Clearly V is a convex subsemigroup of H. If U contains only positive elements then U is also a convex subsemigroup of H. If $c \in U$ is self-negative then for $b \in L$ we find $b < c = b * c$. But this contradicts (4.2.3). Therefore $L = \emptyset$ and $H = U$.

Thus U is also a convex subsemigroup. This shows that H_λ is a convex subsemigroup. We claim that H is the ordinal sum of $(H_\lambda ; \lambda \in \Lambda)$ and that each H_λ is irreducible.

If H_λ is not irreducible but the ordinal sum of its subsemigroups $A_\lambda < B_\lambda$ then $(L \cup A_\lambda , B_\lambda \cup W)$ is a cut between (L,U) and (V,W) contrary to the definition of Λ. Clearly the ordinal sum of the family $(H_\lambda ; \lambda \in \Lambda)$ is well-defined. For $a \in H$ let A denote the union of all *lower classes* L of cuts (L,U) with $a \notin L$ and let B denote the intersection of all lower classes L of cuts with $a \in L$. Then $\lambda := [(A, H \setminus A), (B, H \setminus B)]$ is an essential pair and $a \in H_\lambda$. Therefore H is the ordinal sum of the family $(H_\lambda ; \lambda \in \Lambda)$.

If H contains self-negative elements then due to (4.4.2) we know that Λ has a minimum λ_o and H_{λ_o} contains all self-negative elements.

To prove that each H_λ is a d-subsemigroup of H let now $a,b \in H_\lambda$ with $a < b$. Then there exists $c \in H$ wich $a * c = b$. If $c \notin H_\lambda$ then either $a * c = a$ or $a * c = c$. In both cases $a * c \neq b$. Thus $c \in H_\lambda$.

Finally we show that this representation is unique. Assume that H is the ordinal sum of irreducible T_μ , $\mu \in M$. Then (L_r, U_r)

$$L_r := \bigcup_{\mu < r} T_\mu , \qquad U_r := \bigcup_{\mu \geq r} T_\mu$$

as well as $(L_r \cup T_r, U_r \setminus T_r)$ are cuts. As T_r is irreducible we find $[(L_r, U_r), (L_r \cup T_r, U_r \setminus T_r)] \in \Lambda$. This construction yields a bijection between the family of the H_λ and the family of the T_μ since both families are partitions of H.

∎

The family $(H_\lambda ;\ \lambda \in \Lambda)$ will be called the *ordinal decomposition* of H. From the proof it is clear that any positively, linearly ordered commutative semigroup has a unique ordinal decomposition, too. The only point in the proof using (4.1) was an argument treating the existence of self-negative elements. This case does not appear if H is positively ordered. Then all subsemigroups $(H_\lambda, *_\lambda, \leq_\lambda)$ are positively ordered, too.

If H is a d-semigroup then all irreducible subsemigroups are naturally ordered with the only possible exception of the first subsemigroup H_{λ_o}. An element of a semigroup is called *idempotent* if $a * a = a$.

For example, the neutral element is always idempotent.

(4.6) Proposition

An irreducible, positively ordered d-semigroup H contains at most one idempotent element a. Then $a = \max H$ and $a * b = a$ for all elements $b \in H$.

Proof. Assume the existence of an idempotent element a. Irreducibility implies that the only cuts are $(H,\emptyset)$ and $(\emptyset,H)$. Let $U := \{x > a\}$. Monotonicity implies that $L = H \setminus U$ and U are subsemigroups of H. Now let $x \in L$, $y \in U$. Then $a < y$ and $a * z = y$ for some $z \in H$. $x \leq a$ and H positively ordered imply $a \leq a * x \leq a * a = a$, i.e. $a = a * x$. Hence $x * y = x * a * z = a * z = y$. Therefore (L,U) is a cut. As $a \in L$ we find $U = \emptyset$. Then $a = \max H$ and $a \leq a * b \leq a * a = a$ for all $b \in H$ complete the proof.

■

In the positively ordered case we know that we can assume w.l.o.g. that $(H,*,\leq)$ contains a neutral element. Then $e \leq a$ for all $a \in H$ and thus $(\{e\}, H \setminus \{e\})$ is the immediate successor of $(\emptyset, H)$. This shows that Λ has a minimum λ_o with $H_{\lambda_o} = \{e\}$.
If $(H,*,\leq)$ is a d-semigroup and contains self-negative elements then from (4.5.2) again we know that Λ has a minimum λ_o.
Then let P_- denote the set of all self-negative elements and

$$P_+ := \{c \in H \mid \exists\, a,b \in P_- : a * c = b\} \setminus P_- .$$

Further $M := P_- \dot{\cup} P_+$ and $\bar{M} = H \setminus M$. Clearly, $P_- < P_+$.
The following result is due to LUGOWSKI [1964].

(4.7) <u>Proposition</u>

Let H be a d-semigroup with $P_- \neq \emptyset$. Then $M < \bar{M}$ and M is a linearly ordered commutative group with positive cone P_+.

<u>Proof</u>. At first we prove $M < \bar{M}$. This is trivial if $\bar{M} = \emptyset$. Thus let $\bar{M} \neq \emptyset$. Let $c \in M$ and $d \in \bar{M}$. Then $c \neq d$. If $c \in P_-$ then $c * d \leq d$ (cf. 4.2.3). Further d is positive, and in particular $c \leq c * d$. Thus $c \leq d$. If $c \in P_+$ then there exist $a,b \in P_-$ with $a * c = b$. Assume $d < c$. Then $d * x = c$ for some $x \in H$. Further $a * x \leq$ $\leq a * x * d = a * c = b$. But $a * x \leq b \in P_-$ implies $a * x \in P_-$. Thus $(a * x) * d = b$ implies $d \in P_+$ contrary to the choice of d.

Secondly, we prove that M is a d-subsemigroup of H with positive cone P_+. For example consider $a \in P_-$, $b \in P_+$ with $c * b = d$ for $c,d \in P_-$. Then $c * (a * b) = d * a \in P_-$ implies $a * b \in P_+$. Other cases can be treated with similar arguments. Thus M is a subsemigroup. Now let $a,b \in M$ with $a < b$ and $a * x = b$ for some $x \in H$. For example we consider $a,b \in P_+$. Then let $c * b = d$ with

$c,d \in P_-$. Now $c * a \leq c * a * x = c * b = d \in P_-$ implies $c * a \in P_-$ and therefore $x \in P_+$.

Thirdly, we prove the existence of a unique idempotent element, i.e. of $a \in M$ with $a^2 = a$. For $c \in P_-$ we find $c * c < c$ and therefore for some $x \in P_+$: $c * c * x = c$. Then $(c * x)^2 = c * x$ and $c * x \in P_+$. On the other hand P_+ is a naturally ordered commutative semigroup. Due to (4.5) it has a unique representation as ordinal sum of its irreducible subsemigroups. From (4.6) we know that each of these subsemigroups contains at most one idempotent which then is its maximum. Therefore for any idempotent $a \in P_+$ and all $b \in P_+$ we find

$$
\begin{aligned}
&b < a \quad \Rightarrow \quad b * a = a \\
&a < b \quad \Rightarrow \quad a * b = b .
\end{aligned}
\tag{4.8}
$$

Now let a,b denote two idempotent elements in P_+. W.l.o.g. assume $b < a$. Let $c,d \in P_-$ with $c * a = d$. Then $d = c * a = c * a^2 = d * a$. $P_- < P_+$ implies $d < b$. Therefore $d * x = b$ for some $x \in P_+$. Then

$$b * a = d * x * a = d * x = b < a$$

contrary to (4.8).

The unique idempotent of P_+ is now denoted by e. We show that e is a neutral element of M. If $c \in P_-$ then similarly as in the existence proof of an idempotent element we find $c * c * x = c$ and $c * x = e$. Thus $c * e = c$. If $c \in P_+$ and $e < c$ then (4.8) implies $c * e = c$. Now suppose the existence of $c \in P_+$, $c < e$. For $a \in P_-$ we find

$$a \leq a * c \leq a * e = a,$$

i.e. $a = a * c$. Since $a < c$ let $a * x = c$ for some $x \in P_+$. Then

$$c * c = a * x * c = a * x = c$$

implies $c = e$ contrary to the assumption $c < e$. Hence $e = \min P_+$. Finally we show the existence of inverse elements. Let $a \in P_-$. Then $a * x = e$ for some x and $x = a^{-1}$. For $e < a \in P_+$ let $c * a = d$. Then $a = d * c^{-1}$ and $a^{-1} = d^{-1} * c$ as

$$a * (d^{-1} * c) = d * d^{-1} = e.$$

■

Proposition (4.7) shows in particular that a d-semigroup with $P_- \neq \emptyset$ always contains a neutral element. If e is the neutral element of M then for $a \in H \setminus M$ there exists $x \in H$ with $e * x = a$. Then $a = e * x = e * e * x = e * a$ implies that the neutral element of M is a neutral element of H. The existence of a neutral element in a d-semigroup always implies the existence of a minimum λ_o of Λ. If $P_- = \emptyset$ then we may adjoin a neutral element. Therefore we can assume w.l.o.g. that a d-semigroup is a d-monoid and Λ has a minimum λ_o.

Further $M \subseteq H_{\lambda_o}$, i.e. M is always contained in the first irreducible semigroup H_{λ_o} of the ordinal decomposition of H. Suppose the opposite, i.e. let $b \in M \setminus H_{\lambda_o}$. Then $a * b = b$ for all $a \in M$. For $a \neq e$ this implies $a * b = e * b$. Multiplication with b^{-1} yields $a = e$ contrary to the choice of a.

(4.9) <u>Proposition</u>

Let H be a d-semigroup with $P_- \neq \emptyset$. Then H is a d-monoid and

(1) $M * \bar{M} = M$,

(2) $a \in M$, $b \in \bar{M}$ and $a * c = b$ imply $c \in \bar{M}$,

(3) H is the ordinal sum of M and $\bar{M}$ if and only if $\bar{M}$ is naturally ordered,

(4) if $\bar{M}$ has a minimum then $\bar{M}$ is naturally ordered.

Proof. (1) Let $a \in M$, $b \in \bar{M}$. Assume $a * b \in M$. Then $a^{-1} * (a * b) \in M$ contrary to $b \in \bar{M}$. Thus $M * \bar{M} \subseteq \bar{M}$. $\{e\} * \bar{M} = \bar{M}$ shows the reverse inclusion.

(2) Let $a \in M$, $b \in \bar{M}$ and $a * c = b$. Then $c = a^{-1} * (a * c) = a^{-1} * b \in \bar{M}$.

(3) If H is the ordinal sum of M and $\bar{M}$ then let $a, b \in \bar{M}$ with $a < b$. For some $c \in H$ we find $a * c = b$. If $c \in M$ then $a * c = a$ contrary to $a < b$. Thus $c \in \bar{M}$. As $e < \bar{M}$ we know that $\bar{M}$ is positively ordered. For the reverse direction we assume that $\bar{M}$ is naturally ordered. From (4.7) we know $M < \bar{M}$, M is a group and, as $\bar{M}$ is positively ordered, $\bar{M}$ is a subsemigroup of H. It suffices to show that $a * b = b$ for all $a \in M$, $b \in \bar{M}$. Let $b \in \bar{M}$. For $a = e$ this is trivial. Let $e < a$. Then $b \leq a * b$. Assume $b < a * b$. Then $b * x = a * b$ for some $x \in \bar{M}$. Now $a^{-1} * b * x = b < a * b$ implies by cancellation $a^{-1} * x < a \in M$; in particular $a^{-1} * x \in M$ contrary to (2). Now consider the case $a < e$. Then $e < a^{-1}$ and therefore $a^{-1} * b = b$. Composition with a yields $b = b * a$.

(4) It suffices to show $a * b = b$ for all $a \in M$, $b \in \bar{M}$. Then H is the ordinal sum of M and $\bar{M}$ and we can apply (3). Let $a \in M$, $b \in \bar{M}$. Let d denote the minimum of $\bar{M}$. From (2) we know $a * x = d$ for some $x \in \bar{M}$. Assume $e < a$. Then $x \leq d$ implies $x = d$. In the case $a < e$ we know $e < a^{-1}$ and therefore $a^{-1} * d = d$. Again this implies $d = d * a$. Now let $d * y = b$ for $y \in H$. Then

$$a * b = a * d * y = d * y = b$$

for all $a \in M$.

■

We can now describe the ordinal decomposition of d-semigroups in more detail. W.l.o.g. we assume that a d-semigroup is a d-monoid (possibly after adjoining a neutral element in the case $P_- = \emptyset$).

The following result is due to LUGOWSKI [1964].

(4.10) Theorem (Decomposition of d-monoids)

Let H be a d-monoid. Then:

(1) H has a unique ordinal decomposition $(H_\lambda ; \lambda \in \Lambda)$ and Λ has a minimum λ_o.

(2) $H \setminus H_{\lambda_o}$ and all H_λ, $\lambda \neq \lambda_o$ are naturally ordered.

(3) If $P_- = \emptyset$ then $H_{\lambda_o} = \{e\}$.

(4) If $P_- \neq \emptyset$ and H_{λ_o} is a group then $H_{\lambda_o} = M$.

(5) If $P_- \neq \emptyset$ and H_{λ_o} is not a group then

a) $M \subsetneqq H_{\lambda_o}$ and $M < R := H_{\lambda_o} \setminus M$,

b) R is a positively ordered commutative semigroup, is not naturally ordered, and has no minimum,

c) R is irreducible and contains at most one idempotent element a, which then fulfills $a = \max H_{\lambda_o}$ and $a * b = a$ for all $a \in H_{\lambda_o}$.

Proof. (4.5), (4.7) and (4.9) imply (1) - (4) and (5a). $R = H_{\lambda_o} \cap \bar{M}$ shows that R is a positively ordered commutative semigroup. If R contains a minimum or is naturally ordered then H_{λ_o} is the ordinal sum of M and R. As H_{λ_o} is irreducible we find (5b).

Now we suppose that R is the ordinal sum of S,T with $S < T$.

Then consider $M \cup S, T$. Clearly $M \cup S < T$. $M \cup S$ is a subsemigroup as $a * b \leq b * b \in S$ for $a \in M$, $b \in S$. S is a subsemigroup due to our assumption. Let $a \in M \cup S$, $b \in T$. For $a \in S$ we know $a * b = b$ due to our assumption. Now consider $a \in M$. If $a \in P_+$ then for $c \in S$ we find $a * x = c$ for some $x \in \bar{M}$ (use 4.9.2). $a \in P_+$ implies $x \leq c$ and therefore $x \in S$. Thus $x * b = b$ which implies $b = c * b = a * x * b = a * b$. For $a \in P_-$ we know $a^{-1} \in P_+$. Thus $a^{-1} * b = b$. Multiplication with a yields $b = b * a$. Hence H_{λ_o} is the ordinal sum of $M \cup S$ and T contrary to H_{λ_o} irreducible. Thus R is irreducible.

Now assume that R contains an idempotent element a and consider the partition $S := \{b \in R \mid b \leq a\}$, $T = R \setminus S$. Clearly $S < T$ and S and T are subsemigroups of R. Let $b \in T$. Then $a * x = b$ for some $x \in H_{\lambda_o}$. Therefore $a * b = a * a * x = a * x = b$. Let $c \in S$. Then $a \leq a * c \leq a * a = a$ shows $a = a * c$. Thus we find $c * b = c * a * b = a * b = b$. As R is irreducible and $a \in S$ we know $T = \emptyset$. Therefore $a = \max R = \max H_{\lambda_o}$. For $c \in R$ we have already shown $a * c = a$. Let $a \in P_+$. Then $a \leq a * c \leq a * a = a$. Therefore $a * c = a$. For $c \in P_-$ we know $c^{-1} \in P_+$ and thus $c^{-1} * a = a$. Again multiplication with c yields $a = a * c$.

■

The following example illustrates the situation occurring in (4.10.5). Let $H = \{0,1,...,n\} \times \mathbb{Z} \cup \{(n+1,0)\}$ for $n \geq 1$. Then $(H, *, \preccurlyeq)$ is a d-monoid with internal composition defined by

$$(a,b) * (c,d) := \begin{cases} (a+c, b+d) & \text{if } a+c < n+1, \\ (n+1, 0) & \text{otherwise} \end{cases}$$

and with lexicographical order relation $\preccurlyeq$. We find

$$P_- = \{0\} \times \{z \mid z < 0\}, \qquad P_+ = \{0\} \times \{z \mid z \geq 0\}.$$

Now we will show $H = H_{\lambda_0}$. We know that always $M = P_- \cup P_+ \subseteq H_{\lambda_0}$.
For $(0,z) \in M$ with $z \neq 0$ and $(x,y) \in \bar{M} \setminus \{(n+1,0)\}$ we find

$$(0,z) + (x,y) \neq (x,y).$$

Therefore an upper class U of a cut of H can contain at most one element, i.e. $(n+1,0)$. But then $(1,0)$ is an element of the lower class L and therefore $(1,0)^{n+1} = (n+1,0)$ implies $U = \emptyset$. Hence H is irreducible.

Obviously H is not a group. Further $R = \bar{M}$. All statements in (4.10.5) are easily verified. In particular $(n+1,0)$ is the only idempotent element in R.

Theorem (4.10) shows how we may construct examples from a set of naturally ordered d-semigroups and a first semigroup as described in (4.10.5). In fact, if we can characterize such semigroups in the irreducible case then we have characterized all d-monoids. Unfortunately such a result is not known in general. Nevertheless, if additional properties are assumed, then such characterizations have been developed.

A d-semigroup H is called *cancellative* if it fulfills the cancellation law (2.4), i.e.

$$a * c = b * c \quad \Rightarrow \quad a = b$$

for all $a,b,c \in H$. From theorem (2.8) we know that if H is positively ordered then $H_+ = \{a \mid e \leq a\}$ is isomorphic to the positive cone of a linearly ordered group.

(4.11) Proposition

Let H be a cancellative d-monoid. Then H can be embedded in a linearly ordered commutative group G such that H_+ is isomorphic to the positive cone G_+ of G.

Proof. If H is positively ordered then we refer to proposition (2.8). Otherwise $P_- \neq \emptyset$.

Now H_+ is naturally ordered. Therefore H_+ is isomorphic to the positive cone G_+ of a linearly ordered group G. Let $\varphi: H_+ \to G_+$ denote the isomorphism and extend φ to $\tilde{\varphi}: H \to G$ by $\varphi(a^{-1}) = (\varphi(a))^{-1}$ for all $a^{-1} \in P_-$. As $a^{-1} \in P_-$ iff $a \in P_+$ this is well-defined. $\tilde{\varphi}$ is an embedding of H into G.

■

An ordinal sum H of cancellative d-semigroups fulfills the *weak cancellation law*

$$a * c = b * c \Rightarrow a = b \vee a * c = c,$$

for all $a,b,c \in H$. A d-semigroup fulfilling this law is called *weakly cancellative*. Vice versa we may ask whether all weakly cancellative d-semigoups have an ordinal decomposition in cancellative d-semigroups.

(4.12) Proposition

Let H be an irreducible weakly cancellative d-semigroup. Then H is cancellative.

Proof. If $H = \{a\}$ then H is the trivial group. Now we assume $|H| > 1$. At first let H be naturally ordered. Then H can contain at most one idempotent a and then $a * b = a$ for all $b \in H$ (cf. 4.6) Let $L = H \setminus \{a\}$ and $U = \{a\}$. Then $L < U$ (cf. 4.6) and if L is a subsemigroup then H is the ordinal sum of L and U. Therefore for some $b,c < L$ we find $a * b = b * c = a$. But then weak cancellation implies $a = c$ contrary to the choice of c. Therefore H has no idempotent element.

Now define a binary relation by

$$a \sim b \quad :\Longleftrightarrow \quad \max(a,b) < a * b \,.$$

Clearly τ is reflexive and symmetric. Let $a,b,c \in H$ and assume w.l.o.g. $a < c$. If $\max(a,c) = a * c$ then $a * b = a * b * c$. Weak cancellation implies $a * b = a$ or $b = b * c$. Therefore $\sim$ is transitive. Let L denote an equivalence class of $\sim$. Then for $a,b \in L$ we find $a * b \in L$ since weak cancellation leads to $a \sim a * b$. If $a < c < b$ then $b < a * b \leq a * c$ shows $a \sim c$, i.e. $c \in L$. Thus equivalence classes are convex subsemigroups. Let U be another equivalence class and let $a < d$ for $a \in L$, $d \in U$ w.l.o.g. Then $L < U$ leads to an ordinal decomposition of H. Therefore $\sim$ has only one equivalence class H. Then $a,b \in H$ satisfy $a < a * b$ which implies that H is cancellative.

Secondly, we consider the case that H contains self-negative elements. Then $H = P_- \cup P_+ \cup R$ as given in (4.10.4) or (4.10.5). If $R = \emptyset$ then H is a group and thus cancellative. We will show that the naturally ordered d-subsemigroup $\bar{H} := (P_+ \setminus \{e\}) \cup R$ is irreducible. Suppose $\bar{H}$ is the ordinal sum of $A < B$. As $A \neq \emptyset$ and $a < a * b$ for all $a,b \in P_+$ we find $(P_+ \setminus \{e\}) \subseteq A$. Now consider $A' := P_- \cup \{e\} \cup A$. If $\bar{H}$ is the ordinal sum of A and B then H is the ordinal sum of A' and B. Therefore $\bar{H}$ has to be irreducible. Then $\bar{H}$ is cancellative which implies that H_+ is cancellative. Now let $a,b,c \in H$ and $a * b = a * c$. If $a \notin R$ then composition with a^{-1} shows $b = c$. Assume $a \in R$. If $b,c \in H_+$ then we know $b = c$. Assume w.l.o.g. $b \in P_-$. Then $a = a * c * b^{-1}$. If $c \in R$ then cancellation in H_+ leads to $e = c * b^{-1} \in R$ (see 4.9.1) contrary to $e \notin R$. Assume $c \notin R$. Then $a * c^{-1} = a * b^{-1}$ leads to $c^{-1} = b^{-1}$, i.e. $b = c$. ■

CONRAD [1960] shows that a weakly cancellative, positively ordered d-monoid is the ordinal sum of cancellative, positively ordered d-monoids. Theorem (4.10) and propositions (4.11) and (4.12) imply (1) and (2) in the following more general theorem.

(4.13) Theorem

Let H be a weakly cancellative d-monoid. Then

(1) H has a unique ordinal decomposition $(H_\lambda ; \lambda \in \Lambda)$ and Λ has a minimum λ_o,

(2) all H_λ are cancellative and can be embedded in groups G_λ, and

(3) for $\lambda > \lambda_o$ all H_λ are isomorphic either to the strict positive cone of G_λ or to the trivial group G_λ.

Proof. It remains to show (3). From (4.10) we know that H_λ is positively ordered. If H_λ contains a neutral element e_λ then $e_\lambda \leq a$ for all $a \in H_\lambda$. On the other hand e_λ is idempotent and due to (4.6) we find $a \leq e_\lambda$ for all $a \in H_\lambda$. Then H_λ is isomorphic to the trivial group, i.e. $H_\lambda = \{e_\lambda\}$. Otherwise we adjoin a neutral element to H_λ. Then from (4.11) it follows that H_λ is isomorphic to the strict positive cone $(G_\lambda)_+ \setminus e_\lambda$ where e_λ denotes the neutral element of G_λ. ■

We may assume w.l.o.g. that all G_λ, $\lambda \in \Lambda$ are mutually disjoint. Let $F_{\lambda_o} := G_{\lambda_o}$ and let for $\lambda > \lambda_o$ either $F_\lambda := G_\lambda$ (iff H_λ is isomorphic to the trivial group) or $F_\lambda := (G_\lambda)_+ \setminus \{e_\lambda\}$. Then the ordinal sum F of F_λ , $\lambda \in \Lambda$ is also a weakly cancellative d-monoid. We will assume later on that a weakly cancellative d-monoid is always given in this form. Clearly H can be embedded in F. In the same manner as we previously adjoined a neutral element, here we adjoin the possibly missing inverse element of the first irreducible subsemigroup.

We remark that for the irreducible case a characterization of weakly cancellative d-monoids follows from proposition (4.2) and theorem (4.5), i.e. all cancellative d-monoids can be embedded into a Hahn-group.

CLIFFORD ([1958], [1959]) characterizes another class of irreducible d-monoids. A d-monoid H is called *conditionally complete* if each subset A which is bounded from above has a least upper bound. Then each A which is bounded from below has a greatest lower bound. CLIFFORD considers conditionally complete d-monoids which are positively ordered. Due to the positive order relation the first subsemigroup in the ordinal decomposition is always the trivial group {e}. Therefore we will consider only d-semigroups without neutral element. A positively ordered d-semigroup H is called *Archimedean* if

(4.14) $\qquad \exists\, n \in \mathbb{N} : \qquad a^n \geq b$

for all $a,b \in H$, $a \neq e$. The following results are due to CLIFFORD ([1958], [1959]).

(4.15) Proposition

Let H be an irreducible, conditionally complete positively ordered d-semigroup without neutral element. Then H is Archimedean.

Proof. Let $a,b \in H$. If $a^n = a^{n+1}$ for some $n \in \mathbb{N}$ then a^n is idempotent and due to (3.6) the maximum of H. Therefore $a^n \geq b$.

Otherwise $a^n < a^{n+1}$ for all $n \in \mathbb{N}$. Suppose $a^n < b$ for all $n \in \mathbb{N}$. Then $c := \sup\{a^n \mid n \in \mathbb{N}\}$ exists and $a^n < c$ for all $n \in \mathbb{N}$ ($a^n = c$ implies $a^{n+1} = a^n$). Now $a * d = c$ for some $d \in H$. If $d < c$ then $d < a^n$ for some n and therefore $a * d \leq a^{n+1} < c$. Thus $c = d$, i.e. $a * c = c$.

Let $L := \{g \mid g < c\}$. For $x,y \in L$ we find $\max(x,y) < a^n$ for some $n \in \mathbb{N}$ and therefore $x * y \leq a^{2n}$ implying $x * y \in L$. Further $x * c \leq a^n * c = c$ shows $x * c = c$. Let $U := \{g \mid g \geq c\}$. Clearly positivity implies that U is a subsemigroup. Let $y \in U$, $c < y$. Then $c * z = y$ for some $z \in H$. Suppose $z \in L$. Then $c * z = c$ contrary to $c < y$. Therefore $z \notin L$. Let $x \in U$. Then $x * y = x * c * z = c * z = y$. Thus we have found that H is an ordinal sum of L and U, contrary to our assumption. Therefore H is Archimedean.

■

Now Archimedean, positively ordered d-semigroups have been characterized by HÖLDER [1901] and CLIFFORD [1954]. It should be noted that an Archimedean, linearly ordered divisor semigroup always is commutative (HÖLDER [1901], LUGOWSKI [1964]). Thus our assumption of commutativity is redundant.

The following result is due to CLIFFORD [1954]:

(4.16) Proposition

Let H be an Archimedean, positively ordered d-semigroup.

If H is not cancellative then

(1) H contains a maximum u,

(2) for all $a \neq e$ there exists $n \in \mathbb{N}$ such that $a^n = u$,

(3) if $a * b = a * c < u$ for $a,b,c \in H$ then $b = c$.

Proof. Due to the assumption there exist $a,b,c \in H$ with $a * b = a * c =: u$ and $b < c$. Now $b * x = c$ for some $x \neq e$. Thus $u * x = u$. Suppose $u < y \in H$ then $y \leq x^n$ for some $n \in \mathbb{N}$. Therefore we find the contradiction $u = u * x^n \geq u * y \geq y > u$. This shows (1) and (3). Now for $a \in H$ $(a \neq e)$ we know $a^n \geq u$ for some $n \in \mathbb{N}$, i.e. $a^n = u$.

■

Now the following theorem characterizes Archimedean, positively ordered d-semigroups. The result is due to HÖLDER [1901] and CLIFFORD [1954]. Its proof is drawn from FUCHS [1966].

(4.17) Theorem

Let H be an Archimedean, positively ordered d-semigroup.

(1) If H is cancellative, then it is isomorphic to a subsemigroup of the positive cone $(\mathbb{R}_+, +, \leq)$ of the real numbers. In particular if H has a minimum and contains no neutral element then H is isomorphic to the additive semigroup of natural numbers $(\mathbb{N}, +, \leq)$.

(2) If H has a maximum u and u has no immediate predecessor then H is isomorphic to a subsemigroup of $([0,1], *, \leq)$ with respect to the usual ordering and $a * b := \min(a+b, 1)$. In particular, if H has a minimum and contains no neutral element, then H is isomorphic to $\{k/n \mid 0 < k \leq n\}$ for some $n \in \mathbb{N}$.

(3) If H has a maximum u and u has an immediate predecessor then H is isomorphic to a subsemigroup of $([0,1]\cup\{\infty\},\oplus,\leq)$ with respect to the usual ordering and $a\oplus b := a+b$ if $a+b\leq 1$ and $a\oplus b := \infty$ otherwise.

<u>Proof</u>. We always discard the neutral element e if it exists in H. At first we assume that H contains a minimum a. Let $b\in H$, $b\neq a$. Then $a^k < b \leq a^{k+1}$ for some $k\in\mathbb{N}$ and $a^k * c = b$ for some $c\in H$. Now $a^{k+1} = a^k * a \leq a^k * c = b \leq a^{k+1}$ shows $a^{k+1} = b$. Therefore H is generated by a. If $a^n < a^{n+1}$ for all $n\in\mathbb{N}$ then H is isomorphic to $(\mathbb{N},+,\leq)$. Otherwise $a < a^2 < \ldots < a^n = a^{n+1}$ for some $n\in\mathbb{N}$. Then H is isomorphic to $(\{k/n \mid 1\leq k\leq n\},*,\leq)$.

Secondly, we assume that H does not contain a minimum. For $a\in H$ let $b<a$. Then $b*c = a$ for some $c\in H$. Let $z = \min(b,c)$. Then $z^2\leq a$. Therefore for all $t\in\mathbb{N}$ there exists $z\in H$ such that $z^t\leq a$. Now choose a fixed element $v\in H$ ($v\neq u$ if u exists). In the following we define a function $f: H\to\mathbb{R}_+$. Let $a\in H$ ($a\neq u$) and define two subsets of $\mathbb{N}\times\mathbb{N}$ by

$$L_a := \{(m,n) \mid v\leq x^n \text{ and } x^m\leq a \text{ for some } x\in H\},$$

$$U_a := \{(k,l) \mid y^l\leq v \text{ and } a\leq y^k \text{ for some } y\in H\}.$$

As H is Archimedean $L_a, U_a \neq \emptyset$. We claim that

(I) $\qquad L_a\leq U_a$,

and that for all $t\in\mathbb{N}$ there exist $s,r\geq t$ such that

(II) $\qquad (s,r+1)\in L_a, \qquad (s+1,r)\in U_a$.

For $t\in\mathbb{N}$ let $z^t\leq\min(b,c)$. For $b,c\in H$ we define $r,s\in\mathbb{N}$ by $z^r\leq b<z^{r+1}$ and $z^s\leq c<z^{s+1}$. Then $r,s\geq t$. In particular, for $b:=v$ and $c:=a$ we find (II). Further for $b:=x$ and $c:=y$

in L_a resp. U_a corresponding to (m,n) resp. (k,l) we find

$$z^{rm} \leq a < z^{(s+1)k}, \quad z^{sl} \leq v < z^{(r+1)n} .$$

Therefore rm < (s+1)k and sl < (r+1)n. This implies

$$m/n < (1 + 1/r)(1 + 1/s)\ k/l .$$

As t can be chosen arbitrarily large and $r,s \geq t$ we find (I). Further the distance between the elements in (II) is arbitrarily small for sufficiently large t. Therefore there exists a unique real number α separating L_a and U_a. Now we define $f(a) := \alpha$. In particular $f(v) = 1$.

Further $f(a) > 0$ as $U_a \neq \emptyset$ $(e \neq u \neq a)$. For $a < b$ $(\neq u)$ let $a * c = b$. Then for $t \in \mathbb{N}$ let $z^t \leq \min\{a,b,c,v\}$ and define r,s,μ,λ by $z^r \leq v < z^{r+1}$, $z^s \leq a < z^{s+1}$, $z^{\mu} \leq c < z^{\mu+1}$ and $z^{\lambda} \leq b < z^{\lambda+1}$. Then $t \leq r,s,\mu,\lambda$. Again $(s+1,r) \in U_a$ and $(\lambda,r+1) \in L_b$. Now $z^{\mu+s} \leq a * c = b < z^{\lambda+1}$ implies $\mu + s < \lambda + 1$. Therefore

$$\lambda/(r+1) > (\mu+s)/(r+1) \geq (t+s)/(r+1) .$$

For sufficiently large t we see $(t+s)/(r+1) > (s+1)/r$. Therefore $(\lambda,r+1) \in U_a$. This shows $f(a) < f(b)$. Now (II) shows the existence of sequences in L_a, U_a with limit $f(a)$. Therefore, if $a * b < u$, then we find sequences in L_{a*b}, U_{a*b} with limit $f(a) + f(b)$. Thus $f(a * b) = f(a) + f(b)$. Hence f is an embedding of H (without u) into $\mathbb{R}_+$.

If H is cancellative then from (4.16) we know that H contains no neutral element. Therefore f embeds H into $\mathbb{R}_+$. Otherwise H contains a maximum u. Then f(H) is bounded; in particular, if $a^n = u$, then $f(H) \leq n$. Let $\sup f(H) = \gamma$. If $f(a) < \gamma$ for all $a \in H \setminus \{u\}$ then we define $f(u) := \gamma$. Otherwise we define $f(u) := \infty$.

Then g, defined by $g(a) = f(a)/\gamma$, embeds H into $[0,1]$ or $[0,1] \cup \{\infty\}$ with respect to (2) or (3).

■

In this way all irreducible conditionally complete, positively ordered d-semigroups are characterized. CLIFFORD [1959] gives also an example for an irreducible positively ordered d-semigroup which is not Archimedean and thus not conditionally complete. Further he considers certain completions of such semigroups. Let H be a positively ordered d-semigroup. If H is a subsemigroup of a conditionally complete, positively ordered d-semigroup $\bar{H}$ such that for all $a \in \bar{H}$ there exist subsets A,B of H with $\sup A = a = \inf B$ then $\bar{H}$ is called a *normal completion* of H. If H has a normal completion $\bar{H}$ which is isomorphic to a subsemigroup of one of the semigroups in (4.17) then we say that H can be *normally embedded* into one of these semigroups. Clifford shows that this is possible if and only if H is also Archimedean.

We remark that the five semigroups in theorem (4.17) are conditionally complete and have no conditionally complete subsemigroups with the exception of the trivial group. Each is the ordinal sum of its neutral element and the set of all other elements. We remark that then the obtained discrete semigroup in (4.17.2) for $k = n = 1$ is isomorphic to the trivial group. Therefore each irreducible conditionally complete, positively ordered d-semigroup is isomorphic to one of these five semigroups (after discarding the neutral element), called the *fundamental semigroups*. Thus every conditionally complete, positively ordered d-semigroup is an ordinal sum of fundamental semigroups. CLIFFORD [1958] shows that then the index set Λ of the family is conditionally complete, too. The possible choice of the fundamental semigroups for $\lambda \in \Lambda$ depends on the local structure of Λ; in particular, it is important whether λ has an immediate predecessor or successor (cf. 4.23).

Although we have now quite well characterizations of d-monoids, in particular of cancellative and conditionally complete ones, for practical applications as in part II we need a list of properties which are always fulfilled in d-monoids. Such properties will be used in the discussion of methods proposed for the solution of the optimization problems considered.

For $a \in H$ we consider the following sets

$$N(a) := \{b \in H \mid a > a * b\},$$
$$neg(a) := \{b \in H \mid a \geq a * b\},$$
$$dom(a) := \{b \in H \mid a = a * b\},$$
$$pos(a) := \{b \in H \mid a \leq a * b\},$$
$$P(a) := \{b \in H \mid a < a * b\}.$$

The elements of these sets are called *strictly negative*, *negative*, *dominated*, *positive* and *strictly positive* (*with respect to* a).

(4.18) Proposition

Let H be a d-monoid. Let $a, c \in H$ and $b := a * c$. Then

(1) $N(a) < dom(a) < P(a)$ is an ordered partition of H;

(2) $dom(a) \subseteq dom(b)$, $pos(a) \subseteq pos(b)$, $P(b) \subseteq P(a)$, $neg(a) \subseteq neg(b)$, $N(b) \subseteq N(a)$;

(3) $P(c) \subseteq pos(a)$, $N(c) \subseteq neg(a)$;

(4) dom(a) and pos(a) are d-monoids;

(5) neg(a) is a linearly ordered commutative monoid.

Proof. (1) It suffices to prove the inequalities. Let $x \in dom(a)$ and $y \in P(a)$. Then $a * x = a < a * y$ implies by cancellation $x < y$. $N(a) < dom(a)$ is similarly proved. (2) is an immediate conclusion from the definitions and (1).

(3) Let $x \in P(c)$. Then $c * x > c = c * e$. Therefore $x > e$. Then for $a \in H$ we get $a * x \geq a$, i.e. $x \in pos(a)$. $N(c) \subseteq neg(a)$ is shown in the same way.

(4) At first consider dom(a). Let $x, y \in dom(a)$. Then $a * (x * y) = a * x = a$. Now assume $x < y$ and $x * c = y$ for some $c \in H$. Then $a * c = a * x * c = a * y = a$ shows $c \in dom(a)$. Now let $x, y \in pos(a)$. Then $a \leq a * y \leq a * x * y$. Assume $x < y$ and $x * c = y$ for some $c \in H$. Then $x < x * c$ shows $c \in P(x) \subseteq pos(a)$ (cf. (3)).

(5) Similarly as in (4) the discussion of pos(a).

■

We remark that in general neg(a) is not a d-semigroup. This is due to the asymmetry in the divisor rule. For example assume $P_- \neq \emptyset$ and let $a \in P_-$. Then $a < e$ and for a solution $c \in H$ with $a * c = e$ we know $c = a^{-1} > e$. We may consider linearly ordered commutative semigroups $(H,*,\leq)$ which fulfill

$$(4.1)' \qquad a < b \Rightarrow \exists\, c \in H: \quad a = b * c$$

for all $a, b \in H$. Then $(H,*,\leq)$ is called a *dual d-semigroup*. If $(H,*,\leq)$ is a d-semigroup and a dual d-semigroup then $(H,*,\leq)$ is a linearly ordered commutative group. All results for d-semigroups can be translated into a result for the corresponding dual d-semigroup which we get by reversing the order relation of H.

Now we define for a d-monoid H with ordinal decomposition $(H_\lambda ; \lambda \in \Lambda)$ the *index* $\lambda(a)$ of an element $a \in H$ by $\lambda(a) := \mu$ if a is an element of H_μ.

(4.19) Proposition

Let H be a weakly cancellative d-monoid with ordinal decomposition $(H_\lambda ; \lambda \in \Lambda)$.

(1) If $\lambda(a) \leq \min(\lambda(b),\lambda(c))$ then $a * b = a * c$ implies $b = c$.

(2) If $\lambda(a) \leq \min(\lambda(b),\lambda(c))$ and $d \leq a$ then $a * b \leq d * c$ implies $b \leq c$.

(3) For $a < b$ there exists a unique $c \in H$ with $a * c = b$. Then $\lambda(c) = \lambda(b)$.

(4) If $\lambda(a) = \lambda(b)$ and $a * b = a$ then $b = e$ or $H_{\lambda(a)} = \{a\}$.

(5) If $a^n = b^n$ then $a = b$.

(6) If $b_j \leq a$ for $j = 1,2,\ldots,n$ and $a \leq b := b_1 * b_2 * \ldots * b_n$ then $\lambda(a) = \lambda(b)$.

(7) Let $a := a_1 * a_2 * \ldots * a_n$, $b := b_1 * b_2 * \ldots * b_n$. If $a * c < b * c$ then there exists μ with $a * c < b_\mu * (\ast_{j \neq \mu} a_j) * c$.

Proof.

(1) Assume $b \neq c$. Then $a * b = a = a * c$. Therefore $\lambda(b) \leq \lambda(a)$ and $\lambda(c) \leq \lambda(a)$. Then $\lambda(a) = \lambda(b) = \lambda(c)$ yields a contradiction as the irreducible H_λ are cancellative.

(2) From $a * b \leq d * c$ and $d \leq a$ we find $a * b \leq a * c$. Now (1) implies $b = c$ if equality holds. Otherwise $b < c$.

(3) We know $a * c = b$ for some $c \in H$. Then $\lambda(a),\lambda(c) \leq \lambda(b)$.

Suppose $\lambda(c) < \lambda(b)$. Then $\lambda(a) = \lambda(b)$ and therefore $a * c = a < b$ contrary to the choice of c. Thus $\lambda(c) = \lambda(b)$. If $\lambda(a) < \lambda(b)$ then $c = b$. Otherwise $\lambda(a) = \lambda(b) = \lambda(c)$ and c is uniquely determined as $H_{\lambda(a)}$ is cancellative.

(4) This follows from (4.13.3). If $\lambda(a) = \lambda_o$ then $b = e$. Otherwise suppose that $H_{\lambda(a)}$ is not the trivial group. Then $H_{\lambda(a)}$ is isomorphic to the strict positive cone of a group. Hence $a * b > a$ contrary to the assumption.

(5) $a^n = b^n$ implies $\lambda(a) = \lambda(b)$. Now $H_{\lambda(a)}$ is cancellative and thus $a = b$.

(6) W.l.o.g. assume $b_n = \max\{b_j \mid j = 1,2,\ldots,n\}$. Then $\lambda(b) = \lambda(b_n) \leq \lambda(a)$. On the other hand $a \leq b$ implies $\lambda(a) \leq \lambda(b)$.

(7) Suppose that $a * c \geq b_\mu * (\ast_{j \neq \mu} a_j) * c$ for all $\mu = 1,2,\ldots,n$. Then $a^n * c^n \geq b * a^{n-1} * c^n$. If ">" holds then cancellation leads to the contradiction $a > b$. Otherwise $\lambda(b * c) \leq \lambda(a * c)$. On the other hand $a * c < b * c$ implies $\lambda(a * c) \leq \lambda(b * c)$. Then cancellation in $H_{\lambda(a*c)}$ implies $a * c = b * c$ contrary to the assumption.

■

In the remaining part of this chapter we give examples for different classes of d-monoids. The relationship between these classes is visualized in the following diagram (Figure 5).

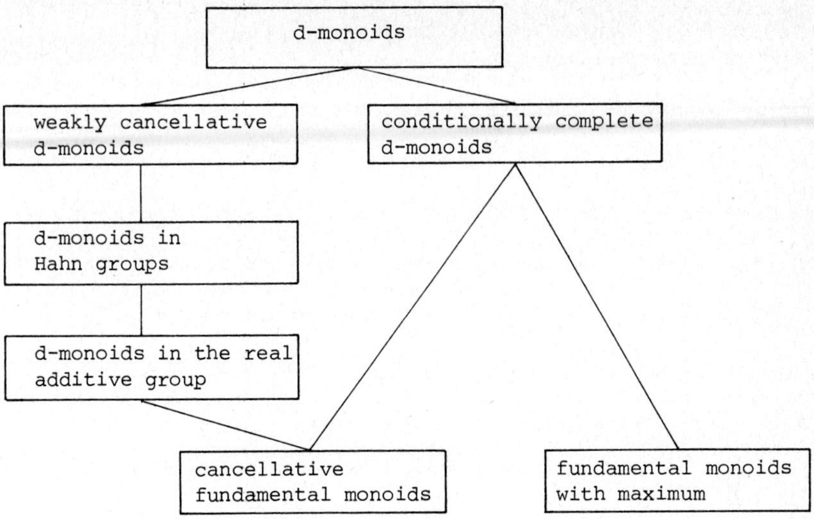

Figure 5. Relationship of classes of d-monoids

A line joining two classes means that the upper class contains the lower one. More details can be found in the following remarks and examples. The *fundamental monoids* are the monoids explicitly mentioned in (4.17).

(4.20) <u>D-monoids in the real additive group</u>

The only conditionally complete subgroups are $(\mathbb{R},+,\leq)$, $(\mathbb{Z},+,\leq)$; positive cones of these are the cancellative fundamental monoids in (4.17). If a d-monoid H contains an element $a \in \mathbb{R}_+ \setminus \mathbb{Z}_+$ then H is *dense* in $\mathbb{R}_+$ (i.e. for $\alpha, \beta \in \mathbb{R}_+$ with $\alpha < \beta$ there exists $\gamma \in H$ such that $\alpha < \gamma < \beta$). A typical example is $(\mathbb{Q}_+,+,\leq)$, the positive cone of the additive group of rational numbers. Another example can be derived from a transcendental number $\alpha \in (0,1)$. Let R denote all real numbers of the form $\Sigma_{j=0}^{r} p_j \cdot \alpha^j$ with

$r \in \mathbb{N} \cup \{0\}$ and $p_j \in \mathbb{Z}$ for each j. Due to the transcendency of α this representation is unique and R is a well-defined dense subgroup of $\mathbb{R}$. Two linearly ordered commutative groups isomorphic to $(\mathbb{R},+,\le)$ are considered in the following. The multiplicative group $(\mathbb{R}_+ \setminus \{0\},\cdot,\le)$ of all positive real numbers is isomorphically mapped onto $\mathbb{R}$ by $x \to \ln x$. Now let $\rho := (-1)^{1/p}$ for $p \in \mathbb{N}$. On $R := \mathbb{R}_+\rho \cup \mathbb{R}_+$ we define an extension of the usual order relation of $\mathbb{R}_+$ by $\mathbb{R}_+\rho \le \mathbb{R}_+$ and $\alpha\rho \le \beta\rho$ iff $\alpha \ge \beta$ for $\alpha,\beta \in \mathbb{R}_+$. With respect to the internal composition $a \oplus b := (a^p+b^p)^{1/p}$ the system $(R,\oplus,\le)$ is a linearly ordered commutative group which is isomorphically mapped onto $(\mathbb{R},+,\le)$ by $x \to x^p$.

(4.21) D-monoids in Hahn-groups

A d-monoid in a group is always cancellative and therefore irreducible. In particular, d-monoids in the real additive group are examples. Further from (4.11) we know that a cancellative d-monoid can always be embedded in a linearly ordered commutative group. Such groups can be embedded into Hahn-groups. Therefore all cancellative d-monoids are isomorphic to a d-monoid in a Hahn-group. A typical example of a Hahn-group is the additive group $(\mathbb{R}^n,+,\preccurlyeq)$ of real n-vectors ordered with respect to the lexicographical order relation. If $n > 1$ then the strict positive cone P of this group is a cancellative d-semigroup which is positively ordered and irreducible but not Archimedean. For example, let $n = 2$. Then $(0,1) \prec (1,0)$ but there exists no $n \in \mathbb{N}$ such that $n(0,1) = (0,n) \succcurlyeq (1,0)$. Thus P is not Archimedean and hence not conditionally complete (cf. 4.15). A more general example is the following.

Let $(\Lambda, \leq)$ be a nonempty, linearly ordered set and let for each $\lambda \in \Lambda$ H_λ be a cancellative d-monoid. We consider the subset H of the Cartesian product $\times_{\lambda \in \Lambda} H_\lambda$ which consists in all elements $(\ldots x_\lambda \ldots)$ such that $S_x := \{\lambda \in \Lambda \mid x_\lambda \neq e_\lambda\}$ is *well-ordered*. (Each subset of S_x contains a minimum.)

On H an internal composition $*$ is defined componentwise. Further H is linearly ordered with respect to the lexicographical order relation, i.e. $x \preccurlyeq y$ if $x = y$ or $x_\lambda < y_\lambda$ for $\lambda = \min\{\mu \in \Lambda \mid x_\mu \neq y_\mu\}$. Then $(H, *, \preccurlyeq)$ is called the *lexicographical product* of the family $(H_\lambda, \lambda \in \Lambda)$ [Denotation: $H = \Pi_{\lambda \in \Lambda} H_\lambda$]. H is a cancellative d-monoid if and only if for all $\lambda > \min \Lambda$ H_λ is a linearly ordered commutative group. Further H is a linearly ordered commutative group if and only if for all λ H_λ is a linearly ordered commutative group.

For example, let H denote the set of all functions $f: [0,1] \to \mathbb{R}$ such that $S_f := \{x \mid f(x) \neq 0\}$ is well-ordered. Together with componentwise addition and lexicographic order relation H is a linearly ordered commutative group. We remark that H is the lexicographic product $\Pi_{\lambda \in [0,1]} H_\lambda$ of real additive groups, i.e. $H_\lambda = \mathbb{R}$ for all $\lambda \in [0,1]$. As all H_λ coincide with $\mathbb{R}$ we call H a *real homogeneous lexicographical product*. H is isomorphic to the Hahn-group $V([0,1])$ (cf. chapter 3) with respect to the inverse ordered set $[0,1]$. In this way each Hahn-group $V(\Lambda)$ is "inverse" isomorphic to the real homogeneous lexicographical product $\Pi_{\lambda \in \Lambda} H_\lambda$.

(4.22) Weakly cancellative d-monoids

All weakly cancellative d-monoids are ordinal sums of cancellative d-semigroups. Therefore all irreducible weakly cancellative d-semigroups can be embedded into Hahn groups.

As first example we consider $([0,1],*,\le)$ with internal composition defined by

$$a * b := a + b - ab$$

and by

$$a * b := (a+b)/(1 + ab).$$

Both are weakly cancellative, positively ordered d-monoids. The irreducible subsemigroups are $\{0\}$, $(0,1)$ and $\{1\}$; in particular the nontrivial subsemigroup can isomorphically be mapped onto $(\mathbb{R},+,\le)$ by $x \to -\ln(1-x)$ and by $x \to -\ln((1-x)/(1+x))$. A further example is the *real bottleneck semigroup* $(\mathbb{R},\max,\le)$ with internal composition defined by $(a,b) \to \max(a,b)$ and with respect to the usual order relation of real numbers. Its ordinal decomposition is $(\{a\};\ a \in \mathbb{R})$, i.e. all irreducible subsemigroups are isomorphic to the trivial group. We may adjoin $-\infty$ as a neutral element. All elements of the bottleneck semigroup are idempotent. This semigroup is not cancellative but conditionally complete. Now we consider $(\mathbb{R} \times \mathbb{R}_+,*,\preccurlyeq)$ with internal composition defined by

$$(a,b) * (c,d) := \begin{cases} (a,b) & \text{if } a > c, \\ (c,d) & \text{if } c > a, \\ (a,b+d) & \text{if } c = a, \end{cases}$$

for all $(a,b),(c,d) \in \mathbb{R} \times \mathbb{R}_+$. Its irreducible subsemigroups are of the form $\{(a,0)\}$ and $\{a\} \times (\mathbb{R}_+ \setminus \{0\})$ for $a \in \mathbb{R}$. We define a suitable index set $\Lambda := \{1,2\} \times \mathbb{R}$. Clearly, $(1,a)$ is the index of $\{(a,0)\}$, and $(2,a)$ is the index of $\{a\} \times (\mathbb{R}_+ \setminus \{0\})$. Λ is linearly ordered by $\preccurlyeq$. We may adjoin a neutral element $(-\infty,0)$ or a first irreducible semigroup $(\{-\infty\} \times \mathbb{R},+,\preccurlyeq)$. Then the new ordinal sum is called *time-cost semigroup*. This semigroup is not conditionally complete. The most general example is the following. Let G_λ be a real homogeneous lexicographic product for each $\lambda \in \Lambda$. Let Λ have a minimum λ_o. Then H_λ is either $(G_\lambda)_+$ or $(G_\lambda)_+ \setminus \{e_\lambda\}$ for $\lambda > \lambda_o$. Let H be the ordinal sum of G_{λ_o} and all H_λ. Then H_λ is called a *homogeneous weakly cancellative d-monoid*. Clearly, all weakly cancellative d-monoids can be embedded into a homogeneous weakly cancellative d-monoid.

(4.23) <u>Conditionally complete d-monoids</u>

All fundamental monoids are conditionally complete. In the positively ordered case all irreducible, conditionally complete d-semigroups are isomorphic to the strict positive cone of a fundamental semigroup. A weakly cancellative example, the bottleneck semigroup, is given in (4.22). Our next example is not weakly cancellative. We consider $(\mathbb{R} \times [0,1],*,\preccurlyeq)$ with internal composition defined by

$$(a,b) * (c,d) := \begin{cases} (a,b) & \text{if } a > c, \\ (c,d) & \text{if } c > a, \\ (a,\min(b+d,1) & \text{if } a = c, \end{cases}$$

for all $(a,b),(c,d) \in \mathbb{R} \times [0,1]$. Similar to the time-cost-semigroup in (4.22) the irreducible subsemigroups are $\{(a,0)\}$ and $\{a\} \times (0,1]$ for $a \in \mathbb{R}$. Again $\Lambda = \{1,2\} \times \mathbb{R}$ is a suitable index set. We may adjoin a neutral element $(-\infty,0)$. The reader will easily construct further examples using fundamental semigroups (without neutral element) from (4.17) in the following way. Let Λ be a conditionally complete, linearly ordered set and let H_λ be a fundamental semigroup for each $\lambda \in \Lambda$ such that the following rules are satisfied (cf. CLIFFORD [1958]).

(1) If $\lambda \in \Lambda$ has no immediate successor, but is not the greatest element of Λ, then H_λ must have a greatest element.

(2) If $\lambda \in \Lambda$ has no immediate predecessor, but is not the least element of Λ, then H_λ must have a least element.

(3) If $\lambda, \mu \in \Lambda$, $\lambda < \mu$ and λ is the immediate predecessor of μ then either H_λ has a greatest or H_μ has a least element.

It is easy to see that these rules are necessary for conditionally complete d-monoids.

If we do not assume that the considered semigroup is positively ordered then we may consider the ordinal sum of an arbitrarily irreducible conditionally complete d-semigroup H_{λ_o} with negative elements and an arbitrary conditionally complete positively ordered d-semigroup H without a neutral element. Again we have to fulfill rules (1) - (3). In this way we get *all* conditionally complete d-monoids. As an example consider

$$(\{-\infty\} \times \mathbb{R} \cup \mathbb{R}_+ \times [0,1], *, \preccurlyeq)$$

which is defined as the ordinal sum of the group $(\{-\infty\} \times \mathbb{R}, +, \preccurlyeq)$ and the subsemigroup $(\mathbb{R}_+ \times [0,1], *, \preccurlyeq)$ of the above example.

(4.24) D-monoids

All examples considered before are d-monoids. An example which is neither weakly cancellative nor conditionally complete can be found from the first example in (4.23) by eliminating all subsemigroups of the form $\{(a,0)\}$ in the ordinal decomposition. Then $(\mathbb{R} \times (0,1], *, \preccurlyeq)$ is no longer conditionally complete as the rules in (4.23) are not satisfied, but it can normally be embedded into the original one. Here the question arises whether a d-monoid exists which is neither weakly cancellative nor conditionally complete and, furthermore cannot be normally embedded into a conditionally complete one. CLIFFORD [1959] gives the following example. Let

$$H = \{(0,n),(n,0) \mid n \in \mathbb{N}\} \cup \{(1,z) \mid z \in \mathbb{Z}\}$$

be lexicographically ordered, i.e.

$$(0,1) \prec (0,2) \prec \ldots \prec (1,-1) \prec (1,0) \prec (1,1) \prec \ldots$$
$$\ldots \prec (2,0) \prec (3,0) \prec \ldots .$$

We define an internal composition by

$$(n,x) * (m,y) := \begin{cases} (n+m,x+y) & \text{if } n+m < 2, \\ (n+m,0) & \text{otherwise.} \end{cases}$$

Then $(H,*,\preccurlyeq)$ is an irreducible, positively ordered d-semigroup without a neutral element. As H is not Archimedean it cannot be conditionally complete. Further H is not cancellative. Now we try to embed H into a normal completion. The only undefined supremum (infimum) is $\alpha := \sup\{(0,n) \mid n \in \mathbb{N}\}$. We consider $H \cup \{\alpha\}$ with the obvious extension of the linear order relation on H. If the extended internal composition $*$ satisfies the monotonicity condition then it is uniquely defined by

$$\alpha * (0,n) := \alpha \quad , \quad \alpha * \alpha := (2,0)$$

$$\alpha * (1,z) := (2,0), \quad \alpha * (n,0) := (n+1,0)$$

for all $n \in \mathbb{N}$ and for all $z \in \mathbb{Z}$. But now $\alpha < (1,0)$ and there exists no $x \in H \cup \{\alpha\}$ such that $\alpha * x = (1,0)$. Therefore $H \cup \{\alpha\}$ is not a d-semigroup. CLIFFORD [1959] shows that, in general, a positively ordered d-semigroup can normally be embedded into a conditionally complete d-semigroup if and only if all its irreducible subsemigroups are Archimedean. Then this normal completion is uniquely determined.

5. Ordered Semimodules

In this chapter we introduce ordered algebraic structures which generalize rings, fields, modules and vectorspaces as known from the theory of algebra. Additionally these structures will be ordered and will satisfy monotonicity conditions similar to the case of ordered semigroups.

Let $(R,\oplus)$ be a commutative monoid with neutral element 0 and let $(R,\otimes)$ be a monoid with neutral element 1 $(0 \neq 1)$. If

$$\begin{aligned} &\alpha \otimes (\beta \oplus \gamma) = (\alpha \otimes \beta) \oplus (\alpha \otimes \gamma), \\ (5.1) \qquad &(\beta \oplus \gamma) \otimes \alpha = (\beta \otimes \alpha) \oplus (\gamma \otimes \alpha), \\ &0 = \alpha \otimes 0 \quad = 0 \otimes \alpha , \end{aligned}$$

for all $\alpha,\beta,\gamma \in R$ then $(R,\oplus,\otimes)$ is called a *semiring with unity 1 and zero 0*. We will shortly speak of a semiring R. (5.1.1) and (5.1.2) are the laws of *distributivity*. We remark that $\otimes$ is not necessarily commutative. Otherwise we call R a *commutative semiring*. If all elements of R are idempotent with respect to $\oplus$ then R is called an *idempotent semiring*.

Now, if $(R,\oplus,\leq)$ is an ordered commutative monoid and

$$\begin{aligned} &\alpha \leq \beta \Rightarrow \alpha \otimes \gamma \leq \beta \otimes \gamma \text{ and } \gamma \otimes \alpha \leq \gamma \otimes \beta, \\ (5.2) \qquad &\alpha \leq \beta \Rightarrow \alpha \otimes \delta \geq \beta \otimes \delta \text{ and } \delta \otimes \alpha \geq \delta \otimes \beta \end{aligned}$$

for all $\alpha,\beta,\gamma,\delta \in R$ with $\gamma \geq 0$, $\delta \leq 0$, then R is called an *ordered semiring*.

(5.3) Proposition

(1) A semiring is idempotent iff $1 \oplus 1 = 1$.

(2) An idempotent semiring is partially ordered with respect to the binary relation defined by $\alpha \leq \beta$ iff $\alpha \oplus \beta = \beta$ for all $\alpha, \beta \in R$.

Proof. (1) The if-part is obvious. Now let $1 \oplus 1 = 1$. Then $\alpha \otimes (1 \oplus 1) = \alpha \otimes 1$ for $\alpha \in R$. Distributivity shows that α is idempotent.

(2) If α is an idempotent element then $\alpha \leq \alpha$. Commutativity implies antisymmetry. Associativity leads to transitivity. $(R, \oplus, \leq)$ is partially ordered as $\alpha \oplus \beta = \beta$ implies $(\alpha \oplus \gamma) \oplus (\beta \oplus \gamma) = \beta \oplus \gamma$ for all $\alpha, \beta, \gamma \in R$. Clearly $0 \leq \gamma$ for all $\gamma \in R$. Then distributivity implies (5.2.1).

■

In particular, if $(R, \oplus)$ is a commutative group then R is called a *ring*. A semiring R with

(5.4) $\alpha \otimes \beta = 0 \quad \Rightarrow \quad \alpha = 0 \quad \text{or} \quad \beta = 0$

for all $\alpha, \beta \in R$ is called a semiring *without (nontrivial) divisors of zero*. A commutative ring without divisors of zero is called an *integral domain*. A ring R with commutative group $(R \setminus \{0\}, \otimes)$ is called a *field*. A well-known theorem from the theory of algebra is the following.

(5.5) Proposition

A (linearly ordered) integral domain R can be embedded in a (linearly ordered) field.

Proof. We give only a sketch of the proof which is similar to the proof of theorem (2.8). An equivalence relation $\sim$ on $R \times (R \setminus \{0\})$ is defined by $(\alpha,\beta) \sim (\gamma,\delta)$ iff $\alpha \otimes \delta = \beta \otimes \gamma$. We denote the equivalence class containing (α,β) by α/β and we identify $\alpha \in R$ with $\alpha/1$. Then the internal composition may be extended by

$$(5.6) \qquad \begin{aligned} \alpha/\beta \oplus \gamma/\delta &:= [(\alpha \otimes \delta) \oplus (\beta \otimes \gamma)]/(\beta \otimes \delta), \\ \alpha/\beta \otimes \gamma/\delta &:= (\alpha \otimes \gamma)/(\beta \otimes \delta) \end{aligned}$$

for all α/β, γ/δ. It can be seen that then the set F of all equivalence classes is a well-defined field. With respect to the linear order relation defined by

$$(5.7) \qquad \alpha/\beta \leq \gamma/\delta \quad :\Longleftrightarrow \quad \alpha \otimes \delta \leq \beta \otimes \gamma$$

for all α/β, γ/δ this field is linearly ordered.

■

The following result for the Archimedean case is similar to (4.17.1).

(5.8) Proposition

An Archimedean linearly ordered field $(F,\oplus,\otimes,\leq)$ is isomorphic to a subfield of the linearly ordered field $(\mathbb{R},+,\cdot,\leq)$ of real numbers.

Proof. We only indicate how this result follows from the proof of (4.17). The details are left to the reader.

Let $\bar{0}$ resp. $\bar{1}$ denote zero resp. unity of F. It suffices to prove that $F' := \{\alpha \in F \mid \alpha > \bar{0}\}$ is isomorphic to the strict positive cone of a subfield of $\mathbb{R}$ $(\mathbb{R}' := \{a \in \mathbb{R} \mid a > 0\})$. F' does not contain a minimum. Therefore using the proof of (4.17) we can choose an isomorphism φ with $\varphi(\bar{1}) = 1$ such that φ maps $(F', \oplus, \leq)$ isomorphically onto a dense subsemigroup of $(\mathbb{R}', +, \leq)$. We find that φ maps $\bar{Q}' := \{\varphi^{-1}(q) \mid q \in \mathbb{Q},\ q > 0\}$ onto $\mathbb{Q}$ such that $\varphi(\alpha \otimes \beta) = \varphi(\alpha)\varphi(\beta)$ for all $\alpha, \beta \in \bar{Q}'$. Then

$$\alpha = \sup\{\beta \in \bar{Q}' \mid \beta < \alpha\} = \inf\{\beta \in \bar{Q}' \mid \beta > \alpha\}$$

for all $\alpha \in F'$ leads to $\varphi(\alpha \otimes \beta) = \varphi(\alpha)\varphi(\beta)$ for all $\alpha, \beta \in F'$.

■

Now let $(H, *)$ be a commutative monoid with neutral element e and let $(R, \oplus, \otimes)$ be a semiring with unity 1 and zero 0. If $\square: R \times H \to H$ is an *external composition* such that

$$
\begin{aligned}
(\alpha \otimes \beta) \square a &= \alpha \square (\beta \square a) \ , \\
(\alpha \oplus \beta) \square a &= (\alpha \square a) * (\beta \square a) \ , \\
\alpha \square (a * b) &= (\alpha \square a) * (\alpha \square b) \ , \\
0 \square a &= e \ , \\
1 \square a &= a \ ,
\end{aligned}
\tag{5.9}
$$

for all $\alpha, \beta \in R$ and for all $a, b \in H$ then $(R, \oplus, \otimes; \square; H, *)$ is called a *semimodule over* R. We remark that the adjective 'external' does not exclude the case R = H. In our definition we only consider a *left* external composition $\square: R \times H \to H$. Similarly, one may consider a *right* external composition $\square: H \times R \to H$. If we want to distinguish between such semimodules then we will use the adjective "left" or "right". Usually we will only treat left semimodules and omit the adjective "left".

A direct consequence of (5.9) is the relation $\alpha \square e = e$ for all $\alpha \in R$. This follows from $e = 0 \square e = (\alpha \otimes 0) \square e = \alpha \square (0 \square e) = \alpha \square e$.

In particular, if R is a ring then H is a commutative group with $a^{-1} = (-1) \square a$ for all $a \in H$ (-1 is the additive inverse of 1 in R). Then H is called a *module* over R. If R is a field then H is called a *vectorspace* over R. We remark that if R is a ring and H a commutative group then $0 \square a = e$ is a redundant assumption in (5.9).

Now let $(H,*,\leq)$ be an ordered commutative monoid with neutral element e and let $(R,\oplus,\otimes,\leq)$ be an ordered semiring with unity 1 and zero 0. Let $\square$: $R \times H \to H$ be an external composition satisfying (5.9),

(5.10.1) $\quad \alpha \leq \beta \;\Rightarrow\; \alpha \square c \leq \beta \square c \;$ and $\; \alpha \square d \geq \beta \square d$

for all $\alpha,\beta \in R$, $c,d \in H$ with $d \leq e \leq c$ and

(5.10.2) $\quad a \leq b \;\Rightarrow\; \gamma \square a \leq \gamma \square b \;$ and $\; \delta \square a \geq \delta \square b$

for all $a,b \in H$, $\gamma \in R_+ := \{\varepsilon \in R \mid \varepsilon \geq 0\}$ and $\delta \in R_- := \{\varepsilon \in R \mid \varepsilon \leq 0\}$. Then $(R,\oplus,\otimes,\leq;\square;H,*,\leq)$ is called an *ordered semimodule* over R. We remark that we will use the same denotation $\leq$ for the different order relations in R and H. From the context it will always be clear which one is meant specifically. If, in particular, $(R,\oplus,\otimes,\leq)$ is the positive cone of a field then H is called an *ordered semivectorspace* over R.

Some direct consequences should be mentioned:

(5.11) <u>Proposition</u>

Let H be an ordered semimodule over R.

(1) If H is a group then

$$\alpha \leq \beta \;\Rightarrow\; \alpha \square c \leq \beta \square c$$

for all $\alpha, \beta \in R$ and $c \in H_+$ implies (5.10.1).

(2) If R is a ring then

$$a \leq b \Rightarrow \gamma \square a \leq \gamma \square b$$

for all $a, b \in H$ and $\gamma \in R_+$ implies (5.10.2).

(3) If R is the positive cone of a linearly ordered ring $\tilde{R}$ and H is the positive cone of a linearly ordered group $\tilde{H}$ then the external composition can be continued in a unique way on $\tilde{R} \times \tilde{H}$ such that $\tilde{H}$ is a linearly ordered modul over $\tilde{R}$.

<u>Proof</u>. (1) In an ordered group $c \leq d$ is equivalent to $c^{-1} \geq d^{-1}$. This follows from composition of $c \leq d$ with $c^{-1} * d^{-1}$. Now let $\alpha, \beta \in R$ with $\alpha \leq \beta$ and let $d \in H_-$. Then $d^{-1} \in H_+$ and, due to our assumption, $\alpha \square d^{-1} \leq \beta \square d^{-1}$. Now $(\alpha \square d^{-1}) * (\alpha \square d) = \alpha \square (d^{-1} * d) = e$ shows that $\alpha \square d^{-1} = (\alpha \square d)^{-1}$. Therefore $\alpha \square d \geq \beta \square d$.

(2) In particular, $(H, *)$ is a group. Now let $a, b \in H$ with $a \leq b$ and let $\delta \in R_-$. Then $-\delta \in R_+$ and, due to our assumption, $(-\delta) \square a \leq (-\delta) \square b$. Now $[(-\delta) \square a] * (\delta \square a) = [(-\delta) \oplus \delta] \square a = e$ shows that $(-\delta) \square a = (\delta \square a)^{-1}$. Therefore $\delta \square a \geq \delta \square b$.

(3) As R and H are linearly ordered we know $\tilde{R} = R \cup \tilde{R}_-$ $(R \cap \tilde{R}_- = \{0\})$ and $\tilde{H} = H \cup \tilde{H}_-$ $(H \cap \tilde{H}_- = \{e\})$. Therefore we may continue $\square$ on $\tilde{R} \times \tilde{H}$ by

$$(5.12) \qquad \begin{aligned} 0 \square b &:= e, \\ \alpha \square b &:= (\alpha \square b^{-1})^{-1}, \\ \beta \square a &:= ((-\beta) \square a)^{-1}, \\ \beta \square b &:= (-\beta) \square b^{-1}, \end{aligned}$$

for all $\alpha \in R \setminus \{0\}$, $\beta \in \tilde{R} \setminus R$, $a \in H$, $b \in \tilde{H} \setminus H$. Then $\tilde{H}$ is a module over $\tilde{R}$. In a module the equations used in (5.12) are necessarily fulfilled. Therefore the extension of $\square$ is uniquely determined. As H is an ordered semimodule over R we know that $\alpha \leq \beta$ implies $\alpha \square c \leq \beta \square c$ for all $\alpha, \beta \in R$ and $c \in H$ and that $a \leq b$ implies $\gamma \square a \leq \gamma \square b$ for all $a,b \in H$ and $\gamma \in R$. It suffices to show that the first implication holds for all $\alpha, \beta \in \tilde{R}$ and that the second implication holds for all $a,b \in \tilde{H}$. Then (1) and (2) imply that $\tilde{H}$ is a linearly ordered module over $\tilde{R}$. At first we remark that for $\alpha \leq \beta$ only the additional cases $\alpha \in \tilde{R}_-$, $\beta \in R$ and $\alpha, \beta \in \tilde{R}_-$ occur. Then $\alpha \square c \leq 0 \square c = e \leq \beta \square c$ resp. $(-\alpha) \square c \geq (-\beta) \square c$ lead to $\alpha \square c \leq \beta \square c$. The second implication follows in a similar way.

■

Proposition (5.11) shows that for special semimodules some of the monotonicity conditions are redundant. Next we consider some examples.

(5.13) Partially ordered semirings

Let $(R,\oplus,\otimes,\leq)$ be a partially ordered semiring. Then $(R,\oplus,\leq)$ is a partially ordered semimodule over $(R,\oplus,\otimes,\leq)$ with respect to the "external" composition defined by $\alpha \square \beta := \alpha \otimes \beta$ for all $\alpha, \beta \in R$. In particular, (5.9) follows from (5.1) and (5.10) follows from (5.2). As an example, $(\mathbb{Z}_+,+,\leq)$ is a linearly ordered commutative semimodule over the linearly ordered commutative semiring $(\mathbb{Z}_+,+,\cdot,\leq)$. As in (5.11.3) this semimodule can be extended to the linearly ordered commutative module $(\mathbb{Z},+,\leq)$ over the linearly ordered commutative ring $(\mathbb{Z},+,\cdot,\leq)$. Further the additive group of a subfield of the real numbers is a vectorspace over this subfield. The positive cone of such an additive group is a semivectorspace over the positive cone of such a field. Finally we remark that $(R^n, \oplus, \leq)$ with respect to componentwise composition and componentwise order relation is again a partially ordered monoid. Therefore $(R^n, \oplus, \leq)$ is a (right as well as left) partially ordered semimodule over $(R, \oplus, \otimes, \leq)$. In particular $\mathbb{Z}^n$ ($\mathbb{Q}^n$ and $\mathbb{R}^n$) are lattice-ordered commutative modules (vectorspaces) over $\mathbb{Z}$ ($\mathbb{Q}$ and $\mathbb{R}$).

(5.14) Semirings

Let $(R, \oplus, \otimes)$ denote a semiring. Then

$$a \leq b \quad :\Longleftrightarrow \quad a \oplus b = b$$

defines a pseudo-ordering on R with minimum 0. In general $(R, \oplus, \leq)$ is not a pseudo-ordered monoid but satisfies

(5.14.1) $\alpha \leq \beta$ and $\gamma \leq \delta \quad \Rightarrow \quad \alpha \oplus \gamma \leq \beta \oplus \delta$,

(5.14.2) $\alpha \leq \beta$ and $\gamma \leq \beta \quad \Rightarrow \quad \alpha \oplus \gamma \leq \beta$

for all $\alpha, \beta, \gamma, \delta \in R$. Further (5.2) is satisfied. If R is an idempotent semiring then R is a partially ordered semiring. From (5.13) we conclude that $(R, \oplus, \leq)$ is a partially ordered semimodule $(R, \oplus, \otimes, \leq)$. In this partial order the least upper bound is well-defined, i.e.

$$\sup(a,b) = a \oplus b$$

but the greatest lower bound may not exist, in general.

(5.15) Residuated lattice-ordered commutative monoids

Let $(R, \otimes, \leq)$ be a residuated lattice-ordered commutative monoid with neutral element 1. At the end of chapter 2 we have mentioned lattice-ordered commutative groups, pseudo-Boolean and, in particular, Boolean lattices as examples for this ordered structure. If not present we adjoin a minimum O and a maximum ∞. The internal composition is extended by $O = O \otimes \alpha = \alpha \otimes O$ for all $\alpha \in R' := R \cup \{O,\infty\}$ and $\infty = \infty \otimes \alpha = \alpha \otimes \infty$ for all $\alpha \in R \cup \{\infty\}$. Let $\alpha \oplus \beta := \alpha \vee \beta$ for all $\alpha, \beta \in R'$. Then (5.1) is satisfied by all $\alpha, \beta, \gamma \in R'$. In particular distributivity follows from (2.13). Therefore $(R', \oplus, \otimes, \leq)$ is a lattice-ordered, idempotent, commutative semiring with unity 1 and zero O. From (5.14) we conclude that $(R', \oplus, \leq)$ is a lattice-ordered commutative semimodule over R'.

(5.16) Linearly ordered semimodules over real numbers

Let R_+ be the positive cone of a subring of the linearly ordered field of real numbers with $\{0,1\} \subseteq R_+$. Let $(H, *, \leq)$ be a linearly ordered commutative monoid and let $\square: R_+ \times H \to H$ be an external composition such that $(R_+, +, \cdot, \leq; \square; H, *, \leq)$ is a linearly ordered commutative semimodule over R_+.
In particular we consider the *discrete semimodule* H over $\mathbb{Z}_+$ with external composition defined by $z \square a := a^z$ for all $z \in \mathbb{Z}_+$ and for all $a \in H$. As we discuss only semimodules over semirings containing distinct zero and unity such a discrete semimodule is contained in each semimodule. Linearly ordered semimodules over real numbers are discussed in chapter 6.

Now we introduce matrices for semirings R and for semimodules H over R. Matrices with entries in R and H will be denoted by $A = (\alpha_{ij})$, B, C and $X = (x_{ij})$, Y, Z. For matrices of suitable size we define certain compositions.

Let A,B and X,Y be two $m \times n$ matrices over R and H. Then $C = A \oplus B$ and $Z = X * Y$ are defined by

(5.17) $$\gamma_{ij} := \alpha_{ij} \oplus \beta_{ij} \quad ,$$

(5.18) $$z_{ij} := x_{ij} * y_{ij} \quad ,$$

for all $i = 1,2,\ldots,m$ and $j = 1,2,\ldots,n$. Let A be an $m \times n$ matrix over R, let B be an $n \times q$ matrix over R and let X be an $n \times q$ matrix over H. Then $C = A \otimes B$ and $Z = A \,\square\, X$ are defined by

(5.19) $$\gamma_{ik} := \bigoplus_{j=1}^{n} (\alpha_{ij} \otimes \beta_{jk}) \quad ,$$

(5.20) $$z_{ik} := \mathop{*}_{j=1}^{n} (\alpha_{ij} \,\square\, x_{jk}) \quad ,$$

for all $i = 1,2,\ldots,m$ and $k = 1,2,\ldots,q$. Properties of these compositions correspond to properties of the underlying semiring and semimodule. If the semiring or semimodule considered is ordered then, for matrices of the same size, an order relation is defined componentwise. A matrix containing only zero-entries is called *zero-matrix* and is denoted by [0]. Similarly the *neutral matrix* [e] contains only neutral elements. A matrix containing unities on the main diagonal and zeros otherwise is called *unity matrix* and is denoted by [1]. Such matrices are considered for any size $m \times n$.

(5.21) <u>Proposition</u>

Let H be an ordered semimodule over R. Let A,B,C,D,E be matrices over R with $D \geq [0]$ and $E \leq [0]$. Let X,Y,Z,U,V be matrices over H with $U \geq [e]$ and $V \leq [e]$. The following properties hold for matrices of suitable size.

(1) $(A \oplus B) \oplus C = A \oplus (B \oplus C)$,

$A \oplus B = B \oplus A$,

$(A \otimes B) \otimes C = A \otimes (B \otimes C)$,

$(A \oplus B) \otimes C = (A \otimes C) \oplus (B \otimes C)$,

$C \otimes (A \oplus B) = (C \otimes A) \oplus (C \otimes B)$,

$[0] \otimes A = [0] = A \otimes [0]$;

(2) $A \leq B \Rightarrow A \oplus C \leq B \oplus C$,

$A \leq B \Rightarrow A \otimes D \leq B \otimes D$ and $D \otimes A \leq D \otimes B$,

$A \leq B \Rightarrow A \otimes E \geq B \otimes E$ and $E \otimes A \geq E \otimes B$;

(3) the set of all $m \times m$ matrices over R is an ordered semiring with unity [1] and zero [0];

(4) $(X * Y) * Z = X * (Y * Z)$,

$X * Y = Y * X$,

$(A \otimes B) \square Z = A \square (B \square Z)$,

$(A \oplus B) \square Z = (A \square Z) * (B \square Z)$,

$[0] \square X = [e]$;

(5) $X \leq Y \Rightarrow X * Z \leq Y * Z$,

$A \leq B \Rightarrow A \square U \leq B \square U$ and $A \square V \geq B \square V$,

$X \leq Y \Rightarrow D \square X \leq D \square Y$ and $E \square X \geq E \square Y$;

(6) the set of all $m \times m$ matrices over H is an ordered semimodule over the set of all $m \times m$ matrices over R.

<u>Proof</u>. These properties are immediate consequences of the definitions of the composition considered. As an example we prove

$$(A \otimes B) \square Z = A \square (B \square Z)$$

for $m \times n$ and $n \times q$ matrices A and B over R and an $q \times r$ matrix Z over H. An entry x_{is} of the left-hand side of the equation is given by

$$x_{is} = \mathop{*}_{k=1}^{q} [(\bigoplus_{j=1}^{n} (a_{ij} \otimes b_{jk})) \square z_{ks}] .$$

Using (5.1) and (5.9) we find

$$x_{is} = \mathop{*}_{k=1}^{q} [\mathop{*}_{j=1}^{n} (a_{ij} \square (b_{jk} \square z_{ks}))]$$

$$= \mathop{*}_{j=1}^{n} [a_{ij} \square (\mathop{*}_{k=1}^{q} (b_{jk} \square z_{ks}))],$$

which is an entry of the right-hand side of the equation considered.

■

For special semimodules a theory similar to the classical theory of linear algebra has been developed. We refer to module theory (cf. BLYTH [1977]) and to min - max algebra (cf. CUNINGHAME-GREEN [1979]). A linear algebra theory in the general case is not known. Further properties will be discussed in the chapters on optimization problems. Here we have described the necessary basic concept for the formulation of equations, inequations and linear mappings over ordered semimodules. Duality principles will be discussed in chapter 10 and chapter 11.

6. Linearly Ordered Semimodules Over Real Numbers

Let $(R,+,\cdot,\leq)$ denote an arbitrary subring of the linearly ordered field of real numbers with zero $0 \in R$ and unity $1 \in R$. Let $(H,*,\leq)$ be a linearly ordered commutative monoid. In this chapter $(R_+,+,\cdot,\leq;\square;H,*,\leq)$ will always be a (linearly ordered) semimodule. As multiplication is commutative the semimodule is commutative, too. Further $(R_+,+,\leq)$ is a d-monoid as for $\alpha,\beta \in R_+$ with $\alpha \leq \beta$ we know $\beta - \alpha \in R_+$. From the proof of (4.17) we conclude that R_+ is either equal to $\mathbb{Z}_+$ or R_+ is a dense subset of $\mathbb{R}_+$. In any case, $\mathbb{Z}_+ \subseteq R_+$ as $\{0,1\} \subseteq R_+$. Therefore the properties of the *discrete semimodule* $(\mathbb{Z}_+,+,\cdot,\leq;\square;H,*,\leq)$ play an important role.

We do not assume in the following that the semimodule considered is ordered. Therefore the monotonicity conditions (cf. 5.10) may be invalid. Nevertheless, for discrete semimodules we find:

(6.1) $\quad e < \alpha \square a \quad$ and $\quad \alpha \square b < e$

for all $\alpha \in \mathbb{Z}_+ \setminus \{0\}$ and all $a,b \in H$ with $e < a$, $b < e$,

(6.2) $\quad \alpha \leq \beta \quad \Rightarrow \quad \alpha \square a \leq \beta \square a \quad$ and $\quad \alpha \square b \geq \beta \square b$

for all $\alpha,\beta \in \mathbb{Z}_+$ and all $a,b \in H$ with $e \leq a$, $b \leq e$, and

(6.3) $\quad a \leq b \quad \Rightarrow \quad \alpha \square a \leq \alpha \square b$

for all $\alpha \in \mathbb{Z}_+$ and all $a,b \in H$. We remark that (6.2) and (6.3) for all $\alpha,\beta \in R_+$ and all $a,b \in H$ are equivalent to (5.10). We will show how these properties are related and that they hold in all *rational* semimodules ($R_+ \subseteq \mathbb{Q}_+$).

(6.4) Proposition

Let H be a linearly ordered commutative monoid which is a semimodule over $\mathbb{R}_+$. If

(1) $a \leq b \Rightarrow \alpha \square a \leq \alpha \square b \quad \forall\, \alpha \in \mathbb{R}_+ \quad \forall\, a,b \in H$

then the following properties are satisfied:

(2) $e < \alpha \square a \quad \forall\, \alpha \in \mathbb{R}_+ \setminus \{0\} \quad \forall\, a \in H$ with $e < a$,

(2') $\alpha \square a < e \quad \forall\, \alpha \in \mathbb{R}_+ \setminus \{0\} \quad \forall\, a \in H$ with $a < e$,

(3) $e \leq \alpha \square a \quad \forall\, \alpha \in \mathbb{R}_+ \quad \forall\, a \in H$ with $e \leq a$,

(3') $\alpha \square a \leq e \quad \forall\, \alpha \in \mathbb{R}_+ \quad \forall\, a \in H$ with $a \leq e$,

(4) $\alpha \leq \beta \Rightarrow \alpha \square a \leq \beta \square a \quad \forall\, \alpha,\beta \in \mathbb{R}_+ \quad \forall\, a \in H$ with $e \leq a$,

(4') $\alpha \leq \beta \Rightarrow \alpha \square a \geq \beta \square a \quad \forall\, \alpha,\beta \in \mathbb{R}_+ \quad \forall\, a \in H$ with $a \leq e$.

Furthermore (2), (3) and (4) as well as (2'), (3') and (4') are equivalent.

Proof. (1) ⇒ (3). Let $a = e$. Then $\alpha \square a = e$ leads to (3). In the same way $b = e$ shows (1) ⇒ (3'). We prove only the equivalence of (2), (3) and (4). The equivalence of (2'), (3') and (4') follows similarly.

(2) ⇒ (3) is obvious.

(3) ⇒ (2). Let $\alpha \in \mathbb{R}_+ \setminus \{0\}$ and let $e < a \in H$. Suppose $e = \alpha \square a$. Now $1 \leq n\alpha$ for some $n \in \mathbb{N}$. Then

$$a = 1 \square a \leq (1 \square a) * [(n\alpha - 1) \square a] = n \square (\alpha \square a) = e$$

contrary to $e < a$.

(4) ⇒ (3). Let $\alpha = 0$ in (4). This leads to (3).

(3) ⇒ (4). Let $\alpha,\beta \in \mathbb{R}_+$ with $\alpha \leq \beta$ and let $e \leq a \in H$. Then $e \leq (\beta - \alpha) \square a$. Therefore $(\alpha \square a) * e \leq (\alpha \square a) * [(\beta - \alpha) \square a]$.

■

(6.5) Theorem

Let H be a linearly ordered commutative monoid which is a semimodule over R_+. If $R_+ \subseteq \mathbb{Q}_+$ then H is a linearly ordered semimodule.

Proof. Due to proposition (6.4) it suffices to prove (6.4.1). Let $\alpha \in R_+$ and let $a, b \in H$ with $a \leq b$. If $a = b$ or $\alpha = 0$ then $\alpha \square a \leq \alpha \square b$. Now assume $\alpha > 0$ and $a < b$. As $R_+ \subseteq \mathbb{Q}_+$ we find $\alpha = n/m$ for some $m, n \in \mathbb{N}$ with greatest common divisor 1. Therefore there exist $p, q \in \mathbb{Z}$ with $pn + qm = 1$. This implies that

$$1/m = p(n/m) + q \in R_+ .$$

Let $a' = (1/m) \square a$ and $b' = (1/m) \square a$. Now $a < b$ implies $a' < b'$. The latter yields $(a')^{mn} \leq (b')^{mn}$, i.e. $\alpha \square a \leq \alpha \square b$.

∎

In the general case R_+ may contain irrational or, in particular, transcendental numbers. Then even in the case $(H, *, \leq) \equiv (\mathbb{R}, +, \leq)$ the monotonicity conditions may be invalid. This is shown by the following example.

(6.6) Semimodule containing transcendental number

Let $\alpha \in (0,1)$ be a transcendental number and consider the subring R generated by $\mathbb{Z} \cup \{\alpha\}$. Then the elements of R have the form (cf. 4.20)

$$\sum_{j=0}^{r} p_j \alpha^j$$

with $p_j \in \mathbb{Z}$, $j = 0, 1, \ldots, r$. Now we define an external composition $\square: R \times \mathbb{R} \to \mathbb{R}$ by

$$\left(\sum_{j=0}^{r} p_j \cdot \alpha^j\right) \,\square\, a := a \cdot \sum_{j=0}^{r} p_j (-\alpha)^j$$

for all $a \in \mathbb{R}$. Then $\mathbb{R}$ is a module over R. But now

$$\alpha \,\square\, 1 = -\alpha < 0$$

shows that (6.4.3) is invalid. Therefore the monotonicity conditions (5.10) are invalid.

On the other hand we will show that for an important class of monoids all monotonicity properties hold.

(6.7) <u>Proposition</u>

Let H be a d-monoid which is a semimodule over R_+. Then (6.4.1), (6.4.2), (6.4.3), and (6.4.4) are equivalent and imply (6.4.2'), (6.4.3'), and (6.4.4').

<u>Proof</u>. Due to (6.4) it suffices to show that (6.4.2) implies (6.4.1). Let $a, b \in H$, $\alpha \in R_+$ and assume (6.4.2). If $a = b$ or $\alpha = 0$ then $\alpha \square a = \alpha \square b$. Otherwise $a * c = b$ for some $c \in H$ with $c > e$. Then $e < \alpha \square c$ which implies $\alpha \square a \leq (\alpha \square a) * (\alpha \square c) = \alpha \square b$.

■

Proposition (6.7) shows that a d-monoid is a linearly ordered commutative semimodule over R_+ if it is a semimodule over R_+ and the external composition restricted to $R_+ \times H_+$ has only values in H_+. Unfortunately, example (6.6) shows that even in the case of the additive group of real numbers we can construct a semimodule in which this condition is invalid.

(6.8) <u>Theorem</u>

Let H be a positively and linearly ordered commutative monoid which is a semimodule over R_+. Then

(1) (6.4.2), (6.4.3), and (6.4.4) hold,

(2) if H is a d-monoid then H is a linearly ordered semimodule over R_+.

<u>Proof</u>. As H is positively ordered we know $H = H_+$. Therefore (6.4.3) is satisfied. Now (1) follows from (6.4) and (2) follows from (6.7).

■

In chapter 4 we discussed monoids which have an ordinal decomposition. If a monoid H is the ordinal sum of a family $(H_\lambda;\ \lambda \in \Lambda)$ of linearly ordered commutative semigroups and if H is a semimodule over R_+ then this semimodule is called *ordinal*. The *index* $\lambda(a)$ for $a \in H$ is defined with respect to the ordinal decomposition by $\lambda(a) := \mu$ if $a \in H_\mu$.

(6.9) <u>Proposition</u>

Let H be an ordinal semimodule over R_+. Then

$$\lambda(\alpha \square a) = \lambda(a)$$

for all $\alpha \in R_+ \setminus \{0\}$ and all $a \in H$.

<u>Proof</u>. Let $\alpha \in R_+ \setminus \{0\}$ and let $a \in H$. Then $\lambda((n+\alpha) \square a) = \max(\lambda(a), \lambda(\alpha \square a))$ for all $n \in \mathbb{N}$. Therefore it is sufficient to consider $\alpha \in (0,1)$. Now $\lambda(\alpha \square a) \leq \max\{\lambda(\alpha \square a), \lambda((1-\alpha) \square a)\} = \lambda(a)$. On the other hand let $n \in \mathbb{N}$ with $1 \leq n\alpha$. Then $\lambda(\alpha \square a) = \lambda((n\alpha) \square a)$ $= \max\{\lambda(a), \lambda((n\alpha - 1) \square a)\} \geq \lambda(a)$.

■

Proposition (6.9) shows that the external composition decomposes into external compositions on $R_+ \setminus \{0\} \times H_\lambda \to H_\lambda$ for $\lambda \in \Lambda$. From (4.9) we know that H_λ is positively ordered with the possible exception of $\lambda = \min \Lambda$. Therefore we find the following result.

(6.10) Proposition

Let H be an ordinal semimodule over R_+. Then $H_\lambda \cup \{e\}$ is a semimodule over R_+ with (6.4.2), (6.4.3), and (6.4.4) for all $\lambda \in \Lambda \setminus \min \Lambda$.

We remark that if $\lambda = \min \Lambda$ then H_λ is a semimodule over R_+ but the monotonicity properties (6.4.2), (6.4.3), and (6.4.4) may be invalid.

A d-monoid H has an ordinal decomposition $(H_\lambda;\ \lambda \in \Lambda)$. If H is positively ordered then $H_{\lambda_o} = \{e\}$ for $\lambda_o = \min \Lambda$. Otherwise the structure of H_{λ_o} is described in theorem (4.10).

(6.11) Theorem

Let H be a d-monoid which is a semimodule over R_+. If the semimodule H_{λ_o} over R_+ satisfies the monotonicity rule (6.4.3) then H is a linearly ordered semimodule over R_+.

Proof. Due to (6.7), (6.8), and (6.10) $H_\lambda \cup \{e\}$ is a linearly ordered semimodule over R_+ for all $\lambda > \lambda_o$. H_{λ_o} is a semimodule over R_+. The monotonicity rule (6.4.3) together with proposition (6.7) shows that H_{λ_o} is a linearly ordered semimodule over R_+. ■

In (4.18) we proved that dom(a) and pos(a) for $a \in H$ are sub-d-monoids of the d-monoid H. A similar result holds with respect to the corresponding semimodules. For convenience we define $R_+ \square A := \{\alpha \square a \mid \alpha \in R_+,\ a \in A\}$ for $A \subseteq H$.

(6.12) <u>Proposition</u>

Let H be a d-monoid which is a semimodule over R_+. Then

(1) $R_+ \square \mathrm{dom}(a) = \mathrm{dom}(a)$ for all $a \in H$,

(2) H is a linearly ordered semimodule over R_+ iff

$R_+ \square \mathrm{pos}(a) = \mathrm{pos}(a)$ for all $a \in H$.

<u>Proof</u>. (1) Let $b \in \mathrm{dom}(a)$. Then $b^n \in \mathrm{dom}(a)$ for all $n \in \mathbb{N}$. Therefore it suffices to consider $\alpha \in R_+ \cap (0,1)$. Now $(\alpha \square b) * a =$ $= (\alpha \square b) * (\alpha \square a) * ((1-\alpha) \square a) = (\alpha \square (a * b)) * ((1-\alpha) \square a) = a$.
(2) Assume that H is a linearly ordered semimodule over R_+. Again it suffices to consider $\alpha \in R_+ \cap (0,1)$ and $b \in \mathrm{pos}(a)$. If $b \in \mathrm{dom}(a)$ then $\alpha \square b \in \mathrm{dom}(a) \subseteq \mathrm{pos}(a)$ follows from (1). Otherwise $e < b$ and therefore $e < (\alpha \square b)$ as (6.4.2) is valid. Then $a \leq a * (\alpha \square b)$. For the reverse implication it suffices to show that (6.4.3) holds in H over R_+. For the special choice of $a = e$ we find $R_+ \square H_+ = H_+$, i.e. (6.4.3).

■

In proposition (4.14) cancellation rules and related properties in weakly cancellative d-monoids are stated. (4.19.1) implies the following cancellation rule.

(6.13) Proposition

Let H be a weakly cancellative d-monoid which is a linearly ordered semimodule over R_+. If $\alpha \Box a \neq \beta \Box a$ or $\lambda(a) \leq \lambda(b)$ then

$$(\alpha \Box a) * b = (\beta \Box a) * c \quad \Rightarrow \quad b = [(\beta - \alpha) \Box a] * c$$

for all $a,b,c \in H$ and all $\alpha, \beta \in R_+$ with $\alpha \leq \beta$.

Proof. The case $a = e$ is trivial. If $a < e$ then the inverse element $a^{-1} > e$ exists. Composition with $\alpha \Box a^{-1}$ leads to $b = (\beta \Box a) * (\alpha \Box a^{-1}) * c$. Then $\beta \Box a = [(\beta - \alpha) \Box a] * (\alpha \Box a)$ implies $b = [(\beta - \alpha) \Box a] * c$. Now let $e < a$. At first we assume $\alpha \Box a \neq \beta \Box a$. Then $\alpha \Box a < \beta \Box a$ and thus $\alpha < \beta$. Suppose $\lambda(a) > \lambda(b)$. Then $\beta \Box a > \alpha \Box a = \beta \Box a * c$ leads to a contradiction. Therefore $\lambda(a) \leq \lambda(b)$. In the second case $\lambda(a) \leq \lambda(b)$ is valid, too. Therefore in both cases $\lambda(\alpha \Box a) \leq \lambda(b)$ and $\lambda(\alpha \Box a) \leq \lambda(((\beta - \alpha) \Box a) * c)$. (4.19.1) implies $b = [(\beta-\alpha) \Box a] * c$.

■

Weakly cancellative d-monoids which are semimodules over R_+ play an important role in the consideration of optimization problems in part II. The ordinal decomposition of such a monoid H may contain H_λ with $|H_\lambda| = 1$. Such *trivial* subsemigroups with an idempotent element ($a \neq e$ for $\lambda \neq \lambda_0$) lead to difficulties in the formulation and validation of algorithms. We avoid these difficulties by a certain extension of the underlying semimodule. Let H be a weakly cancellative d-monoid with ordinal decomposition $(H_\lambda ; \lambda \in \Lambda)$ and assume that H is a linearly ordered semimodule over R_+. Let $H_\mu = \{a\}$ for some $\mu \in \Lambda$. Then it is convenient to extend the trivial semigroup H_μ to

(6.14.1) $\qquad \tilde{H}_\mu := \{(a,r) \mid r \in R_+ \setminus \{0\}\}$

in the case $\mu > \lambda_o = \min \Lambda$ and to

(6.14.2) $\qquad \tilde{H}_\mu := \{(a,r) \mid r \in R\}$

in the case $\mu = \lambda_o$. Then let

$$(a,r) * b := \begin{cases} (a,r) & \text{if } \lambda(b) < \lambda(a), \\ b & \text{if } \lambda(b) > \lambda(a), \end{cases} \tag{6.15}$$

$$(a,r) * (a,r') := (a,\ r + r'),$$

for all $(a,r), (a,r') \in \tilde{H}_\mu$ and all $b \in H \setminus \{a\}$. The external composition is defined by

$$\alpha \square (a,r) := \begin{cases} e & \text{if } \alpha = 0, \\ (a,\ \alpha r) & \text{if } \alpha > 0. \end{cases} \tag{6.16}$$

We identify $a \leftrightarrow (a,1)$ if $\mu > \lambda_o$ and $a \leftrightarrow (a,0)$ if $\mu = \lambda_o$. The order relation on $\tilde{H}_\mu$ is the usual one with respect to the second component. In this way $\tilde{H}_\mu$ replaces H_μ in the ordinal decomposition. If this is done for all trivial semigroups the new semimodule $\tilde{H}$ is called *extended*. This is again a linearly ordered semimodule over R_+. We remark that therefore an extended semimodule is always linearly ordered by definition. A linearly ordered semimodule without trivial semigroups in its ordinal decomposition is clearly extended. In particular, $\tilde{H}$ is a weakly cancellative d-monoid.

The ordinal decomposition of an extended semimodule H has w.l.o.g. (cf. remarks after 4.13) the following form. $H_{\lambda_o} =: G_{\lambda_o}$ is a nontrivial linearly ordered commutative group with neutral element e which is also the neutral element of H. For $\lambda > \lambda_o$ we find that H_λ is the strict positive cone of a nontrivial linearly ordered commutative group G_λ. We may identify the neutral element of G_λ with e. Then $G_\lambda = \{a^{-1} \mid a \in H_\lambda\} \cup \{e\} \cup H_\lambda$. The order relation on G is completely determined by the positive cone $H_\lambda \cup \{e\}$ (cf. remarks after 2.7). Using (5.12) to continue the external composition on $R \times G_\lambda \to G_\lambda$ for $\lambda > \lambda_o$ we find that G_λ is a module over the ring R for all $\lambda \in \Lambda$. We remark that it is not useful to consider the ordinal sum of the groups G_λ. If $|\Lambda| > 1$ then such an ordinal sum is not an ordered semigroup.

Several cancellation properties in extended semimodules will be helpful in part II.

(6.17) Proposition

Let H be a weakly cancellative d-monoid the ordinal decomposition of which contains no trivial subsemigroups. Then

(1) $a < a * b \iff e < b$ and $\lambda(a) \leq \lambda(b)$,

(2) $a = a * b \iff e = b$ or $\lambda(a) > \lambda(b)$,

(3) $a > a * b \iff e > b$ and $\lambda(a) = \lambda_o$,

for all $a,b \in H$. If H is an extended semimodule over R_+ then

(4) $\alpha \leq \beta \iff \alpha \square a \leq \beta \square a$,

(5) $\alpha \leq \beta \iff \alpha \square b \geq \beta \square b$,

for all $\alpha,\beta \in R_+$ and all $a,b \in H$ with $b < e < a$ and

(6) $a \leq b \iff \alpha \square a \leq \alpha \square b$

for all $\alpha \in R_+$, $\alpha > 0$ and all $a,b \in H$. If $\alpha < \beta$ or $\lambda(a) \leq \lambda(b)$ then

(7) $(\alpha \square a) * b = (\beta \square a) * c \Rightarrow b = [(\beta - \alpha) \square a] * c$

for all $a,b,c \in H$ and all $\alpha,\beta \in R_+$ with $\alpha \leq \beta$.

Proof. (1) Clearly, $a < a * b$ implies $e < b$ and $\lambda(a) \leq \lambda(b)$. For the reverse implication assume $e < b$ and $\lambda(a) \leq \lambda(b)$. If $\lambda(a) < \lambda(b)$ then $a < a * b$ follows from the definition of ordinal sums. If $\lambda(a) = \lambda(b)$ then $a < a * b$ follows from (4.19.4) applied to an extended semimodule.

(2) and (3) are proved similarly.

(4) As (5.10) is satisfied it suffices to prove that $\alpha \,\square\, a \leq \beta \,\square\, a$ implies $\alpha \leq \beta$ if $e < a$. If $\alpha \,\square\, a < \beta \,\square\, a$ then $\alpha < \beta$. Now assume $\alpha \,\square\, a = \beta \,\square\, a$. Suppose $\beta < \alpha$. Then $[(\alpha - \beta) \,\square\, a] * (\beta \,\square\, a) = (\beta \,\square\, a)$ together with (2) imply $(\alpha - \beta) \,\square\, a = e$ contrary to (6.4.2) which holds in a linearly ordered semimodule.

(5) and (6) are proved similarly.

(7) If $\lambda(a) \leq \lambda(b)$ then (7) follows from (6.13). Now assume $\lambda(a) > \lambda(b)$ and $\alpha < \beta$. Then $\lambda(a) > \lambda_o$ and therefore $a > e$. In an extended semimodule this implies $\alpha \,\square\, a < \beta \,\square\, a$. Again (6.13) shows (7).

■

From (1) - (3) in proposition (6.17) we conclude the following corollary which describes the sets corresponding to the different definitions of positivity and negativity considered in the discussion of d-monoids in chapter 4.

(6.18) Corollary

Let H be a weakly cancellative d-monoid the ordinal decomposition of which contains no trivial subsemigroups.

(1) The set H_+ of all positive elements is equal to $\{a \mid e \leq a\}$; strictly positive elements exist only if Λ has a maximum

and then the set of all strictly positive elements is $\{a \mid e < a,\ \lambda(a) = \max \Lambda\}$. The set H_- of all negative elements is nonempty only if $|\Lambda| = 1$; then $H_- = \{a \mid a \leq e\}$ and the set of all strictly negative elements is equal to $H_- \setminus \{e\}$.

(2) The set P_- of all self-negative elements is equal to $H \setminus H_+$; the only idempotent is e and the set $\{a \mid a < a * a\}$ of all *self-positive* elements is equal to $H_+ \setminus \{e\}$.

(3) Let $a \in H$. The partition $N(a) < \mathrm{dom}(a) < P(a)$ of H in strictly negative, dominated and strictly positive elements with respect to a is given by

$$H \setminus H_+ < \{e\} < H_+ \setminus \{e\}$$

for $\lambda(a) = \lambda_o$ and is given by $N(a) = \emptyset$ and

$$\{b \mid \lambda(b) < \lambda(a)\} < \{b \mid \lambda(b) \geq \lambda(a)\}$$

for $\lambda(a) > \lambda_o$.

We remark that any weakly cancellative d-monoid can be embedded into a weakly cancellative d-monoid the ordinal decomposition of which does not contain trivial subsemigroups (via the corresponding discrete semimodule).

Finally we discuss the possibility that a given linearly ordered semimodule H over R_+ is contained in a linearly ordered semimodule H' of R'_+ with $H \subsetneq H'$ or $R \subsetneq R'$, i.e. H over R_+ is a *sub-semimodule* of H' over R'_+. At first we consider two examples and show that the semimodules considered are not sub-semimodules of a semimodule with $H = H'$ and $R \subsetneq R'$.

Let $H = \mathbb{Z}_+$ be the additive monoid of the nonnegative integers.

Then H is a linearly ordered semimodule over $\mathbb{Z}_+$ with usual multiplication as external composition. Suppose that H is a sub-semimodule of a linearly ordered semimodule H over R_+ with $\mathbb{Z}_+ \subsetneqq R_+$. Then R_+ is dense in $\mathbb{R}_+$. For $\alpha \in (0,1)$ we try to define $\alpha \,\square\, 1 \in \mathbb{Z}_+$. Monotonicity shows $0 < \alpha \,\square\, 1 \leq 1$ and therefore $\alpha \,\square\, 1 = 1$. For the special choice $\alpha \in (0, 1/2)$ we find the contradiction

$$1 = (2\alpha) \,\square\, 1 = 2 \,\square\, (\alpha \,\square\, 1) = 2 \,\square\, 1 = 1 + 1 = 2.$$

Therefore no such semimodule exists.

Now let $H = \mathbb{Q}_+$ be the additive monoid of nonnegative rationals. Then H is a linearly ordered semimodule over $\mathbb{Q}_+$ with usual multiplication as external composition. Suppose that H is a subsemimodule of a linearly ordered semimodule H over $\mathbb{R}_+$. Let $\beta \in \mathbb{R}_+ \setminus \mathbb{Q}_+$. Then $\alpha \leq \beta \,\square\, 1 \leq \gamma$ for all $\alpha, \gamma \in \mathbb{Q}_+$ with $\alpha < \beta < \gamma$. There exists no rational number $\beta \,\square\, 1$ with this property. Therefore no such semimodule exists.

In both examples the semimodules considered are subsemimodules of the linearly ordered semimodule $(\mathbb{R}_+, +, \leq)$ over $\mathbb{R}_+$ with respect to usual multiplication as external composition. Secondly, the following example shows that a given linearly ordered semimodule H over R_+ with $R \subsetneqq \mathbb{R}$ exists which is not a subsemimodule of any linearly ordered semimodule H' over R'_+ with $H \subseteq H'$ and $R_+ \subsetneqq R'_+$.

Let H be the fundamental monoid $([0,1)], *, \leq)$ with respect to the usual order relation and $a * b := \min(a+b, 1)$. H is a linearly ordered semimodule over $\mathbb{Z}_+$ with external composition $z \,\square\, a = a^z = \min(za, 1)$. Now we try to define $\alpha \,\square\, 1$ for $\alpha \in (0,1)$.

$(\alpha \square 1) * (\alpha \square 1) = \alpha \square (1 * 1) = \alpha \square 1$ shows that $\alpha \square 1$ is an idempotent element of H'. Let $1 \leq n\alpha$ for some $n \in \mathbb{N}$. Then $1 = 1 \square 1 \leq (n\alpha) \square 1 = (\alpha \square 1)^n = \alpha \square 1$. In particular, let $\alpha \in (0, 1/2]$ and $a \in [1/2, 1]$. Then $\alpha \square 1 = \alpha \square (a * a) = \alpha \square (2 \square a) = (2\alpha) \square a \leq 1 \square a = a \leq 1/2$. This contradiction to $1 \leq \alpha \square 1$ shows that no such semimodule exists.

Thirdly, let H be a submonoid of the linearly ordered commutative monoid H'. Then H and H' are linearly ordered semimodules over $\mathbb{Z}_+$ and H is a sub-semimodule of H'. In general, if H is a linearly ordered semimodule over R_+ then it is not known whether the external composition $\square$: $R_+ \times H \to H$ can be continued on $R_+ \times H'$ such that H' is a linearly ordered semimodule over R_+.

Nevertheless, we make the following conjecture. Let H be a weakly cancellative d-monoid which is a linearly ordered semimodule over R_+. Then this semimodule can be embedded in a linearly ordered semimodule H' over $\mathbb{R}_+$ with weakly cancellative d-monoid $H' \supseteq H$. We remark that using the ordinal decomposition of such a semimodule and using the embedding result of HAHN (cf. theorem 3.5) it is possible to provide a proof for the cases $R_+ \subseteq \mathbb{Q}_+$ and $\mathbb{Q}_+ \subseteq R_+$.

PART II LINEAR ALGEBRAIC OPTIMIZATION

7. Linear Algebraic Problems

In part II we discuss optimization problems which are generalizations of linear and combinatorial optimization problems. Differently from part I we assume that the reader is familiar with the basic concepts and results which are necessary for the solution of such problems in the classical case. For *linear programming* we refer to a chapter in BLUM and OETTLI [1975], for *combinatorial optimization* to LAWLER [1976].

Further we will use without explicit explanation the concept of *computational complexity* as described in GAREY and JOHNSON [1979] or in AHO, HOPCROFT and ULLMAN [1974]. The greater part of the problems considered can be solved by a *polynomial time algorithm* in the classical case. In ordered algebraic structures the computational complexity of performing an internal composition or of checking an equation or inequation is usually not known, in general. Therefore we can only give bounds for the number of such algebraic operations.

The generalization of the optimization problems considered consists in replacing classical linear functions by functions which are linear with respect to a semimodule. Let $(H,*)$ be a semimodule over a semiring $(R,\oplus,\otimes)$ with external composition $\square: R \times H \to H$ (cf. chapters 5 and 6). Let $c_i \in H$ for $i \in N := \{1,2,\ldots,n\}$. Then the function $z: R_+^n \to H$ defined by

(7.1) $$z(x) := x^T \square c = (x_1 \square c_1) * (x_2 \square c_2) * \ldots * (x_n \square c_n)$$

is called a *linear algebraic function* with *coefficients* $c_i \in H$ for $i \in N$.

Such a function is not linear in the classical sense. For example, we consider the semimodule $(\mathbb{R} \cup \{\infty\}, \min)$ over the semiring $(\mathbb{R}_+, +, \cdot)$ with external composition $\square: \mathbb{R}_+ \times (\mathbb{R} \cup \{\infty\}) \to \mathbb{R} \cup \{\infty\}$ defined by

$$\alpha \square a := \begin{cases} a & \text{if } \alpha > 0, \\ \infty & \text{otherwise.} \end{cases}$$

Then we find the linear algebraic function

$$z(x) = \min\{c_j \mid x_j > 0\}$$

for $x \in \mathbb{R}_+^n$ with coefficients $c_i \in \mathbb{R} \cup \{\infty\}$, $i \in N$.

In particular, we may consider the semimodule $(R, \oplus)$ over the semiring $(R, \oplus, \otimes)$. Then the linear algebraic function has the form

$$z(x) = x^T \otimes c = (x_1 \otimes c_1) \oplus (x_2 \otimes c_2) \oplus \ldots \oplus (x_n \otimes c_n)$$

for $x \in R^n$ with coefficients $c_i \in R$, $i \in N$. Further, a commutative semigroup $(H, *)$ is a discrete semimodule over the semiring $(\mathbb{Z}_+, +, \cdot)$. Then the linear algebraic function has the form

$$z(x) = x^T \square c = c_1^{x_1} * c_2^{x_2} * \ldots * c_n^{x_n}$$

for $x \in \mathbb{Z}_+^n$ with coefficients $c_i \in H$, $i \in N$.

Now we consider a subset S of R^n. The elements of S are called *feasible* or *feasible solutions*. In ordered semimodules H over R we define the *linear algebraic optimization problem* by

(7.2) $$\tilde{z} := \inf_{x \in S} x^T \square c$$

with *cost coefficients* $c_i \in H$, $i \in N$. In general, existence of an optimal value $\tilde{z} \in H$ is not known; even if $\tilde{z}$ exists then it is possible that $\tilde{z}$ is not attained for any feasible solution. Such questions are discussed for the respective problems in the following chapters. For most of these problems we can give a positive answer.

As for classical linear programs and combinatorial optimization problems a characterization of the set of all feasible solutions by linear equations and inequations is very useful. The necessary basic definitions for matrix compositions over ordered semimodules are introduced at the end of chapter 5. *Linear algebraic equations (inequations)* appear in various forms similarly to linear algebraic functions; for example

$$A \otimes x \leq b$$

for an $m \times n$ matrix A over R and $b \in R^m$. Then $\{x \in R^n \mid A \otimes x \leq b\}$ is the corresponding set of its solutions. In the discussion of duality principles we find linear algebraic equations of the form

$$A^T \square u \geq c$$

for an $m \times n$ matrix A over R and $c \in H^n$.

In view of the discussion of ordered algebraic structures in part I it is highly improbable that one can develop a reasonable solution method for the general problem (7.2). Nevertheless, it is possible to solve (7.2) for several classes of problems; the simpler the underlying combinatorial structure of the problem the more general the algebraic structure in which a solution method can be developed. By analyzing a given classical optimization problem and a given method for its solution we may find an appropriate algebraic structure in which the generalized method remains valid for the solution of the generalized optimization problem. Using this algebraic approach we can give a unifying treatment of different optimization problems and different solution methods; in this way the relationship between various problems and methods becomes more transparent. Further by specializing the general method for the subsumed classical problems it is often possible to find new solution methods; in this way we may develop more efficient algorithms.

We did not try to cover all optimization problems which have been discussed over ordered algebraic structures. We selected two classes of ordered algebraic structures which have been discussed by many authors during the last three decades. As these investigations were made separately many results have been reinvented from time to time. We do not claim that we succeeded in finding all the relevant literature but we hope that the resulting bibliography is helpful to the reader.

In the first class we consider optimization problems over semirings $(R,\oplus,\otimes)$. We discuss *algebraic path problems* (chapter 8) which can be solved via the evaluation of certain matrix series

$$A^* := I \oplus A \oplus A^2 \oplus \ldots$$

or equivalently via the determination of a certain solution Z of the matrix equation

$$Z = (Z \otimes A) \oplus B .$$

Then we investigate *algebraic eigenvalue problems* (chapter 9)

$$A \otimes x = \lambda \otimes x$$

which are closely related to path problems. Finally we consider (7.2) over semirings, i.e. the *extremal linear program* (chapter 10)

$$\tilde{z} = \inf\{c^T \otimes x \mid A \otimes x \geq b,\ x \in R^n\}.$$

In the second class we discuss *algebraic linear programs* in extended semimodules over real numbers, i.e.

$$\tilde{z} = \{x^T \,\square\, a \mid Cx \geq c,\ x \in R^n_+\}$$

for $R \subseteq \mathbb{R}$. We develop a duality theory similarly to the classical case (chapter 11) which can be applied for a solution of combinatorial optimization problems with linear algebraic objective function. We discuss *algebraic network flow problems*, including solution methods for *algebraic transportation* and *assignment problems* (chapter 12). Many combinatorial structures can be described by *independence systems*. The corresponding algebraic optimization problems are considered in chapter 13; in particular, we investigate *algebraic matroid* and *2-matroid intersection problems*. *Algebraic matching problems* can be solved by similar techniques, the difference lies only in the combinatorial structure.

All discussed algebraic optimization problems involve only linear algebraic constraints and linear algebraic objective functions. There is no difficulty in formulating nonlinear algebraic problems in the same way. Only very few results for nonlinear algebraic problems are known. At the time being a theory for such problems comparable to the linear case has not been developed to our knowledge.

8. Algebraic Path Problems

In this chapter we consider paths in a directed graph $G = (N,E)$ with vertex set N and arc set $E \subseteq N \times N$. We assign *weights* a_{ij} to the arcs $(i,j) \in E$. These weights are elements of a semiring $(R,\oplus,\otimes)$ with unity 1 and zero 0. Such a graph is called a *network*. The *length* $l(p)$ of a path $(e_1,e_2,\ldots,e_m)$ is the number of its arcs, i.e. $l(p) = m$; the *weight* $w(p)$ is defined by

$$w(p) := a_{e_1} \otimes a_{e_2} \otimes \ldots \otimes a_{e_m} .$$

Let P denote the set of all paths in G and let P_{ij} denote the set of all paths in G from vertex i to vertex j. The determination of the value (if it exists)

$$(8.1) \qquad a^*_{ij} := \bigoplus_{p \in P_{ij}} w(p)$$

for some $i,j \in N$ (or for all $i,j \in N$) and the determination of a path $p \in P_{ij}$ with $w(p) = a^*_{ij}$ (if it exists) for some $i,j \in N$ (or for all $i,j \in N$) is called the *algebraic path problem*.

For convenience we introduce the *weight function* $\sigma\colon 2^P \to R$ defined by

$$(8.2) \qquad \sigma(S) := \begin{cases} 0 & \text{if } S = \emptyset , \\ \bigoplus_{p \in S} w(p) & \text{if } S \neq \emptyset . \end{cases}$$

This function is not always well-defined as P is often infinite. Later on we discuss assumptions which imply that at least $\sigma(P_{ij})$ is well-defined. Then $a^*_{ij} = \sigma(P_{ij})$.

At first we consider the classical shortest path problem. In this case the weights are elements of $(\mathbb{R} \cup \{\infty\}, \min, +)$ with zero ∞ and unity 0. We may as well consider the linearly ordered commutative monoid $(\mathbb{R} \cup \{\infty\}, +, \leq)$ with absorbing maximum ∞ and neutral element 0. In any case, the special resulting path problem

$$(8.3) \qquad a^*_{ij} := \min_{p \in P_{ij}} \sum_{(\mu,\nu) \in p} a_{\mu\nu} \qquad , \; (i,j \in N)$$

is called the *shortest path problem*. Then $p \in P_{ij}$ with $a^*_{ij} = w(p)$ is called a *shortest path* from i to j. The first discussion of this problem seems to be due to BORŮVKA [1926]. He admits only nonnegative weights. In this case shortest paths from a given vertex i to all remaining vertices $j \neq i$ in G can simultaneously be determined using an elegant method proposed by DIJKSTRA [1959] and DANTZIG [1960].

Let $N = \{1,2,\ldots,n\}$. The weight u_i of a shortest path from vertex 1 to vertex $i = 2,3,\ldots,n$ is computed in the following procedure. W.l.o.g. we assume $a_{ii} = 0$ for $i \in N$ and $a_{ij} := \infty$ for $(i,j) \notin E$.

(8.4) <u>Shortest paths in networks with real nonnegative weights</u>

Step 1. $u_j := a_{1j}$ for $j = 1,2,\ldots,n$; $T := N \setminus \{1\}$

Step 2. Determine $k \in T$ with

$$u_k = \min\{u_j \mid j \in T\};$$

let $T := T \setminus \{k\}$.

Step 3. If $T = \emptyset$ then stop.

Otherwise $u_j := \min\{u_j, u_k + a_{kj}\}$ for all $j \in T$;

go to step 2.

In particular $u_j = \infty$ at the end of the procedure indicates that there exists no path from vertex 1 to vertex j. We will give an inductive proof of the validity of this method.

We claim that at each stage of the algorithm two properties hold ($Q := N \setminus T$ for the current T):

(8.5) *for all* $j \in Q$ u_j is the weight of a shortest path connecting 1 and j,

(8.6) *for all* $j \in T$ a_j is the weight of a shortest path connecting 1 and j subject to the condition that each node in the path belongs to $Q \cup \{j\}$.

Clearly (8.5) and (8.6) are satisfied after step 1. Now assume that the current Q,T,u fulfill (8.5), (8.6) before step 2. Let $k \in T$ be determined in step 2 and suppose the weight of a shortest path p connecting 1 and k is $v < u_k$. Then p contains a first node $j \notin Q \cup \{k\}$. Therefore $v \geq u_k$ contrary to our assumption. Thus (8.5) is satisfied by $Q \cup \{k\}$, too. In particular, if $T = \{k\}$ then the weights of all shortest paths are determined. Otherwise let v denote the weight of a shortest path connecting 1 and $j \in T \setminus \{k\}$ subject to the condition in (8.6) for $Q \cup \{k,j\}$. Let i denote the predecessor of j in such a path. If $j \neq k$ then $v = u_j$. Otherwise $v = u_k + a_{kj}$. Therefore (8.6) is satisfied for $Q \cup \{k\}$, $T \setminus \{k\}$ and the revised weights u_j in step 3.

The performance of this method needs $O(n^2)$ comparisons and $O(n^2)$ additions.

Next we discuss the generalization of this method for weights from ordered commutative monoids $(H,*,\leq)$ with neutral element e. We remark that H is an ordered semimodule over $(\mathbb{Z}_+,+,\cdot,\leq)$ with respect to the trivial external composition $\square$: $\mathbb{Z}_+ \times H \to H$ defined by

$$(8.7) \qquad z \square a := a^z$$

for all $a \in H$, $z \in \mathbb{Z}_+$ (in particular $a^o := e$). A path p in the directed graph G may be represented by its *incidence vector* $x \in \{0,1\}^E$ defined by

$$(8.8) \qquad x_{ij} = \begin{cases} 1 & \text{if } (i,j) \in p, \\ 0 & \text{otherwise.} \end{cases}$$

The set of all incidence vectors of paths in P_{ij} is denoted by S_{ij}. Then we may rewrite the shortest path problem for the case of such a semimodule as

$$(8.9) \qquad a^*_{ij} := \min_{x \in S_{ij}} x^T \square a \qquad , (i,j \in N)$$

with coefficients (weights) $a_{\mu\nu} \in H$. If we assume the existence of a^*_{ij} and the existence of a path connecting i and j with weight a^*_{ij} (an *algebraic shortest path*) then H has to be totally ordered, in general. An algebraic version of (8.4) can be given by simply replacing the equation in step 3 by

$$(8.10) \qquad u_j = \min\{u_j, u_k * a_{kj}\}.$$

A glance through our previous proof of the validity of (8.4) shows that the algebraic version is valid if H is a positively linearly ordered commutative monoid. Commutativity is not necessary for a proof of the validity; but without commutativity H is not a semimodule over $\mathbb{Z}_+$. If we try to avoid the assumption that H is linearly ordered then we cannot be sure that we can perform the steps in the algorithm: in step 2 we may find that $\min\{u_j \mid j \in T\}$ does not exist, in general. All weights have to be nonnegative (i.e. $e \leq a_{ij}$); otherwise even in the classical case counterexamples can easily be given. In terms of a semiring the algebraic version has been discussed by GONDRAN [1975c], MINOUX [1976] and CARRÉ [1979]. GONDRAN mentions the following examples of positively linearly ordered commutative monoids. For an application of the algebraic version of (8.4) let w.l.o.g. $a_{ij} := \infty$ (maximum of H) for $(i,j) \notin E$ and $a_{ii} = e$ (neutral element of H) for all $i \in N$.

(8.11) Examples

(1) $(\mathbb{R}_+ \cup \{\infty\}, +, \leq)$ with neutral element 0 and maximum ∞. Here we get the *classical shortest path* problem.

(2) $([0,1], \cdot, \geq)$ with neutral element 1 and maximum 0. Here w(p) is the product of arc-weights; this is the so-called *reliability* problem.

(3) $(\{0,1\}, \cdot, \geq)$ with neutral element 1 and maximum 0. Here $a^*_{ij} = 1$ iff a path connecting i and j exists (*connectivity* problem).

(4) $(\mathbb{R}_+ \cup \{\infty\}, \min, \geq)$ with neutral element ∞ and maximum 0. Here w(p) is the capacity of a path p and a^*_{ij} the *maximum capacity* of a path connecting i and j.

For additional examples of such an ordered algebraic structure we refer to chapter 4. In particular, the solution of (8.11.4) is related to maximum weighted trees in undirected graphs (cf. KALABA [1960]).

For the remaining part of this chapter we drop the assumption that the underlying algebraic structure is positively linearly ordered. The semiring $(R,\oplus,\otimes)$ is only pseudoordered (cf. 5.14) and, in general, a path with weight a^*_{ij} will not exist. a^*_{ij} is the supremum over the weights of all paths from i to j; clearly, this supremum will not always exist, as P_{ij} may be infinite.

For convenience we introduce some denotations for sets of paths which will appear in the following discussion. Let $i,j \in N$ and $k \in \mathbb{N} \cup \{0\}$. Then $\hat{P}_{ij}$ denotes the set of all elementary paths in P_{ij}, $P^{[k]}_{ij}$ denotes the set of all paths $p \in P_{ij}$ with $l(p) \leq k$ and P^k_{ij} denotes the set of all paths $p \in P_{ij}$ with $l(p) = k$. For technical reasons we assume the existence of a circuit p of length 0 and unity weight containing only vertex i but no arc. Such a circuit is called the null circuit of i. Then $\sigma(P^{[0]}_{ii}) = 1$ for each $i \in N$ and $\sigma(P^{[0]}_{ij}) = 0$ for each (i,j) with $i \neq j$.

As an introductory example we consider the classical shortest path problem with possible negative weights in the semiring $(\mathbb{R} \cup \{\infty\},\min,+)$ with zero ∞ and unity 0. If G contains circuits of negative weight (denotation: *negative circuits*) then (8.3) may be unbounded. If $P_{ij} \neq \emptyset$ and if G does not contain a negative circuit then there exists a shortest path from i to j which is elementary.

We assume $a_{ii} = 0$ for all $i \in N$ and $a_{ij} = \infty$ if $(i,j) \notin E$. Then for $\sigma(P_{ij}^{[k]})$, $k = 0,1,2,\ldots$ we find the following recursion:

(8.12) $$\sigma(P_{ij}^{[k]}) = \min\{\sigma(P_{ik}^{[k-1]}) + a_{kj} \mid k \in N\}$$

for $k = 1,2,\ldots$. For $k = 0$ we know $\sigma(P_{ii}^{[0]}) = 0$ for $i \in N$ and $\sigma(P_{ij}^{[0]}) = \infty$ for $i \neq j$. An inductive proof of this recursion is given in the following.

For $k = 1$ validity of (8.12) is obvious. W.l.o.g. we assume that G is a complete digraph. Then $P_{ij}^{[k]} \neq \emptyset$ for $k \geq 1$. Let $k \geq 2$ and let $p \in P_{ij}^{[k]}$ be a path with weight $\sigma(P_{ij}^{[k]})$. If $l(p) < k$ then $w(p) = \sigma(P_{ij}^{[k-1]}) + a_{jj}$. Otherwise $l(p) = k$. Then let n denote the predecessor of j in p. Now $w(p) = \sigma(P_{ih}^{[k]}) + a_{hj}$. In both cases (8.12) is satisfied.

As G contains no negative circuits we can derive the weights of the shortest paths from (8.12). We find

(8.13) $$a_{ij}^* = \sigma(P_{ij}^{[n-1]})$$

since an elementary path has at most length $n-1$. Let $U^{[k]}$ denote the matrix with entries $\sigma(P_{ij}^{[k]})$. Then (8.12) can be interpreted as matrix-composition over the semiring $(\mathbb{R} \cup \{\infty\}, \min,+)$ (cf. chapter 5). For $A := (a_{ij})$ we find

(8.14) $$A^* = U^{[n-1]} = (((U^{[0]} + A) + A) \ldots) + A.$$

As $U^{[0]}$ is the identity of the semiring of $n \times n$ matrices over $\mathbb{R} \cup \{\infty\}$ this leads to $A^* = A^{n-1}$. Therefore a computation of A^{n-1} solves the shortest path problem. As $A^s = A^{n-1}$ for all $s \geq n-1$ we may determine A^{n-1} by the sequence

$$A, A^2, A^4, \ldots, A^{2^k} ;$$

therefore $O(\log_2 n)$ matrix multiplications are necessary, each of which needs $O(n^3)$ comparisons and $O(n^3)$ additions. In particular, we can determine the weights of all shortest paths from node 1 to node j for all $j \neq 1$ from the first row A_1^* of A^*, i.e.

$$(8.15) \qquad A_1^* = ((\ldots ((0\ \infty \ldots \infty) + A) + A) \ \ldots\ ) + A;$$

this recursion is the classical BELLMAN - FORD method. It needs $O(n^3)$ additions and $O(n^3)$ comparisons. Improvements of this method can be found in YEN [1975].

In the general case, i.e. for weights in arbitrary semirings, a solution of (8.1), if it exists, is determined by computation of A^k and

$$(8.16) \qquad A^{[k]} := I \oplus A \oplus A^2 \oplus \ldots \oplus A^k$$

where I is the unity matrix in the semiring of all $n \times n$ matrices over the underlying semiring $(R, \oplus, \otimes)$. Problems which can be reduced to (8.1) and methods for the computation of these matrices, called *matrix-methods*, have been considered by many authors during the last thirty years. We mention only those which explicitly use some algebraic structures. The following list will hardly be complete.

LUNTS [1950] considers applications to relay contact networks, SHIMBEL ([1951], [1953] and [1954]) to communication nets. KLEENE [1956] bases a theorem in the algebra of regular events on a Gaussian elimination scheme which is separately developed for the transitive closure of a graph by ROY [1959] and

WARSHALL [1962]. MOISIL [1960] discusses certain shortest path problems, YOELI [1961] introduces Q-semirings and PANDIT [1961] considers a matrix calculus. Further results and generalizations are due to GIFFLER ([1963], [1968]). CRUON and HERVÉ [1965], PETEANU ([1967a], [1967b], [1969], [1970], [1972]), TOMESCU [1968], ROBERT and FERLAND [1968], BENZAKEN [1968] and DERNIAME and PAIR [1971]. Methods based on fast matrix multiplication are discussed by ARLAZAROV et al. [1970], FURMAN [1970], FISCHER and MEYER [1971] and MUNRO [1971]. CARRÉ [1971] develops a systematic treatment of several linear algebra procedures. This approach is continued by BRUCKER ([1972], [1974]), SHIER ([1973], [1976]), MINIEKA and SHIER [1973], GONDRAN ([1975], [1975a], [1975b], [1975c], [1975d], [1979] and [1980]), AHO, HOPCROFT and ULLMAN [1974], MARTELLI ([1974], [1975]), ROY [1975], BACKHOUSE and CARRÉ [1975], TARJAN ([1975], [1976]), FRATTA and MONTANARI [1975], WONGSEELASHOTE ([1976], [1979]), LEHMANN [1977], MAHR ([1979], [1980a] and [1980b]), FLETCHER [1980], and ZIMMERMANN, U. [1981]. GONDRAN and MINOUX [1979b] as well as CARRÉ [1979] summarize many of these results.

The following proposition shows the different character of A^k and $A^{[k]}$. The proof is drawn from WONGSEELASHOTE [1979].

(8.17) <u>Proposition</u>

Let A be an $n \times n$ matrix over a semiring $(R,\oplus,\otimes)$. Then

(1) $(A^m)_{ij} = \sigma(P^m_{ij})$,

(2) $(A^{[m]})_{ij} = \sigma(P^{[m]}_{ij})$,

for all $i,j \in N$ and all $m \in \mathbb{N} \cup \{0\}$.

Proof. For paths $p \in P_{ij}$ and $q \in P_{jk}$ let $p \circ q$ denote the path in P_{ik} traversing at first p and then q. For $Q \subseteq P_{ij}$ and $S \subseteq P_{jk}$ let $Q \circ T := \{p \circ q \mid p \in Q, q \in T\}$. Then

$$(3) \qquad \sigma(Q \circ T) = \sigma(Q) \otimes \sigma(T) ,$$

provided that the weight function is well-defined for Q and T. Now for $m \geq 1$

$$(4) \qquad P^m_{ij} = \cup \{P^1_{ir_1} \circ P^1_{r_1 r_2} \circ \ldots \circ P^1_{r_{m-1} j} \mid \begin{matrix} \text{pairwise different} \\ r_1, \ldots, r_m \end{matrix} \}.$$

Therefore (1) is satisfied for $m \geq 1$. For m = O we find $A^O = A^{[O]} = I$. As P^O_{ii} contains the null-circuit with weight equal to the identity 1 of the semiring (3) and (4) show (1). As

$$P^{[m]}_{ij} = \cup \{P^k_{ij} \mid k = O, 1, \ldots, m\},$$

(2) is implied by (1). ∎

In our introductory example $(\mathbb{R} \cup \{\infty\}, \min, +)$ $A^m = A^{[m]}$ and $A^{[s]} = A^{[n-1]}$ for all $s \geq n-1$. Such results do not hold in the general case. If there exists $r \in \mathbb{N} \cup \{O\}$ such that

$$(8.18) \qquad A^{[r]} = A^{[r+1]}$$

then A is called *stable*. The minimum of the integers satisfying (8.18) is called the *stability index* of A. For stable matrices the values a^*_{ij} in (8.1) exist and from (8.17) we find

$$(8.19) \qquad a^*_{ij} = (A^{[r]})_{ij}$$

for all $i,j \in N$. We remark that (8.18) implies $A^{[s]} = A^{[r]}$ for all $s \geq r$. Sufficient conditions for a matrix to be stable are conveniently formulated in terms of the circuits of the network G. Let C denote the *set of all elementary non-null circuits* and $W := \{w(p) \mid p \in C\}$. Let

(8.20) $\alpha^{[m]} := 1 \oplus \alpha \oplus \alpha^2 \oplus \ldots \oplus \alpha^m$

for $\alpha \in R$ and $m \in \mathbb{N} \cup \{0\}$ and let

(8.21) $\alpha[m] := 1 \oplus \alpha_1 \oplus (\alpha_1 \otimes \alpha_2) \oplus \ldots \oplus (\alpha_1 \otimes \alpha_2 \otimes \ldots \otimes \alpha_m)$

for $\alpha \in R^s$, $s \geq m \in \mathbb{N} \cup \{0\}$. The network G is called *absorptive* if

(8.22) $\alpha^{[0]} = \alpha^{[1]}$ for all $\alpha \in W$,

m-regular if

(8.23) $\alpha^{[m]} = \alpha^{[m+1]}$ for all $\alpha \in W$,

and *m-absorptive* if

(8.24) $\alpha[m] = \alpha[m+1]$ for all $\alpha \in W^{m+1}$.

If G is absorptive then G is m-absorptive. If G is m-absorptive then G is m-regular. Absorptive networks over certain semirings were introduced by CARRÉ [1971]; the generalizations (8.23) and (8.24) are due to GONDRAN [1975a] and ROY [1975].
We remark that a *semiring* R is called *absorptive* (*m-regular, m-absorptive*) if the respective property is satisfied for *all* elements of R. An absorptive semiring is idempotent.

(8.25) Proposition

If G is an m-regular network then for all $\alpha \in W$

(1) $\alpha^{[m]} = \alpha^{[s]}$ for all $s \geq m+1$,

(2) $\alpha^{[m]} = \alpha^{[m]} \oplus \alpha^s$ for all $s \geq m+1$;

if G is an m-absorptive network then for all $\alpha \in W^s$

(3) $\alpha[m] = \alpha[k]$ for all $s \geq k \geq m+1$,

(4) $\alpha[m] = \alpha[m] \oplus (\alpha_1 \otimes \ldots \otimes \alpha_s)$ for all $s \geq m+1$.

<u>Proof</u>. We show (3) and (4). A proof of (1) and (2) is similar.

(3). For $s = k = m+1$ this is obvious. Let $s = k > m+1$ and assume

$$\beta[s-2] = \beta[s-1] \qquad \text{for all } \beta \in W^{s-1}.$$

Then we get

$$\begin{aligned}\alpha[s] &= 1 \oplus \alpha_1 \otimes (1 \oplus \alpha_2 \oplus \ldots \oplus (\alpha_2 \otimes \ldots \otimes \alpha_s))\\ &= 1 \oplus \alpha_1 \otimes (1 \oplus \alpha_2 \oplus \ldots \oplus (\alpha_2 \otimes \ldots \otimes \alpha_{s-1}))\\ &= \alpha[s-1] .\end{aligned}$$

Therefore $\alpha[s-1] = \alpha[s]$ for all $\alpha \in W^s$ and for all $s \geq m+1$. This leads to (3) as $\alpha \in W^s$ implies $(\alpha_1, \alpha_2, \ldots, \alpha_k) \in W^k$ for $s \geq k$. In particular, (3) implies

$$\begin{aligned}\alpha[m] = \alpha[s] &= \alpha[s-1] \oplus (\alpha_1 \otimes \alpha_2 \otimes \ldots \otimes \alpha_s)\\ &= \alpha[m] \quad \oplus (\alpha_1 \otimes \alpha_2 \otimes \ldots \otimes \alpha_s)\end{aligned}$$

for all $\alpha \in W^s$, $s \geq m+1$. Thus (4) is satisfied. ■

In (5.14) we introduced the pseudoordering

$$a \leq b \quad :\Longleftrightarrow \quad a \oplus b = b$$

of the semiring $(R, \oplus, \otimes)$. G is absorptive iff

$$\alpha \leq 1$$

for all $\alpha \in W$, G is m-regular iff

$$\alpha^{m+1} \leq \alpha^{[m]}$$

for all $\alpha \in W$ and G is m-absorptive iff

$$\alpha_1 \otimes \alpha_2 \otimes \ldots \otimes \alpha_{m+1} \leq \alpha[m] \qquad \text{for all } \alpha \in W^{m+1}.$$

Similarly, properties (8.25.1 - 4) can be interpreted in terms of this pseudoordering.

Let Q,T be subsets of P with $Q \subseteq T$. Then Q is called *dense in T* if for all $p \in T \setminus Q$ there exists a finite subset H of Q such that $w(p) \leq \sigma(H)$.

(8.26) Proposition

Let Q be a dense subset of T. If $Q \subseteq B \subseteq T$ then B is dense in T. Further, if B is finite, then $\sigma(B) = \sigma(Q)$.

Proof. Let $p \in T \setminus B$. Then $p \in T \setminus Q$ and therefore there exists a finite subset H of Q with $w(P) \leq \sigma(H)$. As $H \subseteq B$ the set B is dense in T. If B is finite then $B \setminus Q$ is finite. Now for $p \in B \setminus Q$ and some $H \subseteq Q$ we know $w(p) \leq \sigma(H)$. Then $0 \leq \sigma(Q \setminus H)$ and (5.14.1) lead to $w(p) \leq \sigma(Q)$. (5.14.2) shows $\sigma(B \setminus Q) \leq \sigma(Q)$, i.e. $\sigma(B) = \sigma(B \setminus Q) \oplus \sigma(Q) = \sigma(Q)$.

■

The set of all paths $p \in P_{ij}$ which *traverse an elementary non-null circuit at most m times* is denoted by $\hat{P}^m_{ij}$. Similarly $\hat{P}^{(m)}_{ij}$ denotes the set of all paths $p \in P_{ij}$ which *traverse at most m elementary non-null circuits*. Let $p \in \hat{P}^m_{ij}$, $q \in \hat{P}^{(m)}_{ij}$. Then

$$(8.27) \qquad l(p) \leq n\,|C|\,m + n - 1 =: s,$$
$$l(q) \leq n\,m + n - 1 =: t\,;$$

therefore $\hat{P}^m_{ij} \subseteq P^{[s]}_{ij}$ and $\hat{P}^{(m)}_{ij} \subseteq P^{[t]}_{ij}$.

(8.28) Proposition

(1) If G is absorptive then $\hat{P}^0_{ij}$ is dense in P_{ij} ;

(2) if G is m-absorptive and R is commutative then $\hat{P}^{(m)}_{ij}$ is dense in P_{ij} ;

(3) if G is m-regular and R is commutative then $\hat{P}^{m}_{ij}$ is dense in P_{ij} .

Proof. (1) Let $p \in P_{ij} \setminus \hat{P}^{o}_{ij}$. Let $p' \in \hat{P}^{o}_{ij}$ denote the elementary path derived from p by eliminating all circuits in p. As p contains at least one elementary non-null circuit q and $w(q) \leq 1$ the monotonicity condition (5.2) (cf. 5.14) implies $w(p) \leq w(p')$. Thus $\hat{P}^{o}_{ij}$ is dense in P_{ij}.

(2) From (8.25.4) we know that for each s-tuple $q = (q_1, q_2, \ldots, q_s) \in C^s$ with $s \geq m+1$ the relation

$$(8.29) \qquad w(q_1) \otimes w(q_2) \otimes \ldots \otimes w(q_s) \leq q[m]$$

is satisfied. Let $p \in P_{ij} \setminus \hat{P}^{(m)}_{ij}$. Then p traverses s elementary non-null circuits $q_1, q_2, \ldots, q_s$ with $s \geq m+1$. Successive elimination of $q_s, q_{s-1}, \ldots, q_1$ leads to the paths $p_1, p_2, \ldots, p_s$. In particular p_s is an elementary path (or a null path with weight $w(p_s) := 1$). As R is commutative we find

$$w(p) = w(p_s) \otimes w(q_1) \otimes w(q_2) \otimes \ldots \otimes w(q_s) .$$

The monotonicity condition (5.2) together with (8.29) shows

$$(8.30) \qquad w(p) \leq w(p_s) \otimes q[m] .$$

Successive "addition" of $q_1, q_2, \ldots, q_m$ to p_s results in $p_{s-1}, p_{s-2}, \ldots, p_{s-m}$. Let $H := \{p_s, p_{s-1}, \ldots, p_{s-m}\}$. Then (8.30) and commutativity imply

$$w(p) \leq \sigma(H) .$$

As $H \subseteq \hat{P}^{(m)}_{ij}$ this shows $\hat{P}^{(m)}_{ij}$ is dense in P_{ij}.

(3) From (8.25.2) we know

$$(8.31) \qquad w(q)^s \leq w(q)^{[m]} \qquad \text{for all } q \in C \text{ and all } s \geq m+1.$$

Let $p \in P_{ij} \setminus \hat{P}^m_{ij}$. Then p traverses at least one $q \in C$ exactly s times with some $s \geq m+1$. Let $p = p_1 \circ q \circ p_2 \circ q \circ \ldots \circ q \circ p_{s+1}$. Again there may be some null paths in this representation with weight 1. Commutativity leads to

$$w(p) = w(p_1) \otimes w(p_2) \otimes \ldots \otimes w(p_{s+1}) \otimes w(q)^s .$$

From (8.31) and monotonicity condition (5.2) we get

$$(8.32) \qquad w(p) \leq w(p_1) \otimes w(p_2) \otimes \ldots \otimes w(p_{s+1}) \otimes w(q)^{[m]} .$$

Let
$$p^{(1)} := p_1 \circ p_2 \circ \ldots \circ p_{s+1} ,$$
$$p^{(2)} := p_1 \circ q \circ p_2 \circ \ldots \circ p_{s+1}, \ldots ,$$
$$p^{(m)} := p_1 \circ q \circ p_2 \circ q \circ \ldots \circ q \circ p_{m-1} \circ p_m \circ \ldots \circ p_{s+1} .$$

(8.32) together with commutativity shows

$$w(p) \leq \sigma(\{p^{(1)}, p^{(2)}, \ldots, p^{(m)}\}).$$

Further each $p^{(i)}$ is a subpath of p which does traverse q at most m times. If p contains further elementary non-null circuits then the procedure is applied to the paths in $H = \{p^{(1)}, p^{(2)}, \ldots\}$. As there is only a finite number of such circuits the process terminates after a finite number of steps. (5.14.1) and (5.14.2) lead to $w(p) < \sigma(\tilde{H})$ with finite $\tilde{H} \subseteq \hat{P}^m_{ij}$. Thus $\hat{P}^m_{ij}$ is dense in P_{ij}.

■

The *matrix* A is called *absorptive (m-regular, m-absorptive)* if G is absorptive (m-regular, m-absorptive). The following theorem summarizes the results of CARRÉ [1971], GONDRAN [1975a] and ROY [1975]. In this form it can be found in WONGSEELASHOTE [1979].

(8.33) Theorem

Let $(R,\oplus,\otimes)$ be a semiring and let A be an $n \times n$ matrix over R. Then A is stable with stability index r if one of the following conditions holds

(1) A is absorptive (then $r \leq n-1$),

(2) A is m-absorptive and R commutative (then $r \leq nm + n-1$),

(3) A is m-regular and R commutative (then $r \leq nm\,|C| + n-1$).

Proof. From (8.28) we know that (1),((2),(3)) imply that $\hat{P}^{o}_{ij}$ ($\hat{P}^{(m)}_{ij}$, $\hat{P}^{m}_{ij}$) is dense in P_{ij}. Therefore the finite set $P^{[s]}_{ij}$ with $s \geq n-1$ ($s \geq nm+n-1$, $s \geq nm|C|+n-1$) is dense in P_{ij} and for all such s we find (cf. 8.26) that $\sigma(P^{[s]}_{ij})$ is equal to $\sigma(\hat{P}^{o}_{ij})$ ($\sigma(\hat{P}^{(m)}_{ij})$, $\sigma(\hat{P}^{m}_{ij})$). Then (8.17) shows that A is stable and has the respective stability index.

■

Theorem (8.33) gives sufficient conditions for a matrix to be stable. For such matrices the algebraic path problem (8.1) can be solved using the equations

(8.34) $$A^* = A^{[r]} = A^{[r+1]} = \ldots = A^{[s]} = \ldots$$

for all $s \geq r$, the stability index of A. By virtue of (8.16) the following recursive formula is valid

(8.35) $$A^{[k]} = (A^{[k-1]} \otimes A) \oplus I \quad , \qquad k = 1,2,\ldots$$

with $A^{[o]} = I$ where I denotes the $n \times n$ unity matrix over the semiring $(R,\oplus,\otimes)$. From (8.34) and (8.35) we conclude

(8.36) $$A^* = (A^* \otimes A) \oplus I \ .$$

As $A^{[k]} \otimes A = A \otimes A^{[k]}$ for all $k \in \mathbb{N} \cup \{o\}$ we find

(8.37) $$A^* = (A \otimes A^*) \oplus I \ .$$

Equations (8.36) and (8.37) show that A^* is the solution of certain matrix equations over R. Let $Y := A^* \otimes B$ and $Z := B \otimes A^*$ for a matrix B over R. Then (8.36) and (8.37) lead to

(8.38) $$Y = (A \otimes Y) \oplus B, \qquad Z = (Z \otimes A) \oplus B \, .$$

In particular, if B is the unit matrix then A^* is a solution of (8.38); if B is a unit vector, then a column resp. a row of A^* are solutions of (8.38). The semiring of all $n \times n$ matrices over R is pseudoordered by

$$C \oplus D = D \iff C \le D \, .$$

The solutions $A^* \otimes B$ resp. $B \otimes A^*$ are minimum solutions of (8.38) with respect to this ordering if R is idempotent. This follows from

(8.39) $$Y = (A^s \otimes Y) \oplus (A^{[s]} \otimes B)$$

for all $s = 0,1,2,\ldots$, if Y is a solution of (8.38). Then for $s \ge r$ we find

(8.40) $$Y \oplus (A^* \otimes B) = (A^s \otimes Y) \oplus (A^* \otimes B) \oplus (A^* \otimes B) \, .$$

Thus, if R is idempotent, then $A^* \otimes B \le Y$. If the underlying graph contains no circuits then A^s is the zero-matrix for $s \ge r = n-1$ and (8.39) shows $Y = A^* \otimes B$ is the unique solution of (8.38). The same remarks hold for the second equation in (8.38).

The solution of such equations can be determined by methods similar to usual methods in linear algebra. A systematical treatment of such an approach was at first developed by CARRÉ [1971] for certain semirings. Then well-known methods from network theory appear to be particular cases of such methods. Several authors (cf. remarks following 8.48) extended this approach to other algebraic structures or derived methods for networks of special structure. BACKHOUSE and CARRÉ [1975] discussed the similarity to the algebra of regular languages, for which the matrix A^* always exists due to a suitable axiomatic system.

Next we consider two solution methods for (8.1) which can be interpreted as certain linear algebra procedures. A straightforward formulation of a *generalized* or *algebraic JACOBI-method* for the second equation in (8.38) is

$$(8.41) \qquad Z^{(k+1)} := (Z^{(k)} \otimes A) \oplus B \;, \qquad k = 0,1,2,\dots$$

with $Z^{(0)} := B$.

(8.42) Proposition

Let A be a $n \times n$-matrix over the semiring $(R,\oplus,\otimes)$ with stability index r. Then (8.41) is a finite recursion for the determination of $B \otimes A^*$; we find

$$Z^{(k)} = B \otimes A^*$$

for all $k \geq r$.

Proof. An inductive argument shows $Z^{(k)} = B \otimes A^{[k]}$ for $k = 0,1,2,\dots$. ∎

We remark that for the classical shortest path problem over the semiring $(\mathbb{R} \cup \{\infty\},\min,+)$ and for $B = (0\ \infty \dots \infty)$ this method reduces to the method of BELLMAN [1958] for the calculation of the weights of all shortest paths from node 1 to all other nodes. In the general case with $B = I$ a complete solution of

(8.1) is determined; the values a^*_{ij} are *not necessarily attained* as weights of certain paths $p \in P_{ij}$. This method needs $O(r)$ iterations and each iteration requires $O(n^2)$ ⊗-compositions and $O(n^3)$ ⊕-compositions. If A is only m-regular $(m \neq 0)$ $r = nm\,|C| + n-1$. The number of elementary non-null circuits $|C|$ is not bounded by a polynomial in n. If A is m-absorptive then $r = O(n)$. In this case the method requires $O(n^4)$ ⊗-compositions and $O(n^4)$ ⊕-compositions.

The JACOBI-method is a so-called *iterative method* in linear algebra. The second class of solution techniques in linear algebra contains *direct* or *elimination methods*. We will consider the *generalized* or *algebraic GAUSS-JORDAN* method. Let $A^{(0)} := A$ and define recursively for $k = 1,2,\dots,n$

$$\text{(8.43)}\quad \begin{aligned} a_{kk}^{(k)} &:= (a_{kk}^{(k-1)})^*, \\ a_{ij}^{(k)} &:= a_{ij}^{(k-1)} \oplus (a_{ik}^{(k-1)} \otimes a_{kk}^{(k)} \otimes a_{kj}^{(k-1)}), \quad (i,j) \neq (k,k) \end{aligned}$$

for all $i,j \in N$. This recursion is well-defined, if the computation of $(a_{kk}^{(k-1)})^*$ is well-defined. For a discussion we introduce the set $T_{ij}^{(k)}$ which consists in all paths $p \in P_{ij}$ such that *p does not contain an intermediate vertex from the set* $\{k+1, k+2, \dots, n\}$. For a circuit $p \in T_{ii}^{(k)}$ with $k < i$ this means that p is a non-null circuit traversing i exactly once.

(8.44) <u>Proposition</u>

(1) If A is absorptive then $\hat{P}_{kk}^{0} \cap T_{kk}^{(k)}$ is dense in $T_{kk}^{(k)}$,

(2) if A is m-regular and R is commutative then $\hat{P}_{kk}^{m} \cap T_{kk}^{(k)}$ is dense in $T_{kk}^{(k)}$,

(3) if A is m-absorptive and R is commutative then

$\hat{P}_{kk}^{(m)} \cap T_{kk}^{(k)}$ is dense in $T_{kk}^{(k)}$.

Proof. Follows from the proof of (8.28) as in each case the constructed set H for $p \in T_{kk}^{(k)}$ satisfies $H \subseteq T_{kk}^{(k)}$. ∎

This proposition is the essential technical tool in the proof of the next theorem.

(8.45) Theorem

Let $(R,\oplus,\otimes)$ be a semiring and let A be an $n \times n$ matrix over R. Then A is stable with stability index r if one of the following conditions holds:

(1) A is absorptive (then $r \leq n-1$),

(2) A is m-absorptive and R commutative (then $r \leq nm+n-1$),

(3) A is m-regular and R commutative (then $r \leq nm|C|+n-1$).

In each case

(4) $$A^* = A^{(n)}$$

with respect to the recursive formula (8.43) and

(5) $$(a_{kk}^{(k-1)})^* = 1 \oplus a_{kk}^{(k-1)} \oplus \dots \oplus (a_{kk}^{(k-1)})^r$$

for all $k = 1,2,\dots,n$.

Proof. We claim that

(8.46) $$a_{ij}^{(k)} = \sigma(T_{ij}^{(k)})$$

for all $i,j \in N$ and all $k \in N \cup \{0\}$. As $T_{ij}^{(0)} = \{(i,j)\}$ we see that (8.46) holds for $k = 0$. Now assume that (8.46) holds for $k-1$ with $k \geq 1$ and for all $i,j \in N$.

At first we consider the case $i = j = k$. In each of the cases (1), (2) and (3) we find (cf. 8.44)

$$T_{kk}^{(k)} \cap P_{kk}^{[s]} \text{ is dense in } T_{kk}^{(k)}$$

for all $s \geq r$ and therefore

$$(8.47) \qquad \sigma(T_{kk}^{(k)}) = \sigma(T_{kk}^{(k)} \cap P_{kk}^{[r]}) .$$

In particular, the existence of $\sigma(T_{kk}^{(k)})$ has been shown. Now $T_{kk}^{(k)}$ consists in the null circuit and all circuits of the form $p = p_1 \circ p_2 \circ \ldots \circ p_s$ with circuits $p_\mu \in T_{kk}^{(k-1)}$ for $\mu = 1,2,\ldots,s$, and $s \in \mathbb{N}$. Let $\alpha := \sigma(T_{kk}^{(k-1)})$. Then

$$\alpha^* = 1 \oplus \alpha \oplus \alpha^2 \oplus \ldots$$

exists as

$$\alpha^s = \sigma(\{p_1 \circ p_2 \circ \ldots \circ p_s \mid p_\mu \in T_{kk}^{(k-1)} \text{ for } \mu = 1,2,\ldots,s$$

and (8.47) lead to

$$\alpha^{[r]} = \alpha^{[r+1]} .$$

Thus we have proved (5) for k and (8.46) for $i = j = k$. Secondly we consider $(i,j) \neq (k,k)$. Then $p \in T_{ij}^{(k)} \setminus T_{ij}^{(k-1)}$ contains vertex k and has the form $p = p_1 \circ p_2 \circ p_3$ with $p_1 \in T_{ik}^{(k-1)}$, $p_2 \in T_{kk}^{(k)}$ and $p_3 \in T_{kj}^{(k-1)}$. Hence

$$\sigma(T_{ij}^{(k)}) = \sigma(T_{ij}^{(k-1)}) \oplus [\sigma(T_{ik}^{(k-1)}) \otimes \sigma(T_{kk}^{(k)}) \otimes \sigma(T_{kj}^{(k-1)})]$$

which shows together with (8.43) that (8.46) is valid for k.

An inductive argument shows now that (8.46) and (5) hold for all $k = 0,1,\ldots,n$. In particular

$$a_{ij}^{(n)} = \sigma(P_{ij})$$

for all $i,j \in N$, i.e. $A^{(n)} = A^*$. ■

Theorem (8.45) shows that the recursion (8.43) can be used if some of our sufficient conditions for stability of A are fulfilled.

Another proof is due to FAUCITANO and NICAUD [1975] and can be found in GONDRAN and MINOUX [1979b]. They show that if $A^{(n)}$ exists then $A^{(n)} = A^*$. It is claimed that $(a_{kk}^{(k-1)})^*$ exists if diagonal pivoting is incorporated in recursion (8.43). Our proof shows that pivoting is not necessary if one of the assumptions (8.45.1), (8.45.2) or (8.45.3) holds. For the absorptive case a proof is also given by CARRÉ ([1971], [1979]).

Recursion (8.43) can be simplified in the following way. If α^* exists for $\alpha \in R$ then

$$\alpha^* = (\alpha^* \otimes \alpha) \oplus 1 = 1 \oplus (\alpha \otimes \alpha^*)$$

is satisfied (cf. 8.36 and 8.37). Therefore, in (8.43) we find for $i \neq j = k$

$$(8.43') \qquad a_{ik}^{(k)} = a_{ik}^{(k-1)} \oplus (a_{ik}^{(k-1)} \otimes a_{kk}^{(k)} \otimes a_{kk}^{(k-1)}) = a_{ik}^{(k-1)} \otimes a_{kk}^{(k)}$$

and similarly for $j \neq i = k$

$$(8.43'') \qquad a_{kj}^{(k)} = a_{kk}^{(k)} \otimes a_{kj}^{(k-1)} \; .$$

If α^* exists for $\alpha \in R$ then $\alpha^* \otimes \alpha^* = \alpha^*$. Together with (8.43') and (8.43") this leads to the following efficient formulation of (8.43).

(8.48) GAUSS-JORDAN ELIMINATION

Step 1. Define the original matrix A; let $k:= 1$.

Step 2. $a_{kk} := (a_{kk})^*$.

Step 3. $a_{ik} := a_{ik} \otimes a_{kk}$ for all $i \neq k$;

$a_{kj} := a_{kk} \otimes a_{kj}$ for all $j \neq k$.

Step 4. $a_{ij} := a_{ij} \oplus (a_{ik} \otimes a_{kj})$ for all $i,j \neq k$.

Step 5. If $k = n$ stop.

Otherwise $k:= k+1$ and go to step 2.

The method requires $O(n)$ iterations. Each iteration needs $O(n^2)$ $\oplus$-compositions, $O(n^2)$ $\otimes$-compositions and the computation of $(a_{kk})^*$, i.e.

$$(a_{kk})^* = 1 \oplus a_{kk} \oplus \ldots \oplus (a_{kk})^r .$$

Here we need $O(r)$ $\oplus$-compositions and $O(r)$ $\otimes$-compositions. Again if A is only m-regular ($m \neq 0$) then this does not yield a polynomial bound. If A is m-absorptive then $r = O(n)$. If A is absorptive then $(a_{kk})^* = 1$ and the method can be simplified. We may w.l.o.g. assume that the main diagonal of A contains only unities. Then step 2 and step 3 can be eliminated from the algorithm.

If A is m-absorptive then the method requires $O(n^3)$ $\oplus$-compositions and $O(n^3)$ $\otimes$-compositions. Therefore, in general, the GAUSS-JORDAN-method is superior to the JACOBI-method. Nevertheless, for special structured networks (as for example in tree-networks, i.e. $|C| = 0$) the JACOBI-method may turn out to be better (cf. SHIER [1973]).

Specific forms of this method have been developed by many authors. ROY [1959] and WARSHALL [1962] applied it to the transitive-closure problem in graphs, FLOYD [1962] to the classical shortest path problem, ROBERT and FERLAND [1968], CARRÉ [1971], BRUCKER ([1972], [1974]), BACKHOUSE and CARRÉ [1975], and GONDRAN [1975a] gave extensions to certain algebraic structures. MÜLLER-MERBACH [1969] used it for the inversion of Gozinto-matrices, MINIEKA and SHIER [1973], MINIEKA [1974] and SHIER [1976] applied it to the k-shortest path problem. MARTELLI ([1974], [1976]) enumerates all minimal cut sets by means of this method.

Other solution methods, as the GAUSS-SEIDEL iteration method or the ESCALATOR elimination method have been generalized also for certain semirings. The interested reader is referred to GONDRAN and MINOUX [1979b] and CARRÉ [1979].

We remind that stability of a matrix A implies the existence of the closure matrix A^*. Theorem (8.45) shows that certain types of stability imply the validity of the GAUSS-JORDAN method for the determination of A^*. In general, it is not known whether the GAUSS-JORDAN method is valid for all stable matrices. Nevertheless, no counterexample is known and we conjecture the validity of the GAUSS-JORDAN method for all stable matrices.

Instead of assuming that the matrix A satisfies certain conditions we may impose certain conditions on the underlying semiring. We remind that P_{ij} is always a countable set. Thus A^* is well defined for all matrices A over R if we assume that infinite sums of the form

$$a_1 \oplus a_2 \oplus a_3 \oplus \ldots$$

are well defined in R for $a_i \in R$, $i \in \mathbb{N}$. We denote such a series by $\Sigma_I a_i$ for a countable set I. In arbitrary semirings $\Sigma_I a_i$ is well defined only if I is a finite set. Besides $\Sigma_I a_i \in R$ we assume distributivity, i.e.

$$b \otimes (\Sigma_I a_i) = \Sigma_I (b \otimes a_i) ,$$

$$(\Sigma_I a_i) \otimes b = \Sigma_I (a_i \otimes b) ,$$

for all $b \in R$ and associativity, i.e.

$$\Sigma_I a_i = \Sigma_J (\Sigma_{I_j} a_i)$$

for all (countable) partitions (I_j, $j \in J$) of I. Then R is called a *(countably) complete semiring.*

In a complete semiring the GAUSS-JORDAN method is well defined for all matrices A over R. Its validity is shown in the same way as in the proof of theorem (8.45) by verifying (8.46). In a complete semiring the proof becomes much easier since the occuring countably infinite sums

$$\alpha^* = 1 \oplus \alpha \oplus \alpha^2 \oplus \ldots = \Sigma_{\mathbb{Z}_+} \alpha^i$$

are well defined.

(8.49) Theorem

Let $(R,\oplus,\otimes)$ be a complete semiring and let A be an $n \times n$ matrix over R. Then A^* is well defined and $A^* = A^{(n)}$.

Theorem (8.49) is proved by AHO, HOPCROFT and ULLMAN [1974] for idempotent complete semirings. FLETCHER [1980] remarks that idempotency is not necessary if the GAUSS-JORDAN method is formulated in the right way (equivalent to (8.43)). We remind that for stable matrices A^* is a solution of the matrix equation

$$X = I \oplus (A \otimes X) \tag{8.50}$$

(cf. 8.37). In a complete semiring $A^* = \Sigma_{\mathbb{Z}_+} A^i$ satisfies (8.50), too.

Next we discuss a weaker condition which does not necessarily imply the existence of A^*; on the other side a solution of (8.50) will exist and can easily be determined using a modification of the GAUSS-JORDAN method (8.43).

We assume the existence of a unary *closure* operation $a \to \bar{a}$ on the semiring R such that

$$\bar{a} = 1 \oplus (a \otimes \bar{a}) \tag{8.51}$$

for all $a \in R$. Then R is called a *closed semiring*. A complete semiring is always a closed semiring with respect to the closure operation defined by $\bar{a} := a^* = \Sigma_{\mathbb{Z}_+} a^i$. On the other side, if R is a closed semiring and if R is m-absorptive for some $m \in \mathbb{N}$, then $\bar{a} = a^*$ for all $a \in R$. In general, the closure operation does not necessarily coincide with the '*'-operation and is not uniquely determined. If a closure operation is defined only on a subset R' of R then LEHMANN [1977] proposes to adjoin an element $\infty \notin R$, to add the definitions

$$\infty \oplus a = a \oplus \infty = a \otimes \infty = \infty \otimes a = \infty$$

and to extend the closure operation by

$$\bar{a} := \infty$$

for all $a \in (R \cup \{\infty\}) \setminus R'$. Then $R \cup \{\infty\}$ satisfies all axioms of a closed semiring with the exception of

(8.52) $\quad a \otimes 0 = 0 \otimes a = 0 \qquad$ for all $a \in R$,

which is invalid for $a = \infty$. We call $R \cup \{\infty\}$ a *closure* of R. In this way an arbitrary semiring can be closed if we drop the assumption that R contains a zero. In particular, this approach is possible for the closure operation $a \to a^* = \Sigma_{\mathbb{Z}_+} a^i$ which is well defined only on a subset of the underlying semiring. Therefore we will avoid to use (8.52) in solving the matrix equation (8.50) in closed semirings.

The following variant of the GAUSS-JORDAN method (8.43) is well defined in a closed semiring. Let $B^{(0)} := A$ and define recursively for $k = 1,2,\ldots,n$:

(8.53) $\quad b_{ij}^{(k)} := b_{ij}^{(k-1)} \oplus (b_{ik}^{(k-1)} \otimes \bar{b}_{kk}^{(k-1)} \otimes b_{kj}^{(k-1)})$

for all $i,j \in N$. The following theorem is implicitly proved in LEHMANN [1977]. Its simple direct proof is drawn from MAHR [1980b].

(8.54) Theorem

Let $(R,\oplus,\otimes)$ be a closed semiring and let A be an $n \times n$ matrix over R. Then $I \oplus B^{(n)}$ is a solution of the matrix equation $X = I \oplus (A \otimes X)$.

Proof. At first we show

(1) $\quad b_{ij}^{(k)} = a_{ij} \oplus \sum_{\nu=1}^{k} (a_{i\nu} \otimes b_{\nu j}^{(k)}) \qquad (i,j \in N)$

for $k = 0,1,\ldots,n$. The case $k = 0$ is trivial. Let $i,j \in N$ and

$1 \leq k < n$. We assume that (1) is satisfied for $k-1$ and we denote $B^{(k-1)}$ and $B^{(k)}$ by B and B'. Since $1 \oplus (b_{kk} \otimes \bar{b}_{kk}) = \bar{b}_{kk}$ we find

$$b'_{kj} = b_{kj} \oplus (b_{kk} \otimes \bar{b}_{kk} \otimes b_{kj}) = \bar{b}_{kk} \otimes b_{kj} =: \beta$$

Now $b'_{\nu j} = b_{\nu j} \oplus (b_{\nu k} \otimes \beta)$ for all $\nu = 1,2,\ldots,k-1$. Therefore the right-hand-side of equation (1) is equal to

$$[a_{ij} \oplus \sum_{\nu=1}^{k-1} (a_{i\nu} \otimes b_{\nu j})] \oplus ([a_{ik} \oplus \sum_{\nu=1}^{k-1} (a_{i\nu} \otimes b_{\nu k})] \otimes \beta).$$

Applying the inductive assumption to the terms in brackets we find equality with

$$b_{ij} \oplus (b_{ik} \otimes \beta);$$

by (8.53) this term is equal to b'_{ij}.

Now let $B := B^{(n)}$, $C := I \oplus B$, and define $\delta_{ij} := 1$ if $i = j$ and $\delta_{ij} = 0$ otherwise. Then

$$c_{ij} = \delta_{ij} \oplus b_{ij} = \delta_{ij} \oplus a_{ij} \oplus \sum_{\nu=1}^{n} (a_{i\nu} \otimes b_{\nu j})$$

$$= \delta_{ij} \oplus \sum_{\nu=1}^{n} (a_{i\nu} \otimes c_{\nu j})$$

for all $i,j \in N$. Therefore C is a solution of the matrix equation $X = I \oplus (A \otimes X)$.

■

We remark that theorem (8.54) is proved without using (8.52). For $k = n$ the equations (1) in the proof show that $B^{(n)}$ is a solution of the matrix equation

(8.55) $\quad X = A \oplus (A \otimes X)$.

In a closed semiring R satisfying the additional equation

(8.56) $\bar{a} = 1 \oplus (\bar{a} \otimes a)$

for all $a \in R$ we may use the GAUSS-JORDAN method (8.43) for the determination of $I \oplus B^{(n)}$ if we replace the '*'-operation in (8.43) by the closure operation in R. Let $C^{(0)} := A$ and define recursively for $k = 1,2,\dots,n$:

(8.57)
$$c_{kk}^{(k)} := \bar{c}_{kk}^{(k-1)},$$
$$c_{ij}^{(k)} := c_{ij}^{(k-1)} \oplus (c_{ik}^{(k-1)} \otimes c_{kk}^{(k)} \otimes c_{kj}^{(k-1)}), \quad (i,j) \neq (k,k)$$

for all $i,j \in N$.

(8.58) Corollary

Let $(R,\oplus,\otimes)$ be a closed semiring satisfying (8.56) and let A be an $n \times n$ matrix over R. Then $C^{(n)} = I \oplus B^{(n)}$ is a solution of the matrix equation $X = I \oplus (A \otimes X)$.

Proof. A comparison of (8.53) and (8.57) shows that

$$c_{ij}^{(k)} = \begin{cases} 1 \oplus b_{ij}^{(k)} & \text{if } i = j \leq k, \\ b_{ij}^{(k)} & \text{otherwise} \end{cases},$$

for all $i,j \in N$ and for all $k = 0,1,\dots,n$. Therefore $C^{(n)} = I \oplus B^{(n)}$ ∎

We remark that Corollary (8.58) is proved without using (8.52). Therefore we may consider the '*'-operation as a particular well-defined example for a closure operation. Thus $A^{(n)}$ is always a solution of the matrix equation $X = I \oplus (A \otimes X)$ in (a closure of) the semiring R. On the other side A^* does not exist, in general (cf. 8.65).

In the remaining part of this chapter we discuss several examples for path problems. The underlying graph is always denoted by G = (N,E) with weights a_{ij} chosen from the respective semirings.

(8.59) Connectivity in graphs

If p is a path in G from i and j then these vertices are called connected. The existence of such paths in (undirected) graphs has first been discussed by LUNTS [1950] and SHIMBEL [1951] with respect to certain applications. They reduce the problem to the determination of the $n-1^{th}$ power of the *adjacency matrix* A defined by

$$a_{ij} := \begin{cases} 1 & \text{if } (i,j) \in E, \\ 0 & \text{otherwise} \end{cases}$$

for all $i,j \in N$ over the semiring ({0,1},max,min). This is a commutative, absorptive semiring with zero 1 and unity 0. Absorptive semirings have been discussed by YOELI [1961]. He called them *Q-semirings*. In particular, YOELI proved that in absorptive semirings

(8.60) $$A^{n-1} = A^*$$

for all matrices A with $A \oplus I = A$. The condition $A \oplus I = A$ is equivalent to $a_{ii} = 1$ (unity) for all $i \in N$. This shows that in our above example $A^{n-1}_{ij} = 1$ if and only if $P_{ij} \neq \emptyset$. In the case of undirected graphs A and A^{n-1} are symmetric matrices. The graph with adjacency matrix $A^* = A^{n-1}$ is called the *transitive closure* of G, and we speak of the transitive closure problem. ROY [1959] and WARSHALL [1962] developed the

GAUSS-JORDAN method (8.48) for the computation of A^{n-1}. Methods based on STRASSEN's fast matrix multiplication are proposed by ARLAZAROV et al. [1970], FURMAN [1970], FISCHER and MEYER [1971] and MUNRO [1971]. The average behavior of such methods is discussed by BLONIARZ, FISCHER and MEYER [1976] and by SCHNORR [1978].

(8.61) Shortest paths

The shortest path problem has at first been investigated by BORŮVKA [1926]. SHIMBEL [1954] reduced this problem to the computation of the $n-1^{th}$ power of the real arc-value matrix A over the semiring $(\mathbb{R}_+ \cup \{\infty\}, \min, +)$. This is a commutative absorptive semiring with zero ∞ and unity 0. SHIMBEL assumed $a_{ii} = 0$ for all $i \in N$. Therefore $A \oplus I = I$ and (8.60) implies again that $A^* = A^{n-1}$. If negative weights a_{ij} are involved then we have to assume that the network is absorptive, i.e. G does not contain a non-null elementary circuit p with weight $w(p) < 0$ (with respect to the usual order relation of the extended real numbers). Then the semiring $(\mathbb{R} \cup \{\infty\}, \min, +)$ is considered. Hence A^k_{ij} is the weight of a shortest path from i to j of length k and $A^{[k]}_{ij}$ is the weight of a shortest path from i to j of length not greater than k. In particular $A^* = A^{[n-1]}$ (cf. 8.33). Based on the paper of WARSHALL [1962] FLOYD [1962] proposed the application of the GAUSS-JORDAN-method for the determination of A^*. In particular, if all weights are nonnegative then algorithm (8.4) can be used (cf. DANTZIG [1960], DIJKSTRA [1959]). Several other solution methods can be found in the standard literature (cf. LAWLER

[1976]). Recent developments can be found in WAGNER [1976], FREDMAN [1976], JOHNSON [1977], BLONIARZ [1980] and HANSEN [1980]. An excellent bibliography is given by DEO and PANG [1980].

If the weights a_{ij} are chosen in $R := \mathbb{R}^k \cup \{(\infty,\infty,\ldots,\infty)\}$ then the weight of a lexicographically shortest path may be determined. Let

$$a \oplus b := \text{lex min}(a,b)$$

for $a,b \in \mathbb{R}^k$ with respect to the lexicographical order relation $\preccurlyeq$ of real vectors. Then $(R,\oplus,+)$ is an idempotent, commutative semiring with zero $(\infty,\infty,\ldots,\infty)$ and unity $(0,0,\ldots,0)$. A^*_{ij} is the weight of a lexicographically shortest path $p \in P_{ij}$. A^* exists if and only if the network is absorptive, i.e. if G contains no circuit p with $w(p) \prec (0,0,\ldots,0)$. Such a problem was first considered by BRUCKER ([1972], [1974]).

(8.62) Most reliable paths

If the weights a_{ij} are chosen from the semiring $([0,1],\max,\cdot)$ with zero 0, unity 1 and usual multiplication of real numbers then the weight

$$w(p) = a_{i_1 j_1} \cdot a_{i_2 j_2} \cdot \ldots \cdot a_{i_s j_s}$$

of a path $p = ((i_1,j_1),(i_2,j_2),\ldots,(i_s,j_s))$ may be interpreted as the *reliability* of the path p. This semiring is again commutative and absorptive. A^*_{ij} is the weight of the most reliable path from node i to node j. This problem was first considered by KALABA [1960].

(8.63) Maximum capacity paths

If the weights a_{ij} are chosen from the semiring $(\mathbb{R}_+ \cup \{\infty\}, \max, \min)$ with zero ∞ and unity 0 then the weight

$$w(p) = \min\{a_{i_1 j_1}, a_{i_2 j_2}, \ldots, a_{i_s j_s}\}$$

of a path $p = ((i_1,j_1),(i_2,j_2),\ldots,(i_s,j_s))$ may be interpreted as the *capacity* of the path p. This semiring is again commutative and absorptive. A^*_{ij} is the weight of a maximum capacity path from vertex i to vertex j. This problem was first considered by HU [1961].

(8.64) Absorptive semirings and absorptive networks

YOELI [1961] discussed absorptive semirings (Q-semirings) and the relevance of A^{n-1} and A^*. The application of the GAUSS-JORDAN-method in absorptive semirings was proposed independently by TOMESCU [1968] and ROBERT and FERLAND [1968]. CARRÉ [1971] considered such algorithms for absorptive networks over idempotent semirings under the assumption that the composition $\otimes$ is cancellative. In particular, distributive lattices are discussed in BACKHOUSE and CARRÉ [1975]. For Boolean lattices (Boolean algebras) we refer to HAMMER and RUDEANU [1968].

(8.65) Inversion of matrices over fields

A field $(F,+,\cdot)$ with zero 0 and unity 1 is, in particular, a semiring. If A is a stable matrix over F with stability index r then $A^{[r]} = A^{[r+1]} = A^{[r]} + A^{r+1}$ implies that A^s is the zero-matrix for all $s \geq r+1$. Let p be a circuit of length $l(p)$. Then

$m \cdot l(p) \geq r+1$ for some $m \in \mathbb{N}$. Therefore $w(p) = 0$. Thus all circuits of G have weight zero. If all circuits in G have weight zero then A is stable with stability index $r \leq n-1$. (8.36) and (8.37) imply

$$A^* = (I - A)^{-1} .$$

In particular, if A is stable then $I - A$ is a regular matrix over F. The algorithms for the computation of A^* given in this section reduce to the conventional linear algebra methods for the inversion of $I - A$. Applications for networks without circuits have been given by MÜLLER-MERBACH [1970] (so-called Gozinto-graphs).

For the closure $\mathbb{R} \cup \{\infty\}$ of the field of real numbers LEHMANN [1977] introduces the closure operation $a \to \bar{a}$ defined by

$$\bar{a} := \begin{cases} \infty & \text{if } a \in \{1,\infty\}, \\ 1/(1-a) & \text{otherwise} \end{cases} .$$

Clearly $\bar{a} = 1 + (a\bar{a}) = 1 + (\bar{a}a)$ for all $a \in \mathbb{R} \cup \{\infty\}$. We remark that $\bar{a} = a^* = 1 + a + a^2 + \ldots$ for all $a \in \mathbb{R}$ with $|a| < 1$.
The closure operation $a \to a^*$ may be extended on $\mathbb{R} \cup \{\infty\}$ by $a^* := \infty$ if $|a| \geq 1$. Since $1 + a + a^2 + \ldots$ is an oscillating divergent series for $a < -1$ the algebraic path problem does not have a solution in $\mathbb{R} \cup \{\infty\}$ if the underlying graph contains a circuit $p = (e_1, e_2, \ldots, e_p)$ with weight

$$w(p) = a_{e_1} \cdot a_{e_2} \cdot \ldots \cdot a_{e_p} < -1 .$$

For the different closure operations algorithm (8.57) leads to different matrices $C^{(n)}$ denoted by A_- and A_* which both solve the matrix equation $X = I + (AX)$ in $\mathbb{R} \cup \{\infty\}$ (cf.

Corollary 8.58). A_* will contain more infinite elements than A_-. If A_- does not contain an infinite element then $A_- = (I-A)^{-1}$. The same result holds for $A*$. For a given nonsingular matrix A we find $A^{-1} = (I - A)_-$ provided that $(I - A)_-$ does not contain an infinite element. If P is a permutation matrix such that $(I - PA)_-$ does not contain an infinite element then $A^{-1} = (I-PA)_- P$. Using well known pivoting rules A^{-1} can always be determined in this way.

GONDRAN [1975] considered the number of paths from node i to node j. For the adjacency matrix A with respect to the field of real numbers we find that A^k_{ij} is the number of paths from i to j of length k, $A^{[k]}_{ij}$ is the number of paths from i to j of length not greater than k and, if G does not contain a circuit, then $A^*_{ij} = A^{[n-1]}_{ij}$ is the number of all paths from i to j. It is always assumed that $A_{ii} = 0$ for all $i \in N$.

(8.66) <u>Regular algebra</u>

BACKHOUSE and CARRÉ considered closed idempotent semirings with a closure operation $\alpha \to \bar{\alpha}$ satisfying

$$\bar{\alpha} = \overline{(1 \oplus \alpha)}$$

(8.67)

$$\psi = (\alpha \otimes \psi) \oplus \beta \;\Rightarrow\; \psi \geq \bar{\alpha} \otimes \beta$$

for all $\alpha, \beta, \psi \in R$ and

(8.68) $$\psi = (\alpha \otimes \psi) \oplus \beta \;\Rightarrow\; \psi = \bar{\alpha} \otimes \beta$$

for all $\alpha, \beta, \psi \in R$ such that

$$\alpha \otimes \lambda = \lambda \;\Rightarrow\; \lambda = 0$$

for all $\lambda \in R$. They discuss only examples in which $\bar{\alpha}$ coincides with the usual closure operation $\alpha \to \alpha^*$, i.e.

$$\alpha^* = 1 \oplus \alpha \oplus \alpha^2 \oplus \dots \quad .$$

The standard example of a regular algebra is the following. Let $V = \{v_1, v_2, \dots, v_n\}$ be a finite nonempty set. V is called an *alphabet* and its elements are called *letters*. A *word* over V is a finite string of zero or more letters from V. In particular the empty string ε is called *empty word*. The *set of all words* is denoted by $\bar{V}$. A *language* is any subset of $\bar{V}$. The *sum* $\alpha + \beta$ *of two languages* $\alpha, \beta \subseteq \bar{V}$ is $\alpha \cup \beta$. For words $w = w_1 w_2 \dots w_r$ and $t = t_1 t_2 \dots t_s$ we define

$$w \circ t := w_1 w_2 \dots w_r t_1 t_2 \dots t_s \quad ;$$

$w \circ t$ is called the *concatenation* of w and t. The complex concatenation of two languages $\alpha, \beta \subseteq \bar{V}$ is denoted by $\alpha \circ \beta$. Then $(\bar{V}, +, \circ)$ is an idempotent semiring with zero $\emptyset$ and unity $\{\varepsilon\}$. The closure

$$\alpha^* = \{\varepsilon\} \oplus \alpha \oplus \alpha^2 \oplus \dots$$

is well-defined; in particular $\emptyset^* = \{\varepsilon\}$. We denote single-letter languages $\{a\}$ shortly by a. Then a *regular expression* over V is any well-formed formula obtained from the elements $V \cup \{\emptyset\}$, the operators $+, \circ$ and $*$, and the parentheses (and). For example

$$\emptyset^* + (v_2 \circ (v_4 + v_1)^* \circ v_3)^*$$

is a regular expression. Each regular expression denotes a language over V which is called a *regular language*. The *set of all regular languages* is denoted by *S(V)*. $(S(V), +, \circ, *)$ is a regular algebra ($\bar{\alpha} = \alpha^*$). The axiomatic system defining regular algebras was put forward for regular languages by SALOMAA [1969]. In a proof of a theorem on languages in automata theory KLEENE [1956] uses a GAUSS-JORDAN method for S(V). In McNAUGHTON and YAMADA [1960] the same method is described explicitly. For the theoretical background in automata theory we refer to EILENBERG ([1974], p. 175).

Another example of a regular algebra is an absorptive semiring $(R,\oplus,\otimes)$ (cf. 8.64). Then $\bar{\alpha} = \alpha^* = 1 \oplus \alpha = 1$ for all $\alpha \in R$. Together with idempotency this shows (8.67). Distributive lattices are included as a particular case.

BACKHOUSE and CARRÉ [1975] discuss several linear algebra methods for solving matrix equations in regular algebra. In particular, theorem (8.54) shows the validity of a variant of the GAUSS-JORDAN method (cf. 8.53). If the considered closure operation coincides with the usual *-operation then corollary (8.58) shows the validity of the GAUSS-JORDAN method (8.57).

(8.69) Path and cut set enumeration

In order to determine the set of all paths from node i to node j we consider the regular algebra S(E) where E denotes the set of all arcs of the underlying graph G. Then the weight a_{ij} is given by

$$(8.70) \qquad a_{ij} = \begin{cases} \{(i,j)\} & \text{if } (i,j) \in E, \\ \emptyset & \text{otherwise} \end{cases},$$

and the weight of a path $p = (e_1, e_2, \ldots, e_s)$ is

$$w(p) = e_1 \circ e_2 \circ \ldots \circ e_s \ .$$

(Here we denote single-letter languages by e_i rather than $\{e_i\}$.)

For $k \in \mathbb{N}$ we get

$$a_{ij}^{k} = \{w(p) \mid p \in P_{ij}^{k}\} \quad ,$$

$$a_{ij}^{[k]} = \{(w(p) \mid p \in P_{ij}^{[k]}\},$$

$$a_{ij}^{*} = \{w(p) \mid p \in P_{ij}\}$$

for all $i,j \in N$. An actual computation of A^* is not possible in the general case, as the cardinality of a_{ij}^* may be infinite. If we are only interested in all elementary paths then a suitable modification of the underlying regular algebra is the following. A word w is called an *abbreviation* of a word t if w can be obtained from t by deletion of some of the letters of t. For example, METAL is an abbreviation of MATHEMATICAL, or if an elementary path p is obtained from a nonelementary path p' by deletion of the elementary circuits in p' then w(p) is an abbreviation of w(p'). For $\alpha \in S(E)$ let $r(\alpha)$ denote the language containing all words of α for which no abbreviation exists in α. Let

$$\tilde{S}(E) := \{\alpha \in S(E) \mid \alpha = r(\alpha)\}$$

and define

$$\alpha \oplus \beta := r(\alpha + \beta)$$

$$\alpha \otimes \beta := r(\alpha \circ \beta)$$

for all $\alpha,\beta \in \tilde{S}(E)$. Then $(\tilde{S}(E),\oplus,\otimes)$ is an absorptive semiring with zero $\emptyset$ and unity $\{\varepsilon\}$. Further, for A as defined by (8.70), we find

$$a_{ij}^{[k]} = \{w(p) \mid p \in P_{ij}^{[k]}, \quad p \text{ elementary}\}$$

for all $k \in \mathbb{N}$ and $A^* = A^{[n-1]}$, i.e. $a_{ij}^* = a_{ij}^{[n-1]}$ contains the weights of all elementary paths from i to j. For the computation of A^* KAUFMANN and MALGRANGE [1963] proposed the sequence

$$M, M^2, M^4, \ldots$$

with $M := I \oplus A$. Then $M^{2^k} = A^*$ for $2^k \geq n-1$ shows that this method, called *Latin multiplication*, needs $O(\log_2 n)$ matrix multiplications over the semiring $\tilde{S}(E)$. MURCHLAND [1965] and BENZAKEN [1968] applied the GAUSS-JORDAN method for a solution of this problem. Improvements can be found in BACKHOUSE and CARRÉ [1975] and FRATTA and MONTANARI [1975].

A related problem is the enumeration of all minimal cut sets of the underlying graph $G = (N,E)$. Let (S,T) denote a partition of N. Then $c = (S \times T) \cap E$ is called a *cut set* of G. If the arcs of a cut set c are removed from G then there exists no path in G from any $i \in S$ to any $j \in T$. Therefore we say that c *separates i from j*. If a cut set c does not contain a proper subset which is a cut set then c is called a *minimal cut set*. In order to determine the set of all minimal cut sets separating $i \in N$ from $j \in N$ the following algebra is introduced by MARTELLI ([1974], [1976]).

Let $P(E)$ denote the set of all subsets of E. For $\alpha \subseteq P(E)$ let $r(\alpha)$ denote the subset of α containing all sets in α for which no proper subset exists in α. Let

$$C(E) := \{\alpha \subseteq P(E) \mid \alpha = r(\alpha)\}$$

and define

$$\alpha \oplus \beta := r\{a \cup b \mid a \in \alpha,\ b \in \alpha\}$$

$$\alpha \otimes \beta := r(\alpha \cup \beta)$$

for all $\alpha, \beta \in C(E)$. Then $(C(E), \oplus, \otimes)$ is an absorptive semiring with zero $\{\emptyset\}$ and unity $\emptyset$.

We assign weights $\hat{a}_{ij} := \{a_{ij}\} \in C(E)$ where a_{ij} is defined by (8.60). The weight $w(p)$ of a path $p = (e_1, e_2, \ldots, e_s)$ with respect to the semiring $C(E)$ is the set of its arcs, i.e.

$$w(p) = \{e_k \mid k = 1, 2, \ldots, s\}.$$

Due to theorem (8.33) we find $A^* = A^{[n-1]}$. Each element $c \in a^*_{ij}$ has the form

$$c = \cup \{e(p) \mid p \in P_{ij}\}$$

where $e(p)$ denotes an arbitrary arc of p. Thus there exists no path from i to j in the graph $G_c = (N, E \setminus c)$. Let S denote the set of all vertices k such that there exists a path from i to j in G_c, and let $T := N \setminus S$. Then $c = (S \times T) \cap E$, i.e. c is a cut set separating i and j in G. Thus a^*_{ij} is the set of all minimal cut sets separating i and j.

Theorem (8.45) shows that $\hat{A}^*$ can be computed using the GAUSS-JORDAN method. This approach is discussed in detail by MARTELLI [1976].

An arc (μ, ν) is called *basic* with respect to $i, j \in E$ if $\{(\mu, \nu)\}$ is a minimal cut separating i from j. CARRÉ [1979] proposes the following semiring for the determination of all basic arcs in G. Let $P(E)$ denote the set of all subsets of E, and let $\omega \notin E$. We define

$$\alpha \oplus \beta = \alpha \cap \beta$$

$$\alpha \otimes \beta = \alpha \cup \beta$$

for all $\alpha,\beta \in P(E)$ and adjoin $\{\omega\}$ as a zero. Let $R = P(E) \cup \{\omega\}$. Then $(R,\oplus,\otimes)$ is an absorptive semiring with zero $\{\omega\}$ and with unity $\emptyset$. We assign weights

$$\tilde{a}_{ij} := \begin{cases} \{(i,j)\} & \text{if } \{i,j\} \in E, \\ \{\omega\} & \text{otherwise.} \end{cases}$$

Again the weight of a path is the set of its arcs. Due to theorem (8.33) we find $\tilde{A}^* = A^{[n-1]}$. Then $\tilde{a}^*_{ij}$ contains all arcs basic with respect to i and j provided $P_{ij} \neq \emptyset$; otherwise $\tilde{a}^*_{ij} = \{\omega\}$. Since we assume the existence of null circuits we find $\tilde{a}^*_{ii} = \emptyset$ for all $i \in N$.

(8.71) Schedule algebra

GIFFLER ([1963], [1968]) considered the determination of all path weights in a network with nonnegative integer weights. His approach has been extended by WONGSEELASHOTE [1976]. Let $(H,+,\leq)$ be a totally-ordered commutative group with neutral element 0. In particular, we are interested in the additive group of real numbers. Let $\bar{H} = \{\bar{x} \mid x \in H\}$ be a copy of H with $\bar{H} \cap H = \emptyset$. For convenience we define $\bar{\bar{x}} = x$ for all $\bar{x} \in \bar{H}$, i.e. we consider $z \to \bar{z}$ as a unitary operation on $Z = H \cup \bar{H}$. Then

$$\|z\| := \begin{cases} z & \text{if } z \in H, \\ \bar{z} & \text{if } z \in \bar{H}, \end{cases}$$

for all $z \in Z$. We continue the internal composition + of H on Z by

$$\bar{x} + \bar{y} := x + y \quad ,$$

$$x + \bar{y} := \bar{x} + y := \overline{x + y} \; ,$$

for all $x, y \in H$ and for all $\bar{x}, \bar{y} \in \bar{H}$. Then $(Z, +)$ is a commutative group with neutral element O. The inverse elements are denoted by $-z$ for $z \in Z$.

A function $a\colon Z \to \mathbb{Z}_+$ is called *countable* if

$$(8.72) \qquad Z(a) = \{z \mid a(z) \neq 0\}$$

is countable and a is called *well-ordered* if

$$(8.73) \qquad H(a) = \{\|z\| \mid a(z) \neq 0\}$$

is well-ordered (each subset of $H(A)$ contains a minimum). Let U denote the *set of all countable and well-ordered functions* $a\colon Z \to \mathbb{Z}_+$. On U we define two internal compositions by

$$(8.74) \qquad (a + b)(z) := a(z) + b(z)$$

for $a, b \in U$ and $z \in Z$ and by

$$(8.75) \qquad (a \circ b)(z) := \sum_{x+y=z} a(x)b(y)$$

for $a, b \in U$ and $z \in Z$. The right-hand sides of (8.74) and (8.75) are defined with respect to the usual addition and multiplication of nonnegative integers. It is easy to show that $a + b$ is a well-defined element of U.

Now let $a, b \in U$ and $z \in Z$. Then we claim that $S = \{(x,y) \in Z^2 \mid x + y = z,\; a(x)b(y) \neq 0\}$ is a finite set. Suppose S is countably infinite. Then $\{x \mid (x,y) \in S\}$ contains an infinite sequence

$$\|x_1\| < \|x_2\| < \ldots < \|x_k\| < \ldots \; .$$

Let y_k denote a corresponding infinite sequence in $\{y \mid (x,y) \in S\}$ with $x_k + y_k = z$ for $k = 1, 2, \ldots$. Then

$$\|y_1\| > \|y_2\| > \ldots > \|y_k\| > \ldots$$

contrary to the assumption that b is well-ordered. Therefore S is finite which shows that $(a \circ b)(z) \in \mathbb{Z}_+$. It is again easy to show that $a \circ b$ is countable and well-ordered. Therefore $a \circ b \in U$.

Now we introduce a mapping $r\colon U \to U$ defined by

(8.76) $\quad r(a)(z) := a(z) - \min(a(z), a(\bar{z}))$

for $a \in U$ and $z \in Z$. This means that a function $r(a)$ has either $r(a)(z) = 0$ or $r(a)(\bar{z}) = 0$ for each $z \in Z$. As in (8.69) we consider only a subset V of the functions in U defined by

(8.77) $\quad V := \{a \in U \mid a = r(a)\}$.

Further we introduce internal compositions on V in the same manner as in (8.69) by

$$a \oplus b := r(a + b)$$
$$a \otimes b := r(a \circ b)$$

for all $a, b \in V$. $(V, \oplus, \otimes)$ is a commutative ring which was discussed by GIFFLER in the particular case mentioned above. Verification of the ring axiomatics is pure routine. We only remark that the zero function 0 $(0(z) \equiv 0)$ is the zero of V and that the function 1 defined by

$$1(z) := \begin{cases} 1 & \text{if } z = 0, \\ 0 & \text{otherwise,} \end{cases}$$

is the unity of V. Further the additive inverse $(-a)$ of $a \in V$ satisfies $(-a) = \bar{1} \otimes a$ with

$$\bar{1}(z) := \begin{cases} 1 & \text{if } z = \bar{0}, \\ 0 & \text{otherwise.} \end{cases}$$

WONGSEELASHOTE [1976] gave the following nice result:

(8.78) Proposition

$(V,\oplus,\otimes)$ is an integral domain.

Proof. It remains to show that $a \otimes b = O$ implies $a = O$ or $b = O$. Suppose $a,b \neq O$ and let $\alpha \in Z(a)$, $\beta \in Z(b)$ such that

$$\|\alpha\| = \min H(a), \qquad \|\beta\| = \min H(b)$$

(cf. 8.72, 8.73). We claim that $(a \otimes b)(\bar{z}) = O$ for $z = \alpha + \beta$. Suppose $\bar{z} = x + y$ and $a(x)b(y) \neq O$. From $\bar{z} = x + y$ we find $\|\alpha\| + \|\beta\| = \|x\| + \|y\|$. The definition of $\|\alpha\|$ and $\|\beta\|$ together with $a(x)b(y) \neq O$ implies $\|\alpha\| \leq \|x\|$, $\|\beta\| \leq \|y\|$. Therefore

$$\|\alpha\| = \|x\|, \qquad \|\beta\| = \|y\| .$$

But then $a(x)b(y) \neq O$ implies $\alpha = x$ and $\beta = y$ contrary to $\alpha + \beta = z \neq \bar{z} = x + y$.

■

Proposition (8.78) shows that $(V,\oplus,\otimes)$ can be embedded into a field $(F,\oplus,\otimes)$ (cf. 5.5) of quotients a/b with $a,b \in V$, $b \neq O$. An element a of V is identified with $a/1$ in F.
Now we consider the application to the problem of finding the weights $w(p)$ of all paths p in a given network with weights in H. Let a_{ij} denote the weight of $(i,j) \in E$.
With respect to the field F let now $\tilde{a}_{ij}$ denote the function defined by

$$(8.79) \qquad \tilde{a}_{ij}(z) = \begin{cases} 1 & \text{if } z = a_{ij} , \\ O & \text{otherwise} , \end{cases}$$

for all $(i,j) \in E$. Let $\tilde{a}_{ij} \equiv O$ if $(i,j) \notin E$. Then the weight of a path $p = (e_1, e_2, \ldots, e_m)$ is given by

$$\tilde{w}(p) = \tilde{a}_{e_1} \otimes \tilde{a}_{e_2} \otimes \ldots \otimes \tilde{a}_{e_m} .$$

Therefore we get

$$\tilde{w}(p)(z) = \begin{cases} 1 & \text{if } z = w(p) , \\ 0 & \text{otherwise} . \end{cases}$$

With respect to the field F the entries $\tilde{a}_{ij}^{[k]}$ of $\tilde{A}^{[k]}$ are functions given by

$$\tilde{a}_{ij}^{[k]}(z) = |\{p \in P_{ij}^{[k]} \mid w(p) = z\}| ,$$

i.e. $\tilde{a}_{ij}^{[k]}(z)$ is the number of paths $p \in P_{ij}^{[k]}$ with weight z. In the field F we know that A is stable if and only if $\tilde{w}(p) \equiv 0$ for all circuits p in the network. This means that the network contains no circuit. Nevertheless the matrix $\tilde{A}^*$ may exist in a network with circuits. In order to characterize the existence of A^* we consider the existence of a^* for $a \in V$. Here we assume that H is Archimedean. From (4.17) we know that in this case H can be embedded into the additive group of real numbers. The following result is due to WONGSEELASHOTE [1976].

(8.80) <u>Proposition</u>

Let H be the additive group of real numbers and let $a \in V$. If $\alpha = \min H(a) > 0$ then a^* exists in V.

<u>Proof</u>. $\alpha = \min H(a)$ implies

$$k \cdot \alpha = \min H(a^k)$$

for all $k \in \mathbb{N}$. Let $k \in \mathbb{N}$. Then

$$a^k(z) = 0$$

for all $z \in Z = \mathbb{R} \cup \overline{\mathbb{R}}$ with $\|z\| < k \cdot \alpha$. Therefore, $a^*(z) = a^{[k]}(z)$ for all $\|z\| < k \cdot \alpha$. Thus $a^*(z) \in \mathbb{Z}_+$ for all $z \in Z$. a^* is countable as $Z(a^*)$ is the union of a countable number of countable sets. Suppose the existence of $S \subseteq H(a^*)$ which contains an infinite strictly decreasing sequence $\beta_1 > \beta_2 > \ldots$. Let $k \cdot \alpha > \beta_1$. Then $\beta_s \in H(a^{[k]}(z))$ for $s = 1,2,\ldots$. This is a contradiction to the fact that $a^{[k]}$ is well-ordered.

■

The existence of $\tilde{A}^*$ can now be shown using the GAUSS-JORDAN method for a constructive proof. The next theorem is taken from WONGSEELASHOTE [1976].

(8.81) Theorem

Let H be the additive group of real numbers and let $\tilde{A}$ be defined by (8.79) with respect to a network G. If $w(p) > 0$ for all non-null elementary circuits in G then

$$\tilde{A}^* = (I - \tilde{A})^{-1}$$

exists, i.e. $\tilde{A}^*$ is the inverse of the matrix $I - \tilde{A}$ in the matrix ring of all $n \times n$ matrices over the field F.

Proof. Let $\tilde{\sigma}$ be defined by (8.2) with respect to V and let C denote the set of all non-null elementary circuits. Then $w(p) > 0$ for all $p \in C$ implies

$$0 < \alpha := \min H(\tilde{\sigma}(C)) \quad .$$

Let D be an arbitrary set of circuits which contains at least one non-null circuit. If $\tilde{\sigma}(D)$ exists then

(8.82) $$\alpha \leq \min H(\tilde{\sigma}(D)) \; .$$

The existence of $\tilde{A}^*$ will be shown by proving the validity of

(8.83) $$\tilde{a}_{ij}^{(k)} = \tilde{\sigma}(T_{ij}^{(k)})$$

for all $i,j \in N$ and $k = 0,1,\ldots,n$. Here, $\tilde{a}_{ij}^{(k)}$ is recursively defined in the same way as, in the GAUSS-JORDAN method (8.43), $a_{ij}^{(k)}$. Obviously (8.83) holds for $k = 0$. Then $T_{11}^{(0)} = \{(1,1)\}$ and therefore (8.82) together with (8.80) implies the existence of $(\tilde{a}_{11}^{(0)})^*$. Hence $\tilde{a}_{ij}^{(1)}$ is well-defined. An argument similar as in the proof of (8.45) shows that (8.83) holds for $k = 1$. The inductive step from $k-1$ with $k \geq 1$ to k is again quite similar. The inductive hypothesis yields $\tilde{\sigma}(T_{kk}^{(k-1)}) = \tilde{a}_{kk}^{(k-1)}$. (8.82) and (8.80) imply the existence of $(\tilde{a}_{kk}^{(k-1)})^*$. It is easy to see that $\tilde{a}_{kk}^{(k)} = \tilde{\sigma}(T_{kk}^{(k)})$.

The remaining part of the proof proceeds in the same way as the proof of (8.45).

■

Theorem (8.81) shows that for the computation of $\tilde{A}^*$ we may use any linear algebra method for the inversion of $I - \tilde{A}$ over the field F. In the GAUSS-JORDAN method the calculation of $(a_{kk}^{(k-1)})^*$ needs, in general, infinitely many steps since

$$\alpha^* = 1 \oplus \alpha \oplus \alpha^2 \oplus \ldots$$

may be an infinite series and since the set $H(\alpha)$ may be infinite, too. If we are only interested in the path weights of elementary paths then we may calculate $\tilde{A}^{[n-1]}$ using the finite JACOBI-iteration or a suitable modification of the GAUSS-JORDAN method.

A particular example is the investigation of *sign-balanced graphs*. A graph G is called *signed* if each weight is $a_{ij} = +1$ or $a_{ij} = -1$. The *sign of a chain* is the usual product of its arc-weights. If the signs of all cycles of G are positive then G is called *sign-balanced*. Such properties have been studied by HARARY, NORMAN and CARTWRIGHT [1965]. HAMMER [1969] developed a procedure for checking whether G is sign-balanced. BRUCKER [1974] proposed the application of the GAUSS-JORDAN method.

Let $H = \{-1,+1\}$. Then $(H,\cdot,\leq)$ is a totally ordered commutative group with neutral element 1. Let $\tilde{b}_{ij} \equiv 0$ if $(i,j) \notin E$ and

$$\tilde{b}_{ij}(z) = \begin{cases} 1 & \text{if } z = a_{ij}\,, \\ 0 & \text{else}\,, \end{cases}$$

for $(i,j) \in E$. Now $\tilde{A} := \tilde{B} \oplus \tilde{B}^T$ is symmetric. W.l.o.g. we assume $(i,i) \notin E$. As the neutral element is the maximum of H we know that $\tilde{A}^*$ will exist if and only if G contains no circuits. Now G is sign-balanced iff all cycles in G have weight 1 or, equivalently, iff all elementary cycles in G have weight 1. Therefore G is sign-balanced if and only if

$$\tilde{a}_{ii}^{[n]}(-1) = 0$$

for all $i \in N$. This is equivalent to

$$Z(\tilde{a}_{ii}^{[n]}) \subseteq \{1\} \tag{8.84}$$

for all $i \in N$. For a computation of the sets $Z(\tilde{a}_{ij}^{[n]})$ we may simplify our method. We consider the semiring $(2^{\{-1,1\}},\oplus,\otimes)$ with zero $\emptyset$, unity $\{1\}$ and $X \oplus Y := X \cup Y$, $X \otimes Y := \{xy \mid x \in X, y \in Y\}$ for all $X,Y \subseteq \{-1,1\}$. This semiring is idempotent, commutative and contains no nontrivial zero-divisors. Furthermore $\emptyset$, $\{1\}$ are absorptive elements and $\{-1\}$, $\{-1,1\}$ are 1-regular elements. Let

$$\bar{a}_{ij} := Z(\tilde{a}_{ij})$$

for all $i,j \in N$. Then $\bar{a}_{ij} \subseteq \{-1,1\}$. Since all elements of the semiring are at least 1-regular the matrix $\bar{A}^*$ exists. We are only interested in $\bar{A}^{[n]}$ as G is balanced (cf. 8.84) iff

$$\bar{a}_{ii}^{[n]} \subseteq \{1\}$$

for all $i \in N$. This semiring was introduced in BRUCKER [1974].

(8.85) k-shortest paths

We consider a network G over the additive group of real numbers, i.e. all weights $a_{ij} \in \mathbb{R}$ for $(i,j) \in E$. We assume that G contains no negative circuits and w.l.o.g. $(i,i) \notin E$. In (8.84) we discussed methods for the determination of all path weights. Let

$$F_{ij} := \{w(p) \mid p \in P_{ij}\};$$

then F_{ij} is countable. As G contains no negative circuits F_{ij} is well-ordered. Let

$$F_{ij} =: \{\alpha_1, \alpha_2, \ldots\}$$

with $\alpha_1 < \alpha_2 < \ldots$. Let U denote the set of all well-ordered and countable subsets of $\mathbb{R}$. We introduce the internal composition

$$X + Y := \{x + y \mid x \in X,\ y \in Y\}$$

for $X,Y \subseteq U$. Then $(U,\cup,+)$ is an idempotent, commutative semiring with zero $\emptyset$ and unity $\{0\}$. $(U,\cup,+)$ corresponds to $(V,\oplus,\otimes)$ in

(8.84). For $a \in V$ with $Z(a) \subseteq \mathbb{R}$ we find $Z(a) \in U$. For such a the mapping $Z: a \to Z(a)$ is a homomorphism. We define now $\tilde{A}$ by

$$\text{(8.86)} \qquad \tilde{a}_{ij} := \begin{cases} \{a_{ij}\} & \text{if } (i,j) \in E, \\ \emptyset & \text{otherwise} \end{cases} ,$$

for all $i,j \in N$. The next theorem shows the existence of $\tilde{A}^*$ over U.

(8.87) Theorem

If G contains no negative circuits (with respect to the usual sum of the arc weights a_{ij}) then $\tilde{A}^*$ exists over U.

Proof. We give a constructive proof using the GAUSS-JORDAN method (8.43). The proof is quite similar to the proof of (8.81). It should be clear that $\alpha^* \in U$ exists for all $\alpha \in U$ with $0 \leq \min \alpha$. Therefore with $\tilde{\sigma}$ defined by (8.2) with respect to U we can prove (8.83) for all $i,j \in N$ and $k = 0,1,\ldots,n$. In particular, the proof uses $0 \leq \min \tilde{a}_{kk}^{(k-1)} = \min(\tilde{\sigma}(T_{kk}^{(k-1)})$ which implies $\tilde{a}_{kk}^{(k)} = (\tilde{a}_{kk}^{(k-1)})^* = \tilde{\sigma}(T_{kk}^{(k)})$ in the inductive step.

■

Theorem (8.87) implies that we may use the GAUSS-JORDAN method for the calculation of the sets F_{ij} , as

$$\tilde{a}_{ij}^* = F_{ij}$$

for all $i,j \in N$. In general, the calculation contains the determination of the possible infinite sets $(\tilde{a}_{kk}^{(k-1)})^*$. The main difference to theorem (8.81) lies in the fact that $\tilde{A}^*$ exists if $w(p) = 0$ for some circuits in G. Then the function $\tilde{a}_{ij}^* \in V$

in (8.81) may have $\tilde{a}^*_{ij}(z) = \infty$ for some $z \in \mathbb{R}$ which is forbidden for functions in the integral domain V (resp. the field F). Over the semiring U this does not lead to any difficulties. Then z is an element of the set $\tilde{a}^*_{ij} \in U$.

Now we may be interested in finding only the k smallest values in F_{ij}. This problem can be solved with a finite method. Let

$$V := \{X \in U \mid X = k\text{-}\min(X)\} ,$$

i.e. V consists in the elements of U with at most k elements. The function $k\text{-}\min: U \to U$ was introduced in chapter 2 (cf. 2.11) and is defined by

$$k\text{-}\min(X) := \{x_1, x_2, \ldots, x_k\}$$

with $x_1 < x_2 < \ldots$ and $|X| \geq k$ and by

$$k\text{-}\min(X) := X$$

if $|X| < k$. Now we define

$$X \oplus Y := k\text{-}\min(X \cup Y)$$

$$X \otimes Y := k\text{-}\min(X + Y)$$

for all $X, Y \in V$. Then $(V, \oplus, \otimes)$ is an idempotent, commutative semiring with zero $\emptyset$ and unity $\{0\}$. If we consider only weights in $\mathbb{Z}_+$ then $(V, *, \leq)$, the lattice-ordered commutative monoid in (2.11), satisfies $X * Y := X \otimes Y$ and $X \leq Y$ iff $X \oplus Y = Y$ for $X, Y \in V$.

Let $\tilde{A}$ be defined by (8.86). Then $\tilde{A}$ is stable and has stability index $r = n(k-1) + n - 1 = n\,k - 1$. For a proof let $p_1, p_2, \ldots, p_k$ be circuits of G with weight

$$\omega_\mu := \tilde{w}(p_\mu) = \{w(p_\mu)\}$$

for $\mu = 1,2,\ldots,k$. ω_μ is the weight of p_μ with respect to V. Then

$$k - \min\{0,\omega_1,\omega_1+\omega_2,\ldots\qquad,\omega_1+\omega_2+\ldots+\omega_{k-1}\}$$

$$= k - \min\{0,\omega_1,\omega_1+\omega_2,\ldots\qquad,\omega_1+\omega_2+\ldots+\omega_{k-1}+\omega_k\}$$

shows that G is (k-1)-absorptive (over V). Then theorem (8.33) shows the stability of $\tilde{A}$. Hence

$$\tilde{A}^* = \tilde{A}^{[s]}$$

for all $s \geq n\,k-1$. Thus the computation of $\tilde{A}^*$ is possible with the aid of the JACOBI- as well as the GAUSS-JORDAN-method. Then

$$a^*_{ij} = \{\alpha_1,\alpha_2,\ldots,\alpha_k\}$$

contains the k smallest values of F_{ij} (if $|F_{ij}| < k$ then $a^*_{ij} = F_{ij}$).

Such semirings as V were introduced by MINIEKA and SHIER [1973] and SHIER [1976]. Methods for solving the k-shortest path problem have also been discussed by HOFFMAN and PAVLEY [1959], BELLMAN and KALABA [1960], YEN ([1971], [1975]), FOX [1973], WONGSEELASHOTE [1976] and ISHII [1978].

Some further examples for algebraic path problems are contained in GONDRAN and MINOUX [1979b] and in CARRÉ [1979].

9. Eigenvalue Problems

In this chapter A denotes an $n \times n$ matrix over a semiring $(R,\oplus,\otimes)$ with zero O and unity 1. From chapter 5 we know that $(R^n,\oplus)$ is a right as well as a left semimodule over R. In particular,

$$(\lambda \otimes x)_i := \lambda \otimes x_i$$

for $i \in N$ defines the left (external) composition $\otimes: R \times R^n \to R^n$. The right (external) composition coincides with the matrix-composition of x and λ. An n-vector $x \in R^n$, $x \neq O$ is called an *eigenvector* of A if there exists $\lambda \in R$ such that

(9.1) $$A \otimes x = \lambda \otimes x \ .$$

Then λ is called an *eigenvalue* of A. The set of all eigenvectors with eigenvalue λ is denoted by $V(\lambda)$. $V_o(\lambda) = V(\lambda) \cup \{O\}$ is a right semimodule over R; i.e.

$$(x \otimes \alpha) \oplus (y \otimes \beta) \in V_o(\lambda)$$

for all $x,y \in V_o(\lambda)$ and for all $\alpha,\beta \in R$. No method is known for the determination of the eigenvalues and eigenvectors of A in the general case. The results in this chapter will focus on the case where the internal composition $\oplus$ reflects more or less a linear order relation of R. The classical case (R^n is a module over a field R) is not covered by the main results. We refer to BLYTH [1977] for a discussion of the classical problem.

Now we introduce a preorder $\leq$ on R by

(9.2) $$a \leq b \quad :\Longleftrightarrow \quad \exists\, c \in R: \quad a \oplus c = b$$

for all $a,b \in R$. Then $(R,\oplus,\leq)$ is a naturally ordered commutative monoid, $(R,\otimes,\leq)$ is a preordered monoid and $(R,\oplus,\otimes)$ is a preordered semiring with least element 0. If $\leq$ is a partial order then R is a partially ordered semiring and

(9.3) $\quad a \oplus b = 0 \quad \Rightarrow \quad a = b = 0$

for all $a,b \in R$. If R is an idempotent semiring then R is partially ordered and

(9.4) $\quad a \leq b \quad \Longleftrightarrow \quad a \oplus b = b$

for all $a,b \in R$. Then

(9.5) $\quad a \oplus b = \sup(a,b)$

for all $a,b \in R$ and R is completely described by the partially ordered monoid $(R,\otimes,\leq)$ and (9.5). If

(9.6) $\quad a \oplus b \in \{a,b\}$

for all $a,b \in R$ then R is linearly ordered and completely described by the linearly ordered monoid $(R,\otimes,\leq)$ and

(9.7) $\quad a \oplus b = \max(a,b)$

for all $a,b \in R$.

The eigenvalue problem in such semirings is discussed by GONDRAN and MINOUX ([1977], [1978], [1979a]. They characterize $V(\lambda)$ in the particular cases that $(R',\otimes)$ with $R' := R \setminus \{0\}$ is a group and that $\otimes$ is an idempotent composition. The group-case is discussed in detail by CUNINGHAME-GREEN [1979]. For the particular case $(\mathbb{R},\max,+)$ the eigenvalue problem is solved in CUNINGHAME-GREEN [1962] and VOROBJEV ([1963], [1967] and [1970]).

The discussion of the eigenvalue problem in such semirings will show its close relationship to certain path problems considered in chapter 8. For an $n \times n$ matrix A let $G_A = G(N,E)$ denote the directed graph with set of vertices $N = \{1,2,\ldots,n\}$ and set of arcs E defined by

$$(i,j) \in E \quad :\Longleftrightarrow \quad a_{ij} \neq 0$$

for $i,j \in N$. The relationship between A and G_A is similar as in chapter 8. Here G_A is derived from A and does not contain arcs with weight 0. The circuits of G_A play an important role in the following discussion. Differently from chapter 8 we exclude circuits of length 0. Therefore the leading term in the matrix

$$A^* = I \oplus A \oplus A^2 \oplus \ldots \quad ,$$

i.e. the unity matrix I, has no longer an interpretation by paths in G_A. Let P_{ij} $(P_{ij}^{[k]})$ denote *the set of all paths from* i *to* j in G_A *(with* $l(p) \leq k$*)*. The length $l(p)$ of a path p, the weight $w(p)$ and the function σ are defined in the same way as at the beginning of chapter 8. Similar to (8.17) we find that the entry $a_{ij}^{(k)}$ in the matrix

$$A^{(k)} := A \oplus A^2 \oplus \ldots \oplus A^k$$

is equal to $\sigma(P_{ij}^{[k]})$ and that if

$$A^+ := A \oplus A^2 \oplus A^3 \oplus \ldots$$

exists then

$$a_{ij}^+ = \sigma(P_{ij}) \tag{9.8}$$

for all $i,j \in N$. We define $\sigma_j := a_{jj}^+$ for $j \in N$. From

(9.9) $A^* = I \oplus A^+ , \quad A^+ = A \otimes A^*$

we conclude that A^* exists if and only if A^+ exists. Therefore all results and methods of chapter 8 with respect to A^* can easily be applied to A^+. Throughout this chapter either we assume the existence of A^* or other assumptions imply the existence of A^*. Only the latter case will explicitly be mentioned.

We continue with the discussion of some sufficient conditions for the existence of solutions of the eigenvalue problem which are drawn from GONDRAN and MINOUX [1977]. This paper has been published in an English version (GONDRAN and MINOUX [1979a]), too.

(9.10) Proposition

Let R be a semiring; let $\mu \in R$, $k \in N$ and $p \in P_{kk}$. Then

(1) $(1 \oplus \sigma_k) \otimes \mu = \sigma_k \otimes \mu$ iff $A_k^* \otimes \mu \in V_o(1)$;

(2) if $(1 \oplus w(p)) \otimes \mu = w(p) \otimes \mu$ then $A_k^* \otimes \mu \in V_o(1)$;

(3) if $I \oplus A = A$ then $A_j^* \in V_o(1)$ for all $j \in N$.

Proof. (1) $A_k^* \otimes \mu \in V_o(1)$ is equivalent to $A \otimes A_k^* \otimes \mu = A_k^* \otimes \mu$. From (9.9) we know $A \otimes A_k^* = A_k^+$. Therefore we find the equivalent equation $A_k^+ \otimes \mu = A_k^* \otimes \mu$. (9.8) leads to the claimed result.

(2) From the definition of σ_k we know $\sigma_k = w(p) \oplus \sigma(P_{kk} \setminus \{p\})$. Therefore $(1 \oplus \sigma_k) \otimes \mu = [(1 \oplus w(p)) \otimes \mu] \oplus [\sigma(P_{kk} \setminus \{p\}) \otimes \mu]$. Then $(1 \oplus w(p)) \otimes \mu = w(p) \otimes \mu$ implies $(1 \oplus \sigma_k) \otimes \mu = \sigma_k \otimes \mu$ and (1) can be applied.

(3) $I \oplus A = A$ shows $a_{jj} \neq 0$ for all $j \in N$. Therefore $(j,j) \in P_{jj}$ for all $j \in N$. Now $1 \oplus w(j,j) = w(j,j)$ and (2) imply $A_j^* \in V_o(1)$.

■

Proposition (9.10) shows the importance of the equations (9.9). The columns of A^* and the columns of A^+ differ only slightly and therefore they are candidates for the eigenvectors of the matrix A with eigenvalue 1. We remark that $A_j^* \otimes \mu \in V(1)$ implies $A_j^+ \otimes \mu = A_j^* \otimes \mu \in V(1)$. This view is quite different from the approach to the classical problem. We know that in the classical problem $A^* = (I - A)^{-1}$ (cf. 8.55). But if $I - A$ is a regular matrix then A cannot have the eigenvalue 1.

Next we apply (9.10) to two particular cases. In the first case $(R,\otimes)$ is an idempotent monoid, i.e. $a \otimes a = a$ for all $a \in R$ and in the second case $(R',\otimes)$ is a group.

(9.11) <u>Proposition</u>

Let R be an idempotent semiring with idempotent monoid $(R,\otimes)$.

(1) $A_j^* \otimes \sigma_j \in V_o(1)$ for all $j \in N$;

(2) if R is commutative (hence A^* exists) then $\lambda \otimes \sigma_j \otimes A_j^* \in V_o(\lambda)$ for all $j \in N$, $\lambda \in R$.

<u>Proof</u>. (1) As both internal compositions are idempotent we find

$$(1 \oplus \sigma_j) \otimes \sigma_j = \sigma_j \otimes \sigma_j$$

and (9.10.1) leads to the claimed result.

(2) In the commutative case the existence of A^* follows from the fact that R is 1-regular ($1 \oplus a \oplus a^2 = 1 \oplus a$; cf. 8.33). Clearly (1) implies (2).

■

Proposition (9.11) shows that if both internal compositions are idempotent then at least 1 is an eigenvalue and that *all* elements of R may be eigenvalues.
In the second case it is possible to reduce the eigenvalue problem (9.1) to the solution of

(9.12) $(\lambda^{-1} \otimes A) \otimes x = x$

provided that we are only interested in eigenvalues $\lambda > 0$.

(9.13) Proposition

Let R be a semiring with group $(R', \otimes)$ and let R be partially ordered by (9.2). If G_A is strongly connected and $x \in V(\lambda)$ then

$$\lambda > 0 \quad \text{and} \quad x > 0.$$

Proof. If $\lambda = 0$ then $A \otimes x = 0$. (9.3) shows $a_{ij} \otimes x_j = 0$ for all $i,j \in N$. As $(R', \otimes)$ is a group we get $a_{ij} = 0$ or $x_j = 0$ for all $i,j \in N$. As $x \in V(\lambda)$ there exists $x_i \neq 0$. Then $a_{ij} = 0$ for all $j \in N$, contrary to G_A strongly connected. Therefore $\lambda > 0$. Now suppose $x_k = 0$ and let $M := \{j \mid x_j \neq 0\}$. Then $M, N \setminus M \neq \emptyset$ and

$$(A_M \otimes x_M)_i = 0$$

for all $i \in N \setminus M$. Therefore $a_{ij} = 0$ for all $(i,j) \in (N \setminus M) \times M$, contrary to G_A strongly connected.

■

In the discussion of the eigenvalue problem in the case that $(R',\otimes)$ is a group we will only consider solutions of (9.1) with $\lambda > 0$, $x > 0$. Such solutions are called *finite* (cf. CUNINGHAME-GREEN [1979]). Proposition (9.13) gives a sufficient condition such that every solution of (9.1) is finite.

Now we apply (9.10) to (9.12). Let $\tilde{A} := \lambda^{-1} \otimes A$. Then G_A and $G_{\tilde{A}}$ differ only in the weights assigned to the arcs (i.e. a_{ij} and $\lambda^{-1} \otimes a_{ij}$). The corresponding weight function and other denotations involving weights are labeled by "$\sim$" $(\tilde{\sigma}, \tilde{w}, \tilde{\sigma}_j, \ldots)$.

(9.14) Proposition

Let R be a semiring with group $(R',\otimes)$. Let $k \in N$.

(1) If $1 \oplus \tilde{\sigma}_k = \tilde{\sigma}_k$ then $\tilde{A}_k^* \in V_o(\lambda)$;

(2) if R is commutative and $1 \oplus \mu_k = \mu_k$ for

$$\mu_k := \bigoplus_{p \in P_{kk}} (w(p) \otimes \lambda^{-l(p)}) \quad \text{then} \quad \tilde{A}_k^* \in V_o(\lambda).$$

Proof. (1) Here (9.10.1) is fulfilled with $\mu = 1$ for $\tilde{A}$. Therefore $\tilde{A}_k^*$ is an eigenvector of (9.12) with eigenvalue 1. Hence $\tilde{A}_k^* \in V_o(\lambda)$.

(2) In the commutative case the weights of a path p with respect to G_A and $G_{\tilde{A}}$ are given by

$$\tilde{w}(p) = \lambda^{-l(p)} \otimes w(p).$$

Therefore $\tilde{\sigma}_k = \mu_k$ and (1) is applicable. ■

These sufficient conditions should be seen together with the following necessary conditions.

(9.15) Proposition

Let R be a semiring with group (R',⊗) and let R be partially ordered by (9.2). Let $x \in V(\lambda)$ be a finite solution.

(1) (If R is commutative then) $\tilde{w}(p) \leq 1$ $(w(p) \leq \lambda^{l(p)})$ for all circuits p in G_A ;

(2) if R is idempotent then $\tilde{A}$ is absorptive (hence $\tilde{A}^*$ exists);

(3) if R is linearly ordered (and commutative) then there exists a circuit p in G_A with $\tilde{w}(p) = 1$ $(w(p) = \lambda^{l(p)})$.

Proof. (1) Let p be a circuit of G_A. Then (with respect to 9.2)

$$(\lambda^{-1} \otimes a_{ij}) \otimes x_j \leq x_i$$

for all arcs (i,j) of p. Together these inequalities imply

$$\tilde{w}(p) \otimes x_k \leq x_k$$

for each vertex k of p. Then $x_k > 0$ shows

(4) $$\tilde{w}(p) \leq 1.$$

In the commutative case $\tilde{w}(p) = w(p) \otimes \lambda^{-l(p)}$.

(2) If R is idempotent then (4) and (9.4) lead to

$$\tilde{w}(p) \oplus 1 = 1 \quad \text{for all circuits p in } G_A ,$$

i.e. $\tilde{A}$ is absorptive and $\tilde{A}^*$ exists (cf. 8.33).

(3) In the linearly ordered case $A \otimes x = \lambda \otimes x$ means

$$\max\{a_{ij} \otimes x_j \mid j \in N\} = \lambda \otimes x_i$$

for all $i \in N$. This maximum is attained for at least one j(i) in row i. The sequence $k, j(k), j^2(k) \ldots$ contains a circuit p_k. Similarly to (4) we find

$$\tilde{w}(p_k) = 1$$

which in the commutative case means $w(p_k) = \lambda^{l(p_k)}$.

■

If R is linearly ordered, commutative and $V(\lambda)$ contains a finite solution then propositions (9.14) and (9.15) show in particular that at least some columns $\tilde{A}_k^*$ are elements of $V_o(\lambda)$.

Now we consider again the more general situation without specific assumption on the internal composition $\otimes$.

(9.16) Proposition

Let R be a semiring.

(1) If R is idempotent then $x \in V(1)$ implies $A^* \otimes x = x$ and $A^+ \otimes x = x$;

(2) if R is absorptive and commutative (hence A^* exists) then $x \in V(\lambda)$ implies $A^* \otimes x = x$;

(3) if R is absorptive and $a_{ii} = 1$ for all $i \in N$ then $V(\lambda) \subseteq V(1)$ for all $\lambda \in R$.

Proof. (1) Let $x \in V(1)$. Then $A^k \otimes x = x$ for all $k \in \mathbb{N}$. Therefore

$$A^{[k]} \otimes x = x \oplus (A \otimes x) \oplus (A^2 \otimes x) \oplus \ldots = x$$

for all $k \in \mathbb{N}$. Then $A^* \otimes x = x$. Similarly $A^+ \otimes x = x$.

(2) Let $x \in V(\lambda)$. Then commutativity implies

$$A^k \otimes x = \lambda^k \otimes x$$

for all $k \in \mathbb{N}$. Due to (8.33) $A^* = A^{[n-1]}$. Now $\alpha = \lambda \oplus \lambda^2 \oplus \ldots$ $\ldots \oplus \lambda^{n-1} \in R$ shows $1 \oplus \alpha = 1$. Therefore $A^* \otimes x = A^{[n-1]} \otimes x$ $= (1 \oplus \alpha) \otimes x = x$.

(3) Obviously $I \oplus A = A$. Let $x \in V(\lambda)$. Then $A \otimes x = (I \oplus A) \otimes x$ $= (1 \oplus \lambda) \otimes x = 1 \otimes x = x$.

■

We remark that, if R is absorptive and $a_{ii} = 1$ for all $i \in N$, then A^* exists and (9.10.3) shows that all A_j^* are elements of $V_o(1)$ which, by (9.16.3), contains all eigenvectors of A.

(9.17) Theorem

Let R be a linearly ordered semiring. If $x \in V(1)$ then

$$x = A_M^* \otimes x_M$$

with $A_\nu^* \otimes x_\nu \in V(1)$ for all $\nu \in M \subseteq N$.

Proof. From (9.16.1) we know $x = A^* \otimes x$. Thus it suffices to find a reduced representation by terms $A_\nu^* \otimes x_\nu \in V(1)$. From (9.16.1) we also know $A^+ \otimes x = x$. We consider the partial digraph $G'_A = (N,E')$ with

$$E' = \{(i,j) \in E \mid a_{ij} \otimes x_j = \lambda \otimes x_i\}$$

with respect to an eigenvalue λ (here $\lambda = 1$). Then we introduce an ordering on N by $i \leq j$ if and only if there exists a path from i to j in G'_A. On the set of maximal vertices this induces an equivalence relation. This follows from the fact that for all $i \in N$ there exists at least one $j \in N$ such that $i \leq j$ (as $A \otimes x = \lambda \otimes x$ and R linearly ordered). Let N_k , $k = 1,2,\ldots,s$ denote the corresponding equivalence classes. Then a similar argument shows that the subgraphs G'_k generated by N_k in G'_A are strongly connected and contain at least one circuit of G'_A.

Now let p denote a path from ν to μ in G'_A. Using the equations $a_{ij} \otimes x_j = x_i$ along p we find

$$w(p) \otimes x_\mu = x_\nu \quad (9.18)$$

Obviously $x_\mu > O$ iff $x_\nu > O$. From (9.8) and the idempotency of $\oplus$ we get

$$(A_\nu^+ \otimes w(p)) \oplus A_\mu^+ = A_\mu^+ .$$

Therefore

$$(9.19) \qquad (A_\nu^+ \otimes x_\nu) \oplus (A_\mu^+ \otimes x_\mu) = A_\mu^+ \otimes x_\mu$$

which leads to

$$x = A^+ \otimes x = A_M^+ \otimes x_M$$

for $M := \{\nu_k \mid A_{\nu_k}^+ \otimes x_{\nu_k} \neq O,\ k = 1,2,\ldots,s\}$ provided that $\nu_k \in N_k$ for all $k = 1,2,\ldots,s$. As each $\nu \in M$ lies on some circuit p of G'_A (9.18) yields

$$w(p) \otimes x_\nu = x_\nu .$$

Idempotency of R implies

$$(w(p) \otimes x_\nu) \oplus x_\nu = x_\nu \oplus x_\nu = w(p) \otimes x_\nu$$

for $\nu \in M$. Then (9.10.2) leads to $A_\nu^* \otimes x_\nu \in V_o(1)$ for $\nu \in M$. Hence

$$A_\nu^+ \otimes x_\nu = A_\nu^* \otimes x_\nu$$

for $\nu \in M$. Thus $x = A_M^* \otimes x_M$ is a representation with $A_\nu^* \otimes x_\nu \in V(1)$ for all $\nu \in M$.

■

This theorem shows that all eigenvectors with eigenvalue 1 are linear combinations of certain columns of A^*. The corresponding vertices lie on circuits of the graph G_A. If R is furthermore absorptive, i.e. 1 is the maximum of R, then (9.16.3) implies that we can determine all eigenvectors in this way provided that the main diagonal of A contains only neutral elements.

Next we consider the special case of a linearly ordered, absorptive (commutative) semiring with idempotent monoid $(R,\otimes)$. We may as well consider a linearly ordered (commutative) monoid $(R,\otimes,\leq)$ with minimum O and maximum 1 which has an idempotent internal composition $\otimes$. The eigenvalue problem (9.1) is thus of the particular form

(9.20) $$\max\{a_{ij} \otimes x_j \mid j \in N\} = \lambda \otimes x_i$$

for $i \in N$.

(9.21) Theorem

Let R be a linearly ordered, absorptive semiring with idempotent monoid $(R,\otimes)$. Hence A^* exists. Let $B_j := A_j^* \otimes \sigma_j$ for $j \in N$ and $J := \{j \in N \mid \sigma_j \neq O\}$. Then

(1) $V_o(1)$ is the right semimodule generated by $\{B_j \mid j \in J\}$;

(2) if R is commutative then $\lambda \otimes V_o(1) = \lambda \otimes V_o(\lambda)$;

(3) if R is commutative and $a_{ii} = 1$ for all $i \in N$ then $\lambda \otimes V_o(1) = V_o(\lambda)$.

<u>Proof</u>. (1) From (9.11.1) we know $B_j \in V_o(1)$ for all $j \in N$. In an absorptive semiring $a_{jj}^* = 1$ for all $j \in N$. Thus $\sigma_j \neq O$ implies $B_j \in V(1)$. Absorptivity and (9.10.1) show that $A_k^* \otimes \mu \in V_o(1)$ iff $\mu = \sigma_k \otimes \mu$. Due to (9.17) $x \in V(1)$ has a representation $x = A_M^* \otimes x_M$ with $A_k^* \otimes x_k \in V(1)$ for all $k \in M$. Therefore $A_k^* \otimes x_k = A_k^* \otimes \sigma_k \otimes x_k$ for all $k \in M$ which shows $x = B_M \otimes x_M$.

(2) (9.11.2) and (1) imply $\lambda \otimes V_o(1) \subseteq V_o(\lambda)$. Then $\lambda \otimes V_o(1) \subseteq \lambda \otimes V_o(\lambda)$. If $x \in V_o(\lambda)$ then $\lambda \otimes x = A \otimes x$. Thus $\lambda \otimes x = \lambda \otimes (\lambda \otimes x) = A \otimes (\lambda \otimes x)$, i.e. $\lambda \otimes x \in V_o(1)$. Therefore $\lambda \otimes V_o(\lambda) \subseteq V_o(1)$ which implies $\lambda \otimes V_o(\lambda) \subseteq \lambda \otimes V_o(1)$.

(3) (9.16.3) implies $\lambda \otimes x = x$ for $x \in V(\lambda)$. Therefore $\lambda \otimes V_o(\lambda) = V_o(\lambda)$. Together with (2) we get $\lambda \otimes V_o(1) = V_o(\lambda)$.

■

Now we consider the special case of a linearly ordered semiring with group $(R',\otimes)$. We may as well consider a linearly ordered group $(R',\otimes)$ with an adjoined least element O acting as a zero. The neutral element of R' is denoted by 1. CUNINGHAME-GREEN [1979] additionally adjoins a greatest element ∞ but mainly considers finite solutions of the eigenvalue problem $(0 < \lambda < \infty,\ 0 < x_i < \infty)$. A vertex $\nu \in N$ is called an *eigenvertex* of A if ν is a node on a circuit of G_A with weight 1. Two eigenvertices are called *equivalent* if they belong to a common circuit in G_A with weight 1. Otherwise two eigenvertices are called *nonequivalent*.

This denotation is motivated by the observation that "i equivalent j" is an equivalence relation $\sim$ on the set of all eigenvertices. Reflexivity and symmetry are obvious. Let $i \sim j$ and $j \sim k$. Let the corresponding circuits be $p = a \circ b$, $q = c \circ d$ with paths a from i to j, b from j to i, c from k to j and d from j to k. Then $a \circ d \circ c \circ b$ is a circuit with vertices i and k. As $w(a) = w(b)^{-1}$ and $w(c) = w(d)^{-1}$ we find $w(a \circ d \circ c \circ b) = 1$, i.e. $i \sim k$.

(9.22) Theorem

Let R be a linearly ordered semiring with group $(R',\otimes)$.

(1) $V_o(1)$ is the right semimodule generated by $\{A_j^* \mid A_j^* \in V(1)\}$.

If the eigenvalue problem has a finite solution with eigenvalue λ (hence $(\lambda^{-1} \otimes A)^*$ exists) then

(2) every finite eigenvector of A has the same finite eigenvalue λ,

(3) $V_o(\lambda)$ is the right semimodule generated by $S := \{(\lambda^{-1} \otimes A)^*_j \mid j \in J\}$ where J denotes a maximal set of nonequivalent eigenvertices of $\tilde{A} := \lambda^{-1} \otimes A$,

(4) S is a minimal generating set of $V_o(\lambda)$ (S is a *base*) and all such S have the same cardinality.

<u>Proof.</u> (1) From (9.17) we know that $x \in V(1)$ has a representation $x = A^*_M \otimes x_M$ with $A^*_j \otimes x_j \in V(1)$ for all $j \in M$. Then $x_j > 0$ and therefore x_j^{-1} exists. Thus $A^*_j \in V(1)$.

(2) Let λ, β be two finite eigenvalues of A for finite solutions of the eigenvalue problem. Then (9.15.2) implies that $\tilde{A} := \lambda^{-1} \otimes A$ and $\bar{A} := \beta^{-1} \otimes A$ are absorptive. Suppose $\lambda \neq \beta$ and w.l.o.g. $\lambda < \beta$. From (9.15.3) we know that there exists a circuit p in G'_A such that $\bar{w}(p) = 1$. Then $\tilde{w}(p) > \bar{w}(p) = 1$ contrary to $\tilde{A}$ absorptive. Thus $\lambda = \beta$.

(3) Let J denote a maximal set of nonequivalent eigenvertices of $G_{\tilde{A}}$. Let $j \in J$. Clearly, (9.15) implies $\tilde{\sigma}_j = 1$. Therefore $1 \oplus \tilde{\sigma}_j = \tilde{\sigma}_j$ and, by (9.10.1), $\tilde{A}^*_j \in \tilde{V}_o(1) = V_o(\lambda)$. As $\tilde{a}^*_{jj} = 1$ we get $\tilde{A}^*_j \in V(\lambda)$. From the proof of (9.17) we know that $x \in \tilde{V}(1)$ has a representation $x = \tilde{A}^*_M \otimes x_M$ such that all vertices $j \in M$ lie on a circuit p of $G'_{\tilde{A}}$ and $\tilde{w}(p) * x_j = x_j$. Thus $\tilde{w}(p) = 1$, i.e. j is an eigenvertex of $\tilde{A}$. The elements of M were chosen in such a way that any two of them do not belong to the same circuit of $G'_{\tilde{A}}$. As any circuit p in $G_{\tilde{A}}$ with weight $\tilde{w}(p) = 1$ is a circuit in $G'_{\tilde{A}}$ we find that M consists in nonequivalent eigenvertices of $\tilde{A}$.

(4) Let $K := J \setminus \{k\}$ for an arbitrary $k \in J$. Suppose $\tilde{A}^*_k = \tilde{A}^*_K \otimes y$ for some $y \in R^K$. Then

$$1 = \tilde{a}^*_{kk} = \tilde{a}^*_{kj} \otimes y_j$$

for some $j \in K$ and as $\tilde{a}^*_{jj} = 1$

$$\tilde{a}^*_{jk} \geq 1 \otimes y_j = y_j .$$

Now $\tilde{a}^*_{jk}$ ($\tilde{a}^*_{kj}$) is the weight of a shortest path p (q) from j to k (k to j) in $G_{\tilde{A}}$ and therefore $w(p \circ q) = \tilde{a}^*_{kj} \otimes \tilde{a}^*_{jk} \geq 1$ and $p \circ q$ is a circuit in $G_{\tilde{A}}$. Absorptivity implies $w(p \circ q) = 1$. Then k and j are equivalent vertices contrary to the choice of k and j. Thus S is a minimal generating set of $V_o(\lambda)$. As $\sim$ is an equivalence relation on the set of all eigennodes all maximal sets of nonequivalent eigenvertices have the same cardinality.

■

Theorem (9.22) shows how we can compute the finite eigenvectors, i.e. $V_o(\bar{\lambda})$, if the unique finite eigenvalue $\bar{\lambda}$ is known. Then we determine $\tilde{A}^*$ for $\tilde{A} := \bar{\lambda}^{-1} \otimes A$ and select columns $\tilde{A}^*_j \in (R')^n$ with $\tilde{a}^*_{jj} = 1$. Let J denote the set of the indices of such finite eigenvectors and choose a maximal set of nonequivalent eigenvertices from J. Such a set can easily be found by comparing the columns $\tilde{A}^*_j$, $j \in J$, pairwise as $j \sim k$ implies that

$$\tilde{A}^*_j = \tilde{A}^*_k \otimes \tilde{a}^*_{kj}$$

(follows from (9.19) on the corresponding circuit in G'_A with $x := \tilde{A}^*_j$).

The remaining problem is the determination of $\bar{\lambda}$ (if it exists). We assume now that R is commutative. Then we may assume w.l.o.g. that $(R', \otimes)$ is a divisible group (cf. 3.2). For $b \in R'$ let

$x = b^{n/m}$ denote the solution of the equation $x^m = b^n$ with $n/m \in \mathbb{Q}$. From (9.15.3) we know that if $\bar{\lambda}$ exists then there exists a circuit p with

$$\bar{\lambda} = w(p)^{1/l(p)} =: \bar{w}(p)$$

and $\bar{\lambda} \geq \bar{w}(p')$ for all circuits p' in G_A. If p is not elementary then $p = p_1 \circ p_2 \circ \ldots \circ p_s$ with elementary circuits p_k, $k = 1,2,\ldots,s$. The cancellation rule in groups implies that

$$\bar{\lambda} = \bar{w}(p_k) \qquad , \; k = 1,2,\ldots,s.$$

If G_A contains at least one circuit then let

(9.23) $\qquad \lambda(A) := \max\{\bar{w}(p) \mid p \text{ elementary circuit in } G_A\}.$

If $\bar{\lambda}$ exists then $\bar{\lambda} = \lambda(A)$. In any case the matrix

$$\bar{B} := \lambda(A)^{-1} \otimes A$$

is absorptive. Hence $\bar{B}^*$ exists. $\lambda(A)$ is the finite eigenvalue of A if and only if $\bar{B}^*$ contains a column $\bar{B}^*_j > 0$ with $\bar{b}^*_{jj} = 1$. Then $\bar{B}^*_j$ is a finite eigenvector. Otherwise the eigenvalue problem has no finite solution.

A direct calculation of $\lambda(A)$ from (9.23) by enumeration of all elementary circuits does not look very promising. For the special case of the semiring $(\mathbb{R} \cup \{-\infty\}, \max, +)$ LAWLER [1967] proposed a method which has a straightforward generalization to semirings derived from linearly ordered, divisible and commutative groups $(R', \otimes)$. Let $A^{\{k\}}$ be defined by

$$a_{ij}^{\{k\}} := (a_{ij}^k)^{1/k}$$

with $A^k = (a_{ij}^k)$ for all $i,j,k \in N$ (cf. chapter 8). Then

$$\lambda(A) = \max\{a_{ii}^{\{k\}} \mid i,k \in N\}.$$

The computation of A^k for $k \in N$ needs $O(n^4)$ $\otimes$-compositions and $O(n^4)$ $\oplus$-compositions ($\leq$ comparisons). Then $a_{ii}^{\{k\}}$ for $i,k \in N$ can be determined by solving $O(n^2)$ equations of the form $\alpha^k = \beta$, $1 \leq k \leq n$. Further $O(n^2)$ $\oplus$-compositions ($\leq$ comparisons) lead to the value $\lambda(A)$.

For the same special case another method was proposed by DANTZIG, BLATTNER and RAO [1967]. Its generalization is an algebraic linear program over a certain module (cf. chapter 11). Its solution $\tilde{\lambda}$ can be found by a generalization of the usual simplex method (cf. chapter 11). It can be shown (cf. 11.62) that $\tilde{\lambda} \geq \lambda(A)$ and, in particular, $\tilde{\lambda} = \lambda(A)$ if $\bar{\lambda}$ exists. After determination of $\tilde{\lambda}$ using methods discussed in chapter 11 we compute the matrix $(\tilde{\lambda}^{-1} \otimes A)^*$ which exists since $\tilde{\lambda} \geq \lambda(A)$. If this matrix contains a finite column with main diagonal entry 1 then $\tilde{\lambda} = \lambda(A)$ and the column is a finite eigenvector. Otherwise the finite eigenvalue problem has no solution. In particular, if $\tilde{\lambda} > \lambda(A)$ then all main diagonal elements of $(\tilde{\lambda}^{-1} \otimes A)^*$ are strictly less than 1. For the above mentioned special case this method is discussed in CUNINGHAME-GREEN [1979].

We remark that the above results for linearly ordered, commutative groups have some implications for lattice-ordered commutative groups. From theorem (3.4) we know that a lattice-ordered, commutative group $(R', \otimes, \leq)$ can be embedded in the lattice-ordered, commutative and divisible group $V(\tilde{\Lambda})$ with a root system $\tilde{\Lambda}$. Two different finite eigenvalues with finite eigenvectors are incomparable and therefore the number of such finite eigenvalues is bounded by the number of mutually incomparable w(p) for elementary circuits p of G_A. This number is bounded by the number of the elementary circuits in G_A and the number of maximal linearly ordered subsets in $\tilde{\Lambda}$. In particular, if $\tilde{\Lambda}$ consists in a finite number of mutually disjoint maximal linearly ordered subsets $C_1, C_2, \ldots, C_s$ then we can solve the finite eigenvalue problem by solving the finite eigenvalue problems restricted to $V(C_k)$ for k = 1,2,...,s.

In the remaining part of this chapter we discuss some examples for eigenvalue problems. As mentioned previously the approach described in this chapter has been developed in view of ⊕-compositions which correspond to linear (or, at least partial) order relations. Therefore only such examples will be considered although the eigenvalue problem may be formulated for all the examples given in chapter 8.

(9.24) Periodic schedules

CUNINGHAME-GREEN [1960] considers the semiring $(\mathbb{R} \cup (-\infty), \max, +)$ and proves the existence of a unique eigenvalue λ for matrices A over ℝ. In this example λ(A) is the maximum of the *circuit means*

$$\bar{w}(p) := w(p)/l(p)$$

for elementary circuits p in G_A. An interpretation of the eigenvalue problem as described by CUNINGHAME-GREEN ([1960], [1979]) is the following. Let $x_j(r)$ denote the starting-time for machine j in some cyclic process in the rth cycle and let $a_{ij}(r)$ denote the time that this machine j has to work until machine i can start its r+1st cycle. Then

$$x_i(r+1) \geq \max\{a_{ij}(r) + x_j(r) \mid j \in N\}$$

for all $i \in N$. Hence the fastest possible cyclic process is described by

(9.25) $$x(r+1) = A(r) \otimes x(r)$$

for $r = 1,2,\ldots$. The finite eigenvalue λ is then a constant time that passes between starting two subsequent cycles, i.e.

$$x(r+1) = \lambda + x(r) = A(r) \otimes x(r), \qquad r = 1,2,\ldots .$$

In this way the cyclic process moves forward in regular steps, i.e. the cycle time for each machine is constant and minimal with respect to the whole process. The right semimodule $V_o(\lambda)$ contains the finite eigenvectors which define possible starting times for such regular cyclic processes. An example is given in CUNINGHAME-GREEN [1979].

The solution of the equivalent eigenvalue problem in the semiring $(\mathbb{R}_+, \max, \cdot)$ is developed in VOROBJEV ([1963], [1967] and [1970]).

Methods for the determination of the finite eigenvalue $\lambda(A)$ and further applications have been discussed in DANTZIG, BLATTNER and RAO [1967], and LAWLER [1967].

(9.26) Maximum capacity circuits

GONDRAN [1977] considers the semiring $(\mathbb{R}_+ \cup \{\infty\}, \max, \min)$. Let A be an $n \times n$ matrix with $a_{ii} = \infty$ for all $i \in N$. Let σ_j denote the maximum capacity of a circuit with node j. Then (9.21) implies that the vector B_j with components

$$\min\{a^*_{ij}, \lambda, \sigma_j\}$$

for $i \in N$ is an eigenvector for each $\lambda \in \mathbb{R}_+ \cup \{\infty\}$ and that $\{B_j \mid j \in N\}$ generates $V_o(\lambda)$. An application to hierarchical classifications are given in GONDRAN [1977].

(9.27) Connectivity

GONDRAN and MINOUX [1977] consider the semiring ({0,1},max,min) with the only possible finite eigenvalue 1. If G_A is strongly connected then $a^*_{ij} = 1$ for all $i,j \in N$ and the only finite eigenvector is x with $x_i = 1$ for $i \in N$. In general, if G_A decomposes into s strongly connected components with vertex sets $I_1, I_2, \ldots, I_s$ then the eigenvectors with eigenvalue 1 are $x^1, x^2, \ldots, x^s$ defined by $x^i_j = 1$ if $j \in I_i$ and $x^i_j = 0$ otherwise.

More examples are considered by GONDRAN and MINOUX [1977] and can easily be derived from the examples in chapter 8.
For a discussion of linear independence of vectors over certain semirings and related topics we refer to GONDRAN and MINOUX [1978] and CUNINGHAME-GREEN [1979]. CUNINGHAME-GREEN [1979] considers further related eigenvalue problems.

10. Extremal Linear Programs

In this chapter we consider certain linear optimization problems over semirings $(R',\oplus,\otimes)$ with minimum O and maximum ∞ derived from residuated, lattice-ordered, commutative monoids $(R,\otimes,\leq)$ (cf. 5.15). As in chapter 9 the correspondence is given by

$$a \leq b \iff a \oplus b = b$$

or, equivalently, by

(10.1) $$a \oplus b = \sup(a,b) \;.$$

Due to (2.13) we know that the least upper bound is distributive over $\otimes$. The *extremal linear program* is defined by

(10.2) $$z := \inf\{c^T \otimes x \mid A \otimes x \geq b, \;\; x \in R^n\}$$

with given *coefficient vector* $c \in R^n$, $m \times n$ matrix A over R and m-vector b over R. Extremal linear programs are linear algebraic optimization problems (cf. chapter 7). The adjective 'extremal' is motivated by the linearly ordered case. Then the internal composition $\oplus$ attains only extremal values, i.e. $a \oplus b \in \{a,b\}$. In general, linear functions of the form

$$c^T \otimes x = \sup\{c_j \otimes x_j \mid x_j > O\}$$

are considered.

We define a *dual extremal linear program* by

(10.3) $$t := \sup\{y^T \otimes b \mid y^T \otimes A \leq c^T, \;\; y \in R^m\}.$$

As we assume that the reader is familiar with the usual linear programming concepts we may remark that this kind of dualization is obviously motivated by the classical formulation of a dual program. It is, in fact, a special case of the abstract dual program which HOFFMAN [1963] considered for partially ordered semirings. He showed that the weak duality theorem of linear programming can easily be extended to such programs. On the other hand it is very difficult to prove strong duality results.

At first we will generalize this weak duality theorem to arbitrary ordered semimodules $(R,\oplus,\otimes,\leq;\square;H,*,\leq)$ (cf. chapter 5).

The *algebraic linear program*

$$(10.4) \qquad z = \inf\{x^T \square c \mid A \otimes x \geq b,\ x \geq 0,\ x \in R^n\}$$

has the dual

$$(10.5) \qquad t = \sup\{b^T \square y \mid A^T \square y \leq c,\ y \geq e,\ y \in H^m\}$$

The feasible solutions of (10.4) and (10.5) are called *primal* and *dual feasible*.

(10.6) Weak duality theorem

Let H be an ordered semimodule over R. Let x be a primal feasible solution of (10.4) and let y be a dual feasible solution of (10.5). Then

$$b^T \square y \leq x^T \square c .$$

<u>Proof</u>. Using some properties from proposition (5.21) we find

$$b^T \square y \leq (A \otimes x)^T \square y = x^T \square (A^T \square y) \leq x^T \square c.$$

■

For semimodules over real numbers such weak duality theorems have been considered by BURKARD [1975a], DERIGS and ZIMMERMANN,U.

[1978a] and ZIMMERMANN, U. [1980a]. The value of the weak duality theorem lies in the fact that, if we succeed in finding a primal feasible solution x and a dual feasible solution y with $b^T \square y = x^T \square c$, then x is an optimal solution of (10.4), y an optimal solution of (10.5) and we get

$$t = b^T \square y = x^T \square c = z \ ,$$

i.e. a strong duality result. HOFFMAN [1963] gives examples for strong duality results which will be covered by the following discussion. With respect to semimodules over real numbers many examples are known and will be considered in chapters 11 - 13.

The weak duality theorem is valid for (10.2) and (10.3), even in the case $0 \notin R$. Next we will solve (10.3) explicitly. As R is a residuated semigroup (cf. 2.12) we know that for all $a,b \in R$ there exists $c \in R$ with

$$a \otimes x \leq b \iff x \leq c$$

for all $x \in R$. This uniquely determined element c is denoted by b:a and is called residual. This result can be generalized to matrix inequalities. In the following let $N := \{1,2,...,n\}$ and $M := \{1,2,...,m\}$.

(10.7) Proposition

Let R be a residuated, lattice-ordered commutative monoid. Let A and B be $m \times n$ and $m \times r$ matrices over R and define B:A by $(B:A)_{jk} := \inf\{b_{ik} : a_{ij} \mid i \in M\}$. Then

$$A \otimes X \leq B \iff X \leq B:A$$

for all $n \times r$ matrices X over R.

Proof. $A \otimes X \leq B$ is equivalent to $a_{ij} \otimes x_{jk} \leq b_{ik}$ for all i,j,k. These inequalities have the residuals $b_{ik}:a_{ij}$. Therefore we find equivalence to $X \leq B:A$.

■

Proposition (10.7) shows that the dual problem (10.3) has a maximum solution $\tilde{y} := c : A^T$. Hence $\tilde{y}$ is the optimal solution of (10.3) and

$$(10.8) \qquad t = b^T \otimes (c:A^T) .$$

Let $\tilde{x}$ denote the maximum solution of

$$c^T \otimes x \leq t$$

i.e. $\tilde{x} = t:c^T$. If $\tilde{x}$ is primal feasible then

$$t \leq c^T \otimes \tilde{x} \leq t$$

implies $z = t$ and therefore $\tilde{x}$ is optimal. We can prove feasibility of $\tilde{x}$ for lattice-ordered groups, Boolean lattices and bounded linearly ordered sets (cf. chapter 1). For the discussion we introduce the set $R(v,w)$ with respect to $v,w \in R^k$ which consists of all solutions $b \in R$ of the inequality

$$(10.9) \qquad \sup_{\mu=1,2,\ldots,k} v_\mu \otimes [(b \otimes \inf_{\lambda=1,2,\ldots,k} (w_\lambda : v_\lambda)):w_\mu] \geq b.$$

The relevance of this strange-looking inequality is shown by the following result.

(10.10) Theorem

Let R be a residuated lattice-ordered commutative monoid. Let A_i denote the rows of the matrix A. If

$$b_i \in R(A_i, c)$$

for all $i \in M$ then $\tilde{x}$ is an optimal solution of (10.2).

Proof. It suffices to prove primal feasibility of x. As $e \leq f$ implies $e:g \leq f:g$ for all $e,f,g \in R$ we find

$$(b_k \otimes \tilde{y}_k):c_j \leq t:c_j = \tilde{x}_j$$

for all k,j. For $i \in M$ this leads to

$$\sup_{j \in N} a_{ij} \otimes [(b_i \otimes \tilde{y}_i):c_j] \leq A_i \otimes \tilde{x}$$

As $\tilde{y}_i = \inf\{c_\lambda : a_{i\lambda} \mid \lambda \in N\}$ the assumption $b_i \in R(A_i, c)$ implies $b_i \leq A_i \otimes \tilde{x}$.

■

We do not know how to characterize the sets $R(A_i, c)$ in the general case. However, for lattice-ordered groups ($R = R' \setminus \{0, \infty\}$), Boolean lattices ($R' = R$) and bounded linearly ordered sets ($R' = R$) the following result gives a sufficient description.

(10.11) Proposition

(1) Let R be a lattice-ordered commutative group. Then $R(v,w) \equiv R$ for all $v, w \in R^k$.

(2) Let R be a Boolean lattice and let $a \otimes b := \inf(a,b)$. Then $R(v) := \{b \mid b \leq \sup_{\mu=1,2,\ldots,k} v_\mu\} \subseteq R(v,w)$ for all $v, w \in R^k$.

(3) Let R be a bounded linearly ordered set with maximum 1 and let $a \otimes b := \min(a,b)$. For $v, w \in R^k$ let

$$\beta := \min\{w_\mu : v_\mu \mid \mu = 1,2,\ldots,k\}$$

$$\alpha := \max\{v_\mu \mid w_\mu \leq \beta,\ \mu = 1,2,\ldots,k\}.$$

Then $R(v,w) = \{b \mid b \leq \alpha\}$.

<u>Proof</u>. We will denote the left-hand-side of the inquality (10.9) as a function f of the parameter $b \in R$.

(1) For lattice-ordered groups $b:a = b \otimes a^{-1}$ and the greatest lower bound is also distributive over $\otimes$ (cf. 2.14). Therefore f has the form

$$f(b) = \sup_\mu v_\mu \otimes [(b \otimes \inf_\lambda (w_\lambda \otimes v_\lambda^{-1})) \otimes w_\mu^{-1}]$$

$$= b \otimes \sup_\mu (v_\mu \otimes w_\mu^{-1}) \otimes \inf_\lambda (v_\lambda^{-1} \otimes w_\lambda).$$

As in lattice-ordered groups $\sup_\mu \alpha_\mu = (\inf_\mu \alpha_\mu^{-1})^{-1}$ we find

$$f(b) = b$$

which implies $R(v,w) = R$.

(2) For calculations in Boolean lattices it is more convenient to use the denotations $\cap$ and $\cup$ instead of sup and inf. Let

$$\beta := \cap_\mu (w_\mu \cup v_\mu^*)$$

where v_μ^* denotes the complement of v_μ. In a Boolean lattice $b:a = b \cup a^*$ (cf. 1.16). We remark that

$$\beta^* = \cup_\mu (w_\mu^* \cap v_\mu).$$

Then f has the form

$$f(b) = \cup_\mu v_\mu \cap [(b \cap \beta) \cup w_u^*]$$

As a Boolean lattice is in particular a distributive lattice we find

$$f(b) = [\cup_\mu (v_\mu \cap b \cap \beta)] \cup [\cup_\mu (v_\mu \cap w_\mu^*)]$$

$$= [(\cup_\mu v_\mu) \cap b \cap \beta] \cup \beta^*$$

Now let $b \in R(v)$. Then $b \leq \cup_\mu v_\mu$ implies

$$f(b) = [b \cap \beta] \cup \beta^* = b \cup \beta^* \geq b ,$$

i.e. $b \in R(v,w)$.

(3) Bounded linearly ordered sets are introduced in an example for pseudo-Boolean lattices after proposition (1.19). Here

$$b:a = \begin{cases} 1 & \text{if } b \geq a , \\ b & \text{if } b < a . \end{cases}$$

With β and α as defined in (3), f has the form

$$f(b) = \max_\mu \min[v_\mu, \min(b,\beta):w_\mu].$$

If $b \in R(v,w)$ and $b' \leq b$ then $b' \in R(v,w)$. Therefore it suffices to show $f(b) = \alpha$ for all $b \in R$ with $b \geq \alpha$. Let

$$K_- := \{\mu \mid w_\mu < v_\mu\}$$

Assume $K_- = \emptyset$. Then $\beta = 1$ and for $b \geq \alpha$ we find

$$f(b) = \max_\mu \min[v_\mu , b:w_\mu] = \max_\mu v_\mu = \alpha.$$

Otherwise $K_- \neq \emptyset$ and $\beta = \min\{w_\mu \mid w_\mu < v_\mu\}$. Further $\alpha = \max\{v_\mu \mid w_\mu = \beta, \mu \in K_-\}$ and thus $\alpha > \beta$. For $b \geq \alpha$ we find

$$f(b) = \max_\mu \min[v_\mu, \beta:w_\mu] = \alpha .$$

■

This characterization of R(v,w) shows immediately that for lattice-ordered commutative groups $\tilde{x}$ is the optimal solution of (10.2) and that a strong duality result holds for (10.2) and (10.3).

For Boolean lattices the assumption $b_i \leq \sup\{a_{ij} \mid j \in N\}$ is necessary for the existence of primal feasible solutions. Therefore $\tilde{x}$ is an optimal solution provided that there exists any feasible solution of (10.2).

For bounded linearly ordered sets the situation is not so transparent. We will transform the coefficients of such a problem in order to simplify the problem. In a first step let

$$(10.12) \qquad a'_{ij} := \begin{cases} 0 & \text{if } a_{ij} < b_i \,, \\ 1 & \text{otherwise} \,. \end{cases}$$

Then the set of primal feasible solutions remains unchanged.

(10.13) Proposition

Let R be a bounded linearly ordered set with minimum 0 and maximum 1 and $a \otimes b := \min(a,b)$ for all $a,b \in R$. Then

$$A \otimes x \geq b \iff A' \otimes x \geq b$$

for all $x \in R^n$.

Proof. Let A_i (A'_i) denote the i-th row of A (A') for $i \in M$. The following statements are equivalent to $A_i \otimes x \geq b_i$:

$$\exists\, j \in N: \quad \min(a_{ij}, x_j) \geq b_i \,,$$

$$\exists\, j \in N: \quad a_{ij} \geq b_i \quad \text{and} \quad x_j \geq b_i \,,$$

$$\exists\, j \in N: \quad a'_{ij} = 1 \quad \text{and} \quad x_j \geq b_i \,,$$

and $A'_i \otimes x \geq b_i$.

■

Due to the reduction step (10.12) and proposition (10.13) we may assume in the following w.l.o.g. that A is a matrix over {0,1}. Further we exclude trivial problems and nonbinding constraints. If $b_i = 0$ then we delete the i-th constraint. Thus w.l.o.g. let $b > 0$. Then $A_i = (0,0,...,0)$ implies that no feasible solution exists. Therefore we assume in the following

w.l.o.g. that A contains no zero-rows. If A contains a zero-column j or $c_j = 0$ for some $j \in N$ then an optimal value of the corresponding variable is $x_j = 0$. Thus we assume w.l.o.g. that A contains no zero-columns and $c > 0$.
Summarizing our reductions we have to consider in the following only problems with 0 - 1 matrix A containing no 0 - column or 0 - row and with $b,c > 0$. Such a problem is called *reduced*.

(10.14) Proposition

Let (10.2) be a reduced problem over the bounded linearly ordered set R with minimum 0 and maximum 1 and $a \otimes b := \min(a,b)$ for all $a,b \in R$. Then

$$R(A_i,c) \equiv R$$

for all $i \in M$.

Proof. Due to proposition (10.11.3) we know that for $i \in M$ $R(A_i,c) = \{b \mid 0 \leq b \leq \alpha_i\}$ with

$$\beta_i := \min\{c_j : a_{ij} \mid j \in N\} \qquad ,$$

$$\alpha_i := \max\{a_{ij} \mid c_j \leq \beta_i \ , \ j \in N\}.$$

For a reduced problem this leads to

$$\beta_i = \max\{c_j \mid a_{ij} = 1, \ j \in N\}$$

and therefore to $\alpha_i = 1$.

■

Proposition (10.14) shows that for a reduced problem over a bounded linearly ordered set $\tilde{x}$ is an optimal solution of (10.2). We summarize these explicit solutions of (10.2) in the following theorem.

(10.15) Theorem

(1) Let $(R,\otimes,\leq)$ be a lattice-ordered commutative group. Then the optimal value of (10.2) is

$$z = \sup_{i\in M}[b_i \otimes \inf_{j\in N}(c_j \otimes a_{ij}^{-1})]$$

and an optimal solution $\tilde{x}$ is given by

$$\tilde{x}_j := z \otimes c_j^{-1}$$

for all $j \in N$.

(2) Let $(R,\leq)$ be a Boolean lattice with join (meet) $\cup$ ($\cap$), complement a^* of a, and $a \otimes b = a \cap b$ for all $a,b \in R$. Then the value

$$z = \cup_{i\in M}[b_i \cap \cap_{j\in N}(c_j \cup a_{ij}^*)]$$

and the solution $\tilde{x}$ is given by

$$\tilde{x}_j := z \cup c_j^* ,$$

for all $j \in N$ are optimal for (10.2) provided that $\tilde{x}$ is primal feasible. Otherwise there exists no feasible solution of (10.2).

(3) Let $(R,\leq)$ be a bounded linearly ordered set with minimum 0, maximum 1 and $a \otimes b = \min(a,b)$ for all $a,b \in R$. If (10.2) is a reduced problem then its optimal value is

$$z = \max_{i\in M} \min[b_i, \min\{c_j \mid a_{ij} = 1\}]$$

and an optimal solution $\tilde{x}$ is given by

$$\tilde{x}_j := \begin{cases} 1 & \text{if } c_j \leq z , \\ z & \text{otherwise} , \end{cases}$$

for all $j \in N$.

It should be noted that HOFFMAN [1963] gave the optimal solution in (10.15.1) for the additive group of real numbers $(\mathbb{R},+,\leq)$ and a solution of (10.2) in Boolean lattices. Theorems (10.10) and (10.15) show that such problems can be treated and solved in a unified way. Further they describe in axiomatic terms a class of optimization problems for which strong duality results hold.

On the other hand we have to admit that for general residuated, lattice-ordered, commutative monoids no solution method is known for (10.2). As we did not assume that R is conditionally complete it is even unknown whether an optimal value exists.

A solution of

$$(10.16) \qquad z := \inf\{c^T \otimes x \mid A \otimes x \leq b, \quad x \in R^n\}$$

is trivial if we assume $0 \in R$, a solution of

$$(10.17) \qquad z := \sup\{c^T \otimes x \mid A \otimes x \leq b, \quad x \in R^n\}$$

is obviously $x := b{:}A$ which is the maximum solution of $A \otimes x \leq b$. The solution of

$$(10.18) \qquad z := \sup\{c^T \otimes x \mid A \otimes x \geq b, \quad x \in R^n\}$$

is trivial if we assume $\infty \in R$.

If we consider such problems as (10.2) or (10.17) with equality constraints then we can only derive some lower and upper bounds on the optimal value. Let

$$(10.19) \qquad z := \inf\{c^T \otimes x \mid A \otimes x = b, \quad x \in R^n\}$$

and assume the existence of z.

(10.20) Proposition

Let R be a residuated, lattice-ordered, commutative monoid. If $S := \{x \in R^n \mid A \otimes x = b\} \neq \emptyset$ then $\tilde{x} := b:A$ is feasible and a maximum element of S.

Proof. $\tilde{x}$ is the greatest solution of $A \otimes x \leq b$. In particular, if $A \otimes x = b$ then $x \leq \tilde{x}$. This implies $b = A \otimes x \leq A \otimes \tilde{x}$. Thus $A \otimes \tilde{x} = b$.

■

Obviously $\tilde{x}$ yields an upper bound on z. On the other hand (10.3) leads to a lower bound. We find

$$\text{(10.23)} \qquad \sup_{i \in M}[b_i \otimes \inf_{j \in N}(c_j : a_{ij})] \leq z \leq \sup_{j \in N}[c_j \otimes \inf_{i \in M}(b_i : a_{ij})].$$

If the greatest lower bound is distributive over $\otimes$ then these bounds are certain row- and column-infima of matrices with entries $b_i \otimes (c_j : a_{ij})$ resp. $c_j \otimes (b_i : a_{ij})$.
As in chapter 9 we have to assume that R is linearly ordered if we want to develop a finite and efficient solution method. Thus in the following R is a residuated linearly ordered commutative monoid. Further we will assume that (10.2) and (10.19) have optimal solutions if a feasible solution exists. In particular, for one-dimensional inequations (equations) this assumption means for all $a, b \in R$:

(10.24) if $a \otimes x \geq b$ $(a \otimes x = b)$ has a feasible solution x then it has a minimum solution.

Such minimum solutions are uniquely determined and will be denoted by b/a $(b/\!/a)$. If $b/\!/a$ exists then b/a exists, too, and we find $b/a = b/\!/a$. If $b/\!/a$ exists for all a,b then R is a residuated and dually residuated d-monoid.

In a linearly ordered monoid linear algebraic inequations (equations) can be interpreted in the following way. $A \otimes x \geq b$ means that for all rows $i \in M$ there exists at least one column $j(i)$ such that

$$(10.25) \qquad a_{ij(i)} \otimes x_{j(i)} \geq b_i \, .$$

$A \otimes x = b$ means that for all $i \in M$ there exists at least one column $j(i)$ such that

$$(10.26) \qquad a_{ij(i)} \otimes x_{j(i)} = b_i$$

and that

$$(10.27) \qquad a_{ij} \otimes x_j \leq b_i$$

for all $i \in M$, $j \in N$. This interpretation is extensively used in chapter 9 for the solution of eigenvalue problems in certain linearly ordered semirings. Let G (G') denote the set of all ordered pairs $(i,j) \in M \times N$ such that the set $\{x \mid a_{ij} \otimes x_j \geq b_i\}$ ($\{x \mid a_{ij} \otimes x_j = b_i\}$) is nonempty (and $a_{kj} \otimes (b_i \,/\!/\, a_{ij}) \leq b_k$ for all $k \in M$). Then we may reduce the set of feasible solutions to a set containing only solutions x with

$$(10.28) \qquad x_j \in \{b_i/a_{ij} \mid (i,j) \in G\}$$

in the case of inequations and with

$$(10.28') \qquad x_j \in \{b_i /\!/ a_{ij} \mid (i,j) \in G'\}$$

in the case of equations. This leads to the following theorems.

(10.29) Theorem

Let R be a residuated linearly ordered commutative monoid satisfying (10.24). With respect to the extremal linear program (10.2), let $x_j(\alpha) := \max\{b_i/a_{ij} \mid (i,j) \in G,\ \alpha \geq c_j \otimes (b_i/a_{ij})\}$ for all $j \in N$ and for all $\alpha \in R$. If (10.2) has an optimal solution with optimal value z then

(1) $x(z)$ is an optimal solution of (10.2) and

(2) $z \in Z := \{c_j \otimes (b_i/a_{ij}) \mid (i,j) \in G\}$.

Proof. (1) $c^T \otimes x(\alpha) \leq \alpha$ for all $\alpha \in R$ and therefore it suffices to show that $x(z)$ is feasible. Let y be an optimal solution and let $i \in M$. Then there exists $j = j(i)$ with $a_{ij} \otimes y_j \geq b_i$. Therefore $y_j \geq b_i/a_{ij}$ and $z \geq c_j \otimes y_j \geq c_j \otimes (b_i/a_{ij})$. Thus $x_j(z) \geq b_i/a_{ij}$ which implies $a_{ij} \otimes x_j \geq b_i$. Hence $x(z)$ is feasible.
(2) Follows from $z \in \{c^T \otimes x(\alpha) \mid \alpha \in R\}$.

∎

(10.30) Theorem

Let R be a residuated linearly ordered commutative monoid satisfying (10.24). With respect to the extremal linear program (10.19) let $x'_j(\alpha) := \max\{b_i /\!/ a_{ij} \mid (i,j) \in G',\ \alpha \geq c_j \otimes (b_i /\!/ a_{ij})\}$ for all $j \in N$ and for all $\alpha \in R$. If there exists an optimal solution with optimal value z then

(1) $x'(z)$ is an optimal solution of (10.19) and

(2) $z \in Z' := \{c_j \otimes (b_i /\!/ a_{ij}) \mid (i,j) \in G'\}$.

Proof. (1) We get $c^T \otimes x'(\alpha) \leq \alpha$ and from the definition of G' $A \otimes x'(\alpha) \leq b$. Therefore it suffices to show that $x'(z)$ satisfies $A \otimes x(z) \geq b$. Let y be an optimal solution and let $i \in M$.

Then there exists $j = j(i) \in N$ with $a_{ij} \otimes y_j = b_i$ and therefore $y_j \geq b_i // a_{ij}$. This implies $z \geq c_j \otimes y_j \geq c_j \otimes (b_i // a_{ij})$. Further $a_{\mu\nu} \otimes y_\nu \leq b_\mu$ for all $\mu \in M$, $\nu \in N$ shows $(i,j) \in G'$. Therefore $x'_j(z) \geq b_i // a_{ij}$. Thus $A \otimes x'(z) \geq b$.

(2) Follows from $z \in \{c^T \otimes x(\alpha) \mid \alpha \in R\}$.

■

Theorems (10.29) and (10.30) reduce the extremal linear programs (10.2) and (10.19) to

(10.31) $\min\{\alpha \in Z \mid A \otimes x(\alpha) \geq b\}$

and to

(10.32) $\min\{\alpha \in Z' \mid A \otimes x'(\alpha) = b\}$.

For the determination of Z (Z') it is necessary to compute the minimal solutions of O(n m) inequations $a_{ij} \otimes \beta \geq b_i$ (equations $a_{ij} \otimes \beta = b_i$ and $O(n\,m^2)$ comparisons) and to perform O(n m) $\otimes$-compositions. As Z and Z' are linearly ordered an optimal solution can be determined using a binary search strategy together with the threshold method of EDMONDS and FULKERSON [1970]. Here we need $O(\log_2(n\,m))$ steps in which the determination of $x(\alpha)$ consists in O(n m) comparisons and in which the feasibility check consists in O(n m) $\otimes$-compositions and O(n m) comparisons.

This approach can easily be extended to residuated lattice-ordered commutative monoids satisfying (10.24) which are finite products of linearly ordered ones. Thus, in particular, extremal linear programs over the lattice-ordered group of real k-vectors $(\mathbb{R}^k, +, \leq)$, finite Boolean lattices $(2^K, \cap, \subseteq)$ for $K := \{1,2,\ldots,k\}$, and of finite products of bounded linearly ordered sets (R,min,$\leq$) can be solved using a similar reduction. As for such monoids the resulting sets Z resp. Z' are in general not totally ordered it will be necessary to perform enumeration. A complexity bound for the solution of such problems will contain the cardinality of the finite product considered (in the above examples: k).

Similar results as given in this chapter for residuated, lattice-ordered, commutative monoids can be given for dually residuated, lattice-ordered, commutative monoids by exchanging the role of sup and inf. These results follow from the given results by replacing $(R,\otimes,\leq)$ by its lattice-dual $(R,\otimes,\geq)$. We remark in this context that lattice-ordered commutative groups are residuated and dually residuated.

Many authors have considered extremal linear equations, inequations and extremal linear programs. The solution of linear equations and inequations is closely related to the eigenvalue problem discussed in chapter 9. The solution of such equations and inequations has been considered by VOROBJEV ([1963], [1967]) in the special case $(\mathbb{R}_+,\cdot,\leq)$; in a series of papers ZIMMERMANN, K. ([1973a], [1973b], [1974a], [1974b], [1976a] and [1976b]) discussed solution methods for extremal equations, inequations and extremal linear programs in $(\mathbb{R}_+,\cdot,\leq)$. A compact solution of all different possible extremal linear programs in this monoid is given in ZIMMERMANN, U. [1979c] (cf. 10.34). Extremal linear equations and inequations in linearly ordered commutative monoids are discussed by ZIMMERMANN, K. and JUHNKE [1979].

For lattice-ordered groups and related algebraic structures such equations and inequations are considered in detail in the book of CUNINGHAME-GREEN [1979]. For Boolean lattices a discussion is given in the book of HAMMER and RUDEANU [1968] and in the survey article of RUDEANU [1972]. For residuated lattice-ordered commutative monoids the solution of extremal linear programs is previously described in an extended abstract of ZIMMERMANN, U. [1980c].
The discussion of such problems in $(\mathbb{R}_+, \cdot, \leq)$ leads in a natural way to a generalized convexity concept which enables a geometrical description. Such concepts were investigated by ZIMMERMANN, K. ([1977], [1979a], [1979b]) and in a more general context by KORBUT [1965], TESCHKE and HEIDEKRÜGER ([1976a], [1976b]), and TESCHKE and ZIMMERMANN, K. [1979]. ZIMMERMANN, K. [1980] discusses the relationship of certain semimodules and join geometries as defined by JANTOSCIAK and PRENOWITZ [1979]. A unified treatment of the linear algebra viewpoint is developed in CUNINGHAME-GREEN ([1976], [1979]) for lattice-ordered groups and related algebraic structures. Further, CUNINGHAME-GREEN [1979] described several applications. As an example we consider a machine scheduling problem.

(10.33) Machine Scheduling

We assume that a production process is described as in (9.24) by starting times x which lead to finishing times $A \otimes x$ over the monoid $(\mathbb{R} \cup \{-\infty\}, +, \leq)$. If we have prescribed finishing times b then a solution of the system of extremal equations $A \otimes x = b$ is a vector x of feasible starting times. Assume that a preparation procedure is necessary to start machine j which takes time c_j. This means that after starting machine j the next start of machine j is possible not before time $c_j + x_j$. In order to be able to start the process as early as possible a solution of

$$\min\{c^T \otimes x \mid A \otimes x = b\}$$

is required. If the time constraint is not absolutely binding then situations may occur in which we consider inequations instead of equations. It is also possible that $A \otimes x = b$ does not have any feasible solution. Then a reasonable solution is given by minimizing

$$\max\{b_i - (A \otimes x)_i \mid i \in M\}$$

subject to $A \otimes x \leq b$. Its solution is $\tilde{x} = b:A$ since we know that $A \otimes x \leq A \otimes \tilde{x}$ for all x with $A \otimes x \leq b$. Here $\tilde{x}$ is given by

$$\tilde{x}_j = \min\{b_i - a_{ij} \mid i \in M\}$$

for $j \in N$ (cf. 10.7).

(10.34) Bounded Linearly Ordered Commutative Groups

In the following we assume that $(R,\otimes,\leq)$ is a linearly ordered commutative group with $|R| > 1$. R is residuated and dually residuated; the residual b:a, the dual residual b/a, and the minimum solution b/a (b//a) in (10.24) coincide with $b \otimes a^{-1}$. We consider several different types of extremal linear programs denoted by $(\circ,\square,\Delta,\rho)$ with $\circ,\square,\Delta \in \{\max,\min\}$ and with $\rho \in \{\leq,\geq,=\}$. An extremal linear program $(\circ,\square,\Delta,\rho)$ is given by

(10.35) $\quad o\{z(x) \mid x \in P\}$

with objective function defined by

$$z(x) := \square\{c_j \otimes x_j \mid j \in N\}$$

and with the set $P \subseteq R^n$ of feasible solutions defined by

$$\Delta\{a_{ij} \otimes x_j \mid j \in N\} \, \rho \, b_i$$

for all $i \in M$. For example, (10.1) corresponds to (min,max,max,$\geq$). Since $(R,\otimes,\geq)$ is a linearly ordered group, too, each program has a dual version in the order-theoretic sense which is obtained by exchanging max $\leftrightarrow$ min and $\leq \leftrightarrow \geq$. Thus it suffices to solve one problem from each of these pairs. For example, the dual version of (10.1) is (max,min,min,$\leq$).

W.l.o.g. we assume $b_i = 1$ for all $i \in M$. In table 1 we list the explicit solutions to all different programs with the exception of two programs with equality constraints which have to be solved by a threshold method (indicated by "T"). Some trivial cases are excluded by stating an assumption which can easily be checked for given coefficients. For the special group $(\mathbb{R}_+ \setminus \{0\},\cdot,\leq)$ such a table is given in ZIMMERMANN, U. [1979c]. All nontrivial explicit solutions can be described by $\bar{x} = \bar{x}(\Delta)$ with

(10.36) $\quad \bar{x}_j := [\Delta\{a_{ij} \mid i \in M\}]^{-1}$

for all $j \in N$ or by $x^* = x^*(o,\Delta)$ with $x_j^* := z^* \otimes c_j^{-1}$ for all $j \in N$ and with $z^* = z^*(o,\Delta)$ defined by

(10.37) $\quad z^* := \Delta\{o\{c_j \otimes a_{ij}^{-1} \mid j \in N\} \mid i \in M\}.$

Then the optimal values are $\bar{z} := z(\bar{x})$ and $z^* = z(x^*)$, respectively. At first, we verify the results in table 1 for inequality constraints.

ρ \ o	Δ \ □	max: max	max: min	min: max	min: min
$\leq$	max	$\bar{Z}$	$\bar{Z}$	o	o
	min	∞ $(n > 1)$	Z^*	o	o
$\geq$	max	∞	∞	Z^*	o $(n>1)$
	min	∞	∞	$\bar{Z}$	$\bar{Z}$
$=$ $(P \neq \emptyset)$	max	$\bar{Z}$	$\bar{Z}$	T	o $(\lvert P \rvert > 1)$
	min	∞ $(\lvert P \rvert > 1)$	T	$\bar{Z}$	$\bar{Z}$

Table 1 : Optimal value of extremal linear programs (o,□,Δ,ρ).

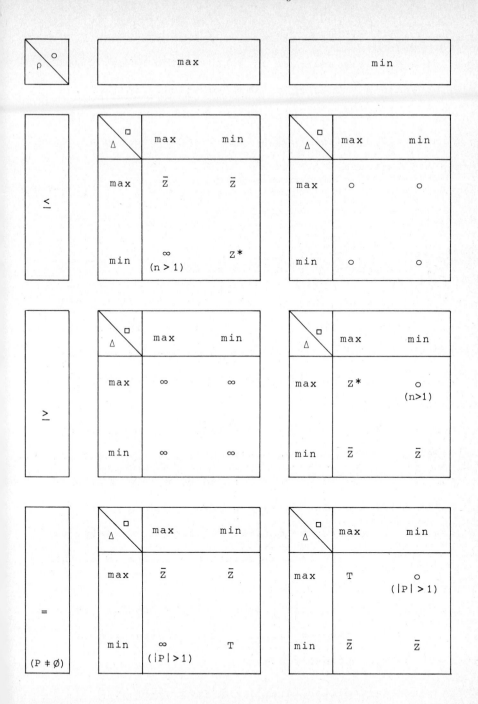

For each problem of the form $(min,\square,\Delta,\leq)$ no component x_j $(j \in N)$ of a feasible solution is bounded from below. Clearly, $P \neq \emptyset$. We admit $x \equiv 0$ as an optimal solution. Similarly, $x \equiv \infty$ is considered as an optimal solution of problems of the form $(max,\square,\Delta,\geq)$. We remind that 0 and ∞ are the adjoined minimum and maximum.

$(max,max,max,\leq)$ has the optimal solution $x = b:A$ (cf. 10.17), i.e.

$$\begin{aligned} x_j &= \min\{a_{ij}^{-1} \mid i \in M\} \\ &= [\max\{a_{ij} \mid i \in M\}]^{-1} = \bar{x}_j(\max) \end{aligned}$$

for all $j \in N$. Clearly, the dual of this problem in the order-theoretic sense is $(min,min,min,\geq)$ and has the optimal solution $\bar{x}(\min)$. Similarly, we find that $\bar{x}(\max)$ and $\bar{x}(\min)$ are the optimal solutions of $(max,min,max,\leq)$ and $(min,max,min,\geq)$.

For $(min,min,max,\geq)$ the case $n = 1$ has to be treated separately. Then $\hat{x}_1 = \min\{b_i \otimes a_{i1}^{-1} \mid i \in M\} = b:A$ is the optimal solution. Otherwise the objective function is not bounded from below. We admit the optimal solution x defined by $x_1 := \hat{x}_1$, and by $x_j := 0$ for all $j \neq 1$. Then $z(x) = 0$. For the dual version $(max,max,min,\leq)$ we find the optimal solution $\hat{x}_1 = \max\{b_i \otimes a_{i1}^{-1} \mid i \in M\}$ $(n = 1)$ and the optimal solution x with $x_1 := \hat{x}_1$ and with $x_j := \infty$ for all $j \neq 1$.

$(min,max,max,\geq)$ is a special instance of the problem (10.1) which is solved in theorem (10.15.1) for lattice-ordered groups. Thus, its optimal value is

$$z = \max_{i \in M} \min_{j \in N} (c_j \otimes a_{ij}^{-1}) = z^*(\min,\max),$$

and an optimal solution is $x^*(\min,\max)$. Then the dual version *(max,min,min,≤)* has the optimal value $z^*(\max,\min)$ and the optimal solution $x^*(\max,\min)$.

Secondly, we discuss problems with equality constraints. Then the set $P(\Delta)$ of feasible solutions is defined by

$$\Delta_{j\in N}(a_{ij} \otimes x_j) = b_i$$

for all $i \in M$. Proposition (10.20) shows that $\bar{x}(\max)$ is the maximum feasible solution of $P(\max)$ if and only if $P(\max) \neq \emptyset$. Similarly, $\bar{x}(\min)$ is the minimum feasible solution of $P(\min)$ if and only if $P(\min) \neq \emptyset$. In the following we assume that $P(\Delta) \neq \emptyset$. This assumption can easily be checked for a given problem by testing $\bar{x}(\Delta) \in P(\Delta)$.

A maximum (minimum) feasible solution is optimal for maximization (minimization) problems. Therefore, $\bar{x}(\Delta)$ is an optimal solution for each of the problems *(max,max,max,=)*, *(max,min,max,=)*, *(min,min,min,=)*, and *(min,max,min,=)* provided that $P(\Delta) \neq \emptyset$.

Next we consider *(min,min,max,=)*. If $|P(\max)| = 1$ then $\bar{x}(\max)$ is optimal. Otherwise, let $x \neq \bar{x}(\max)$ be another feasible solution. Since $\bar{x}(\max)$ is the maximum feasible solution there exists some $\mu \in N$ with $x_\mu < \bar{x}_\mu$. Thus $a_{i\mu} \otimes x_\mu < 1$ for all $i \in M$. Then $\tilde{x}$ with $\tilde{x}_j := \bar{x}_j(\max)$ for $j \neq \mu$ and with $\tilde{x}_\mu := \alpha$ is a feasible solution for all $\alpha \leq \bar{x}_\mu(\max)$. Since α is not bounded from below in R, the objective function is not bounded from below in R. We admit $\alpha = 0$. Then $\tilde{x}$ is an optimal solution with optimal value $z(\tilde{x}) = 0$. Similarly, the optimal solution of the dual version *(max,max,min,=)* is $\bar{x}(\min)$, if $|P(\min)| = 1$, or there exists an optimal solution $\tilde{x}$ with $\tilde{x}_j = \bar{x}_j(\min)$ for $j \neq \mu$ and with

$\tilde{x}_\mu = \infty$ for some $\mu \in N$. Obviously, the condition $|P(\Delta)| > 1$ can easily be checked.

If $P(\max) \neq \emptyset$ then the optimal value z of *(min,max,max,=)* satisfies the inequalities $z^*(\min,\max) \leq z \leq \bar{z}(\max)$, i.e.

$$\max_{i \in M} \min_{j \in N} (c_j \otimes a_{ij}^{-1}) \leq z \leq \max_{j \in N} \min_{i \in M} (c_j \otimes a_{ij}^{-1})$$

(cf. 10.23). If $z^*(\min,\max) < \bar{z}(\max)$ then we apply theorem (10.30). With $\bar{x} = \bar{x}(\max)$ we find

$$Z' = \{c_j \otimes \bar{x}_j \mid j \in N\},$$

$$x'_j(\alpha) = \begin{cases} \bar{x}_j & \text{if } \alpha \geq c_j \otimes \bar{x}_j , \\ 0 & \text{otherwise,} \end{cases}$$

for all $\alpha \in R$. Again, $x'_j(\alpha) = 0$ expresses the fact that this component of x' is not bounded from below. Theorem (10.30) implies that an optimal solution can be determined using one of the following threshold methods. W.l.o.g. we assume

$$c_1 \otimes \bar{x}_1 \leq c_2 \otimes \bar{x}_2 \leq \ldots \leq c_n \otimes \bar{x}_n .$$

(10.38) <u>Threshold method for (min,max,max,=)</u>

Step 1 $k := \min\{j \mid c_j \otimes \bar{x}_j \geq z^*(\min,\max)\}$;

$x'_j := \bar{x}_j$ for $j \leq k$ and $x'_j := 0$ otherwise.

Step 2 If x' is feasible then stop (x' optimal);

$k := k + 1$.

Step 3 $x'_k := \bar{x}_k$ and go to step 2.

(10.39) Primal threshold method for (min,max,max,=)

Step 1 $x' := \bar{x}$; $k := n$.

Step 2 $x'_k := 0$;

if x' is not feasible then go to step 4.

Step 3 $k := k - 1$ and go to step 3.

Step 4 $x'_k := \bar{x}_k$ and stop (x' optimal).

Since x' in method (10.39) is feasible throughout the performance of the algorithm, we call this method primal. To improve the complexity bound of threshold methods it is possible to combine (10.38) and (10.39) and to use a binary search strategy. For example for assignment problems such an approach is described in SLOMINSKI [1979]. Then $O(\log_2 n)$ steps are sufficient. The dual version *(max,min,min,=)* can be solved in the same manner.

We remark that the literature on extremal linear programs over the multiplication group of positive real numbers is discussed above (preceding 10.33).

11. Algebraic Linear Programs

In this chapter $(H,*,\leq)$ denotes a (weakly cancellative) d-monoid which is a linearly ordered or an extended semimodule over the positive cone R_+ of a subring $(R,+,\cdot,\leq)$ of the linearly ordered field of real numbers. Nevertheless, we will mention some results which hold in more general semimodules over real numbers. For a discussion of the relationship of the different types of semimodules and for properties which hold in such semimodules we refer to chapters 5 and 6. We assume that the reader is familiar with basic results for the usual linear programming problem (cf. for example chapter 2 in BLUM and OETTLI [1975]). We consider the algebraic linear program

(11.1) $$\tilde{z} := \inf_{x \in P} x^T \mathbin{\square} a$$

for given coefficients $a_j \in H$, $j \in J := \{1,2,\ldots,n\}$ and

(11.2) $$P := \{x \in R_+^n \mid Cx \geq c\}$$

with an $m \times n$ matrix C and a vector c with entries in R. We assume $\tilde{z} \in H$ and for simplicity of the discussion that P is nonempty and bounded. For constraints with positive entries such a problem was discussed in DERIGS and ZIMMERMANN, U. [1978a]; a great part of this chapter can be found in ZIMMERMANN, U. [1980a]. The case $R = \mathbb{R}$ is also discussed in FRIEZE [1978].

It is possible to introduce slack variables with cost coefficients e (neutral element of H) to transform inequalities to equations. We prefer to develop results explicitly only for inequalities of the form $Cx \geq c$. This choice yields the shortest formulation, and as in classical linear programming, it is sufficient to consider constraints of this type. Equality constraints $Bx = b$ are replaced by $Bx \geq b$, $-Bx \geq -b$; and $Dx \leq d$ is replaced by $-Dx \geq -d$. Then all results of this chapter can easily be carried over. Solution procedures will be given in terms of equality constraints which admit a convenient use of basic solutions within these methods.

On the other hand, the problem

$$(11.3) \qquad \tilde{q} := \sup_{x \in P} \; x^T \,\square\, a$$

is not a special case of (11.1). This is due to the fact that we cannot replace the cost coefficients by their inverses in a semigroup. If H is a weakly cancellative linearly ordered commutative semigroup in which axiom (4.1) is replaced by

$$a > b \quad \Rightarrow \quad \exists\, c \in H: \quad a * c = b$$

for all $a,b \in H$ then a similar theory as given in this chapter for (11.1) can be given for (11.3) simply by reversing the order relation in H. A solution of (11.3) in a semimodule as considered in this chapter can be found using a reduction discussed at the end of this chapter. In fact, it will turn out that in a certain sense (11.3) is a simpler problem than (11.1).

We remark that for the classical case , i.e. for the vectorspace $(\mathbb{R}, +, \leq)$ over the field $\mathbb{R}$, a recent result of HAČIJAN [1979] shows that (11.1) can be solved by a method with polynomial complexity bound. Such a result is not known for (11.1), in general. We will not try to generalize the ellipsoid algorithm described in HAČIJAN [1979] which is based on results of SHOR ([1970], [1977]). We prefer to develop a duality theory which leads to a generalization of the simplex method for (11.1) and which is applicable to combinatorial optimization problems (cf. subsequent chapters).

The first results on dual linear problems due to HOFFMAN [1963] and BURKARD [1975a] have been mentioned in chapter 10 as special cases of the weak duality theorem (10.6) which holds in ordered semimodules.

If H is a linearly ordered commutative group then we assume w.l.o.g. that H is a module over R (via embedding as described in proposition 5.11.3). Then a reasonable dualization of (11.1) is

$$(11.4) \qquad \tilde{g} := \sup_{s \in D} \; c^T \Box s$$

with

$$(11.5) \qquad D := \{ s \in H_+^m \mid C^T \Box s \leq a \} .$$

This is the same dualization as used in the weak duality theorem (10.6). Unfortunately such a dualization fails even for simple examples if H is not a module. Then $C^T \Box s$ is not defined if negative entries occur in C (or c). Such difficulties will appear in the discussion of equality constraints even if all entries in the original constraints are positive.
In order to find a proper dualization it is helpful to split

matrices and vectors into "positive" and "negative" parts. For $\alpha \in R$ let $\alpha_+ := \alpha$ if $\alpha \geq 0$ and $\alpha_+ = 0$ otherwise. Then $\alpha_- := \alpha_+ - \alpha$. For matrices (vectors) A over R we define A_+ and A_- in the same manner componentwise.

Now we call $s \in H_+^m$ *dual feasible* if

$$(11.6)\qquad C_+^T \square s \leq a * C_-^T \square s\ .$$

If $(H,*,\leq)$ is a module then (11.6) is equivalent to $C^T \square s \leq a$. As usual a solution $x \in P$ is called *primal feasible* and a pair $(x;s)$ of a feasible primal and a feasible dual solution is called a *feasible pair*. Now let

$$\alpha(x;s) := (C_-x)^T \square s$$

$$\beta(x;s) := (C_+x)^T \square s$$

for a feasible pair. Further we define *reduced cost coefficients* $\bar{a}$ by

$$C_+^T \square s * \bar{a} := a * C_-^T \square s$$

with $\bar{a}_j := e$ if equality holds in the j-th row of (11.6). Collecting all equations after composition with x_j yields

$$(11.7)\qquad \alpha(x;s) * x^T \square a = x^T \square \bar{a} * \beta(x;s).$$

As $\bar{a} \in H_+^n$ we find the *weak duality theorem*

$$(11.8)\qquad \alpha(x;s) * x^T \square a \geq \beta(x;s).$$

We remark that if C does not contain negative entries then $C_- = 0$ and then $\alpha(x;s) = e$ which implies the usual weak duality theorem, as $x^T \square a \geq (Cx)^T \square s \geq c^T \square s$.

If H is a module then (11.7) is equivalent to

(11.9) $x^T \Box a = x^T \Box \bar{a} * (Cx - c)^T \Box s * c^T \Box s$

which yields $x^T \Box a \geq c^T \Box s$, and $x^T \Box a = c^T \Box s$ if and only if

(11.10) $x^T \Box \bar{a} = (Cx - c)^T \Box s = e.$

In the general case, we consider the ordinal decomposition of the d-monoid H, i.e. the family $(H_\lambda ; \lambda \in \Lambda)$ as described in chapter 4. Such a decomposition leads to a decomposition of the semimodule (cf. chapter 6). Then the *index condition*

(11.11) $\lambda(\alpha(x;s)) \leq \lambda(\tilde{z})$

plays an important role. If $\alpha(x;s) = e$ then (11.11) is satisfied as $\lambda(e) = \lambda_o = \min \Lambda$. In a group (11.11) is always satisfied as $\Lambda = \{\lambda_o\}$. The importance of (11.11) can be seen in the following optimality criterion.

(11.12) Theorem (Dual optimality criterion)

Let H be a weakly cancellative d-monoid which is a linearly ordered semimodule over R_+. Let $s \in H_+^m$ dual feasible and let $\alpha(y) := (C_-y)^T \Box s$, $\beta(y) := (C_+y)^T \Box s$ for all $y \in R_+^n$. If $x \in P$, $\lambda(\alpha(x)) \leq \lambda(\tilde{z})$ and

(1) $\alpha(x) \geq \alpha(y)$ $\forall\, y \in P$,

(2) $\beta(x) \leq \beta(y)$ $\forall\, y \in P$,

(3) $\beta(x) = \beta(x) * x^T \Box \bar{a}$,

then x is an optimal solution of (11.1) and

$$\beta(x) = \alpha(x) * x^T \Box a.$$

Proof. Let $y \in P$. Then

$$\begin{aligned}\alpha(x) * x^T \square a = \beta(x) * x^T \square \bar{a} = \beta(x) &\leq \beta(y)\\ &\leq \beta(y) * y^T \square \bar{a} = \alpha(y) * y^T \square a\\ &\leq \alpha(x) * y^T \square a.\end{aligned}$$

Due to the index condition (11.11) and $\lambda(\tilde{z}) \leq \min(\lambda(x^T \square a), \lambda(y^T \square a))$, cancellation of $\alpha(x)$ yields optimality. (11.12.4) follows from (11.12.3) and (11.7).

■

Theorem (11.12) is a generalization of the usual complementarity criterion. (11.12.1) - (11.12.3) correspond to (11.10) and the index condition (11.11) is added to assure that no "dominating" elements occur in the inequalities.

(11.13) Corollary

Let H be a d-monoid which is a linearly ordered semimodule over R_+. Let C,c have only entries in R_+. If $x \in P$ and

(1) $$c^T \square s = c^T \square s * (Cx - c)^T \square s * x^T \square \bar{a}$$

then x is an optimal solution and

(2) $$c^T \square s = x^T \square a.$$

(11.14) Corollary

Let H be a linearly ordered module over R. If $x \in P$ and

(1) $$(Cx - c)^T \square s = x^T \square \bar{a} = e$$

then x is an optimal solution and

(2) $$c^T \square s = x^T \square a.$$

Special cases of theorem (11.12) and its corollaries are used in solution methods for combinatorial optimization problems in subsequent chapters. A sequence of pairs (x;s) is constructed in such a manner that each pair fulfills some of the primal feasibility conditions, some of the "weak complementarity" conditions (11.12.1) - (11.12.3) and the index condition (11.11). In each step a new dual feasible solution is determined. Therefore it is possible to define reduced cost coefficients $\bar{a}$ (sometimes called "transformed costs"). Then a "transformation" $a \to \bar{a}$ is considered with respect to the respective class of pairs. A sequence of such pairs is determined in order to find a pair fulfilling all these conditions. The construction of such a sequence is rather involved for combinatorial and linear integer problems. This reflects the complicated combinatorial (integer) structure of the underlying primal problems. On the other hand optimality criteria, i.e. feasibility and "weak complementarity" are comparably easy to check for the problems considered. The most difficult condition is the index condition (11.11). This difficulty is due to the occurence of the unknown optimal value $\tilde{z}$ in (11.11). Therefore lower bounds on $\tilde{z}$ play an important role in these primal-dual or dual methods.

In as much as we can interpret "transformation" methods by certain duality principles, it is possible to use theorem (11.12) without an explicit knowledge of duality. Let $T: H^n \to H^n$ denote a "transformation" called *admissible* with respect to (11.1) if there exist functions $\alpha: P \times H^n \to H$, $\beta: P \times H^n \to H$ such that

$$(11.15) \qquad \alpha(y,a) * y^T \square a = y^T \square \bar{a} * \beta(y,a) \qquad \forall\, y \in P.$$

Then $x \in P$ is optimal if (11.12.1) - (11.12.3) and the index condition in theorem (11.12) are satisfied by $\alpha(y) := \alpha(y,a)$ and $\beta(y) := \beta(y,a)$ for all $y \in P$. In the same way corollaries (11.13) and (11.14) can be viewed in terms of a transformation concept. Such an approach was necessary since in the early discussion of algebraic optimization problems a duality theory as described in this chapter was not known. Nevertheless, the transformations were always based on the solution of certain inequalities which could be derived from the primal linear descriptions of the respective problem. Till now it has not been possible to develop transformations for problems the linear description of which is not known.

In an extended semimodule the "weak complementarity" conditions in theorem (11.12) can be simplified. If the index condition (11.11) is fulfilled then (11.7) is equivalent to

$$(11.16) \qquad c_-^T \square s * x^T \square a = x^T \square \bar{a} * (Cx-c)^T \square s * c_+^T \square s .$$

(11.16) follows from the cancellation rule (6.17.7). This is quite similar to the case that H is a module over R. The difference lies only in the splitting of the dual objective $c^T \square s$ into positive and negative parts. In order to obtain the equation (11.16) for all $x \in P$ we assume the *strong index condition*

$$(11.17) \qquad \max\{\lambda(s_k) \mid k = 1,2,\ldots,m\} \leq \lambda(\tilde{z}).$$

Clearly, the strong index condition implies

$$\lambda(\alpha(y;s)) \leq \lambda(\tilde{z})$$

for all $y \in P$. We call $s \in H_+^m$ *strongly dual feasible* if s is dual feasible and satisfies the strong index condition. The set of all strongly dual feasible vectors is denoted by $\tilde{D}$. Then the equation (11.16) holds for all $x \in P$ and $s \in \tilde{D}$.

(11.18) Theorem (Dual optimality criterion)

Let H be a weakly cancellative d-monoid which is an extended semimodule over R_+. Let $s \in \tilde{D}$. If $x \in P$ and

(1) $$c_+^T \square s = c_+^T \square s * (Cx - c)^T \square s * x^T \square \bar{a}$$

then x is an optimal solution of (11.1) and

(2) $$c_+^T \square s = x^T \square a * c_-^T \square s.$$

Proof. In the same way as in the proof of theorem (11.12) we find

$$c_-^T \square s * x^T \square a \leq c_-^T \square s * y^T \square a$$

for all $y \in P$. Cancellation of $c_-^T \square s$ is possible due to the strong index condition (11.17). (11.18.2) is a direct consequence of equation (11.18.1) and equation (11.16).

■

Similar remarks as for theorem (11.12) hold for theorem (11.18). The usual complementarity condition (11.14.1) in corollary (11.14) is replaced by the weak complementarity condition (11.18.1). The complementarity condition is only a sufficient condition, but the weak complementarity condition is necessary and sufficient for the existence of a pair (x;s) fulfilling a "duality theorem" (11.18.2).

Till now we have described the use of certain duality principles for optimality criteria. In order to define a dual program with respect to (11.1) we have to introduce a dual objective. If

$c_- \neq 0$ then $c^T \square s$ is not defined. Therefore let f: $\tilde{D} \to H$ be defined by

$$(11.19) \qquad c_-^T \square s * f(s) := c_+^T \square s$$

for all $s \in \tilde{D}$ with $c_-^T \square s \leq c_+^T \square s$ or $\lambda(c_+^T \square s) = \lambda(c_-^T \square s) = \lambda_o$. Otherwise let $f(s) := e$. As $c_-^T \square s * x^T \square a \geq c_+^T \square s$ for all $x \in P$ and $s \in \tilde{D}$, we find

$$(11.20) \qquad \tilde{f} := \sup_{s \in \tilde{D}} f(s) \leq \inf_{x \in P} x^T \square a .$$

Equality in (11.20) can occur only if there are $s \in \tilde{D}$ with $\lambda(f(s)) = \tilde{\lambda} := \lambda(\tilde{z})$. Therefore it seems to be useful to restrict consideration to values of the objective function which are elements of $H_{\tilde{\lambda}}$. This yields another approach. It is easy to see that $\tilde{\lambda}$ is the optimal value of the *index bottleneck problem*

$$(11.21) \qquad \tilde{\lambda} = \inf_{x \in P} \lambda(x^T \square a) .$$

Its denotation is motivated by

$$\lambda(x^T \square a) = \max(\{\lambda(a_j) \mid x_j > 0\} \cup \{\lambda_o\})$$

and was introduced in ZIMMERMANN, U. ([1976], [1979a]) for the solution of algebraic matroid intersection problems (cf. chapter 13).

A determination of $\tilde{\lambda}$ is possible via a combination of the threshold method of EDMONDS and FULKERSON [1970] and usual linear programs. In fact, (11.21) is a linear bottleneck program as considered by FRIEZE [1975] and GARFINKEL and RAO [1976] and thus is a particular example of an algebraic linear program.

If $\tilde{\lambda}$ is known then the algebraic linear program (11.1) can be reduced to an equivalent problem in a group. Let $M := \{j \mid \lambda(a_j) \leq \tilde{\lambda}\}$. For a vector $u \in H^m$ and $\bar{\lambda} \in \Lambda$ let $u(\bar{\lambda})$ denote the vector with

(11.22) $$u(\bar{\lambda})_i := \begin{cases} u_i & \text{if } \lambda(u_i) \geq \bar{\lambda}, \\ e & \text{otherwise} \end{cases},$$

for $i = 1,2,\ldots,m$. We consider the *reduced algebraic linear program*

(11.23) $$\tilde{t} := \inf_{y \in P_M} y^T \square a(\tilde{\lambda})_M$$

with $P_M := \{y \in R_+^{|M|} \mid C_M y \geq c\}$. This is an optimization problem over the module $G_{\tilde{\lambda}}$ over R (cf. chapter 6). Thus its dual is

(11.24) $$\tilde{g} := \sup_{s \in D} c^T \square s$$

with $D := \{s \in (G_{\tilde{\lambda}})_+^m \mid C_M^T \square s \leq a(\tilde{\lambda})_M\}$. Weak duality shows $\tilde{g} \leq \tilde{t}$. The relationship of these different algebraic linear programs is given in the following theorem.

(11.25) <u>Theorem</u>

Let H be a weakly cancellative d-monoid which is an extended semimodule over R_+. Let $P \neq \emptyset$. Hence $\tilde{\lambda}$ exists.

(1) (11.1) and (11.23) are equivalent. $\tilde{z} = \tilde{t}$. If $x \in P$ is an optimal solution for (11.1) then x_M is an optimal solution for (11.23). If y is an optimal solution for (11.23) then $(y,0)$ is an optimal solution for (11.1).

(2) If the weak complementarity condition (11.18.1) is satisfied for some $x \in P$, $s \in \tilde{D}$ then

$$\tilde{f} = \tilde{g} = c^T \square s(\tilde{\lambda}) = x^T \square a = \tilde{z}.$$

<u>Proof</u>. (1) If $y \in P_M$ then x defined by $x_j := y_j$ for $j \in M$ and $x_j = 0$ otherwise is primal feasible. Therefore $x_j > 0$ for some j with $\lambda(a_j) = \tilde{\lambda}$. Thus $y^T \square a(\tilde{\lambda}) = x^T \square a$ which leads to $\tilde{t} \geq \tilde{z}$. If $x \in P$

is an optimal solution of (11.1) then $\lambda(x^T \square a) = \tilde{\lambda}$ implies $x_j = 0$ for all $j \notin M$. Thus $x_M \in P_M$ and $x_M \square a(\tilde{\lambda})_M = x^T \square a$.

Let $x \in P$. If $x_j > 0$ for some $j \notin M$ then $x^T \square a > \tilde{x}^T \square a$ for some $\tilde{x} \in P$ with $\lambda(\tilde{x}^T \square a) = \tilde{\lambda}$. Otherwise $x_M \in P_M$ and $x_M^T \square a(\tilde{\lambda})_M = x^T \square a$. Therefore $\tilde{t} \leq \tilde{z}$. Together with (1) we find $\tilde{t} = \tilde{z}$. If $y \in P_M$ is an optimal solution of (11.23) then x as defined in the proof of (1) is an optimal solution of (11.1).

(2) Due to theorem (11.18) we know $c_+^T \square s = x^T \square a * c_-^T \square s$. Thus $f(s) = x^T \square a = \tilde{z}$. Then $s(\tilde{\lambda}) \in D$ and $c^T \square s(\tilde{\lambda}) = \tilde{z}$ complete the proof.

■

Naturally we cannot prove an existence theorem for optimal pairs or show the necessity of the assumptions in theorem (11.18), as a duality gap can occur in combinatorial and integer problems even in the classical case. On the other hand, we will now develop a primal-dual simplex method in order to determine such a pair provided that R is a subfield of the field of real numbers. Then H is an extended semivectorspace over R_+.
For the development of a solution method we consider the algebraic linear program

$$(11.26) \qquad \tilde{z} := \inf_{x \in P} x^T \square a$$

with $P := \{x \in R_+^n \mid Ax = b\}$ with $m \times n$ matrix A of rank m over R and $b \in R_+^m$. Every linear algebraic program of the form (11.1) can be transformed to such a linear algebraic program. We have chosen this form for convenience in the description of basic solutions. Another convenient assumption is the nonnegativity of the cost coefficients a_j, i.e. $a_j \geq e$ for all $j \in J$. This allows an obvious starting solution. Self-negative cost coefficients will play a role in the solution of (11.26) only if $\lambda(\tilde{z}) = \lambda_o$; then it suffices to solve a reduced problem similar to (11.23) in the vectorspace G_{λ_o} over the field R. Such a problem can be solved by the usual primal-dual (or other variants of the) simplex method which can easily be seen to be valid in vectorspaces.

In the following method a sequence of strongly feasible vectors $u, v \in H_+^m$ is determined corresponding to the inequality constraints $Ax \geq b$ and $-Ax \geq -b$. From (11.6) we derive

$$(11.27) \qquad A_+^T \square u * A_-^T \square v \leq a * A_-^T \square u * A_+^T \square v$$

and the strong index condition (11.17) reads $\lambda(u_i), \lambda(v_i) \leq \lambda(\tilde{z})$ for all $i \in I := \{1,2,\ldots,m\}$. Reduced cost coefficients $\bar{a} \in H_+^n$ are defined accordingly. A corresponding sequence of vectors $x \in R_+^n$ will satisfy $x^T \square \bar{a} = e$. The method terminates when the current vector x is primal feasible. Then the complementarity condition (11.10) is fulfilled which implies (11.18.1) and hence x is optimal.

Partitions of vectors and matrices are denoted in the usual way. In particular A_j is the j-th column of A. E denotes the $m \times m$ unit matrix and (O 1) is a row vector with n zero-entries followed by m one-entries.

(11.28) <u>Primal dual solution method</u>

Step 1 Let $u_i = v_i = e$ for all $i \in I$, $x_j = 0$ for all $j \in J$.

Step 2 Let $K := \{j \in J \mid \bar{a}_j = e\}$; determine an optimal basis B of the usual linear program

$$\varepsilon := \min\{\Sigma y_i \mid A_K x_K + y = b;\ x_K \geq 0;\ y \geq 0\};$$

let $\pi \in R^m$ be defined by

$$\pi^T := (0\ 1)[A\ E]_B^{-1}\ .$$

Step 3 If $\varepsilon = 0$ then stop (x is optimal).

Step 4 Let $\tilde{K} := \{j \in J \setminus K \mid \pi^T A_j > 0\}$;

if $\tilde{K} = \emptyset$ then stop ($P = \emptyset$).

Step 5 Let $d := \min\{(1/\pi^T A_j) \square \bar{a}_j \mid j \in \tilde{K}\}$;

let $u_i := u_i * (\pi_+)_i \square d$

$v_i := v_i * (\pi_-)_i \square d$

for all $i \in I$ and go to step 2.

Provided that (11.27) is always fulfilled we see that all steps of the method are well-defined in the underlying algebraic structure. In particular in the definition of π in step 2 and the definition of d in step 5 we observe the necessity of the assumption that $(R,+,\cdot)$ is a field. If we replace the compositions "$*$" and "$\square$" by the usual addition and multiplication of real numbers then the above method is the usual primal-dual one for linear programming. For the minimization of *bottleneck objectives*

$$z(x) := \max\{a_j \mid x_j > 0\} \qquad (x \in P)$$

with real cost coefficients $a_j \in \mathbb{R}$, $j \in J$, the method reduces to a method proposed by FRIEZE [1975] and GARFINKEL and RAO [1976].

(11.28) is closely related to the primal-dual method in FRIEZE [1978]. Both methods reduce to the same methods for the usual linear program and the *bottleneck program*

$$\tilde{z} := \min_{x \in P} \max\{a_j \mid x_j > 0\}.$$

The main difference lies in the introduction of a further vector $w \in H^m$ in the constraints replacing (11.27). This leads to a more complicated optimality criterion and a more complicated method in FRIEZE [1978].

In the following we give a proof of the validity and finiteness of (11.28). Due to the assumption $a_j \geq e$ for all $j \in J$ the objective function of the ALP is bounded from below by e. Therefore there are only two possible terminations of the method. In the same way as for usual linear programming we see that $\tilde{K} = \emptyset$ in step 4 means that the current value $\varepsilon > 0$ is the optimal value of the linear program

$$(11.29) \qquad \min\{\Sigma y_i \mid Ax + y = b;\ x \geq 0,\ y \geq 0\} \ ;$$

this implies $P = \emptyset$. Finiteness will follow from the fact that in step 2 a sequence of bases B for (11.29) is considered. It suffices to show that the value ε strictly decreases in an iteration. For an iteration we always assume $\tilde{K} \neq \emptyset$. As mentioned after (11.28) we have at first to show that (11.27) holds after an iteration. We denote the redefined vectors in step 2 by $\bar{u}, \bar{v}$.

(11.30) Proposition

If $u, v \in H^m_+$ satisfy (11.27) then $\bar{u}, \bar{v} \in H^m_+$ and $\bar{u}, \bar{v}$ satisfy

$$A^T_+ \square \bar{u} * A^T_- \square \bar{v} \leq A^T_- \square \bar{u} * A^T_+ \square \bar{v} * a \ .$$

Proof. For convenience denote the left-hand-side of (11.27) by $\beta(u,v)$ and the respective part on the right-hand-side by $\alpha(u,v)$. Furthermore let $\delta^+ := A_+^T\pi_+ + A_-^T\pi_-$, $\delta^- := A_-^T\pi_+ + A_+^T\pi_-$. As $\bar{a} \in H_+^n$ implies $d \in H_+$ we find $\bar{u},\bar{v} \in H_+^m$ from the definition in step 5 . This definition also leads to

$$\text{(11.31)} \quad \begin{aligned} \alpha_j(\bar{u},\bar{v}) &= \alpha_j(u,v) * \delta_j^- \,\Box\, d \\ \beta_j(\bar{u},\bar{v}) &= \beta_j(u,v) * \delta_j^+ \,\Box\, d \end{aligned}$$

for all $j \in J$. Here the distributivity axioms in the semi-vectorspace play an important role. At first we consider $j \in J$ with $\pi^T A_j \le 0$. Then $\delta_j^+ \le \delta_j^-$. As $\beta_j(u,v) \le \alpha_j(u,v) * a_j$ is assumed we find $\beta_j(\bar{u},\bar{v}) \le \beta_j(\bar{u},\bar{v}) * a_j$ using the monotonicity of the external composition and (11.31). Secondly, let $j \in J$ with $\pi^T A_j > 0$. Then $j \notin K$ follows from the optimality of B in step 2. Therefore

$$d \le (1/\pi^T A_j) \,\Box\, \bar{a}_j \ .$$

Using distributivity and monotonicity this leads to

$$\delta_j^+ \,\Box\, d \le \bar{a}_j * \delta_j^- \,\Box\, d \ .$$

Due to our assumptions we know

$$\beta_j(u,v) * \bar{a}_j = \alpha_j(u,v) * a_j$$

and therefore using (11.31) we find

$$\text{(11.32)} \qquad \beta_j(\bar{u},\bar{v}) \le \beta_j(u,v) * \bar{a}_j * \delta_j^- \,\Box\, d = \alpha_j(\bar{u},\bar{v}) \,\Box\, a_j \ .$$

■

In particular the proof of (11.30) shows that for a defining row $k \in \tilde{K}$ of $d := (1/\pi^T A_k) \,\Box\, \bar{a}_k$ we find equality in (11.32). As

$k \notin K$ and $\pi^T A_k > 0$ we know that in the next performance of step 2 the value of ε can strictly be decreased by introducing x_k into the basis provided that the considered polyhedron in (11.29) is nondegenerate. As $(R,+,\cdot)$ is a subfield of the real field the set R is dense in $\mathbb{R}$. Therefore any perturbation method can be used and we may assume nondegeneracy w.l.o.g.. As the primal dual method can be interpreted as special method for solving (11.29) we can of course apply other finiteness rules as for example the usual lexicographic criterion (cf. BLUM and OETTLI [1975]) or the rule of BLAND [1977].

Now if $\varepsilon = 0$ in step 3 then x is primal feasible. Furthermore we know that $x^T \square a = e$. Therefore we have only to prove that $u, v \in H_+^m$ is strongly feasible. Again the iterated vectors in step 5 are denoted by $\bar{u}, \bar{v}$.

(11.33) Proposition

If $u, v \in H_+^m$ are strongly dual feasible then $\bar{u}, \bar{v} \in H_+^m$ are strongly dual feasible.

Proof. The first part has been shown in proposition (11.30). Therefore it suffices to show $\lambda(\bar{u}_i), \lambda(\bar{v}_i) \leq \lambda(\tilde{z})$. Now from the definition of d in step 5 we conclude that $\lambda(d) = \lambda(\bar{a}_k)$ for some $k \in \tilde{K}$. From the definition of the reduced cost coefficients we find $\lambda(\bar{a}_k) \leq \max\{\lambda(u_i), \lambda(v_i), \lambda(a_k) \mid i \in I\}$. Clearly, $\lambda(a_k) \leq \lambda(\tilde{z})$. As the strong index condition holds for u,v this implies $\lambda(d) \leq \lambda(\tilde{z})$. Finally this and the definition of $\bar{u}, \bar{v}$ in step 5 leads to $\lambda(\bar{u}_i), \lambda(\bar{v}_i) \leq \lambda(\tilde{z})$.

■

As the starting solution fulfills the index condition we conclude from proposition (11.33) that in the primal dual method a sequence of strongly dual feasible solutions u,v is determined. Therefore if the method terminates in step 3 we have found an optimal solution x.

The primal dual solution method yields a constructive proof for the following strong duality theorem.

(11.34) <u>Theorem</u> (Linear programming duality theorem)

Let H be a weakly cancellative d-monoid which is an extended semivectorspace over R_+. If P is bounded and nonempty then there exists an optimal basic solution $\tilde{x} \in P$ with basis B for the linear algebraic program (11.26) and there exists a strongly dual feasible solution $\tilde{u}, \tilde{v} \in H_+^m$ such that $\tilde{x}^T \square \bar{a} = e$ and

(1) $$b^T \square \tilde{u} = \tilde{x}^T \square a * b^T \square \tilde{v} ,$$

(2) $$b^T \square \tilde{u} * (-b)^T \square \bar{v}(\tilde{\lambda}) = \tilde{x}^T \square a \qquad \text{for } \tilde{\lambda} = \lambda(\tilde{x}^T \square a).$$

<u>Proof</u>. We remark that the dual objective function $f: \tilde{D} \to H$ on the set of all strongly dual feasible solutions $(u,v) \in \tilde{D}$ is given by

$$b^T \square u =: f(u,v) * b^T \square v$$

in the same way as f(s) in (11.19). The claimed result now follows from theorems (11.18) and (11.25) interpreted for linear programs with equality constraints and applied to the feasible pair $(\tilde{x};\tilde{u},\tilde{v})$ which is determined by the primal dual solution method (11.28).

∎

We remark that $(\tilde{x};\tilde{u},\tilde{v})$ in theorem (11.34) satisfies the assumptions of (1) and (2) in theorem (11.25); therefore the corresponding algebraic linear programs have identical optimal values $(\tilde{t} = \tilde{z} = \tilde{f} = \tilde{g})$.

Next we will state a strong duality theorem from matching theory which has been given in ZIMMERMANN, U. [1980a] based on the results in DERIGS [1979a] and EDMONDS and PULLEYBLANK [1974].

Let $G = (V,N)$ denote a complete graph with even number $|V|$ of vertices. Then a subset $M \subseteq N := \cup \{e_{ij} \mid i,j \in V,\ i \neq j\}$ is called a *perfect matching* if for all $i \in V$ there exists one and only one edge $e_{\mu\nu} \in M$ with $\mu = i$ or $\nu = i$. We consider subsets $T \subset V$ with odd cardinality $2t+1$, $t \in \mathbb{N}$. Let FT denote the set of these subsets. Let $N_T := \cup \{e_{ij} \mid i,j \in T,\ i \neq j\}$ for $T \in FT$ and denote the set of all N_T by FN. Now let B denote the vertex-edge incidence matrix of G and C the set-edge incidence matrix of sets $N_T \in FN$ with the edges of G. Let C_{ij} denote the column of C corresponding to edge e_{ij}. Then the vector with components $c_T := t$ for $T \in FT$ is denoted by c.

EDMONDS and PULLEYBLANK [1974] have shown that the vertices of

$$(11.35) \qquad P := \{x \in \mathbb{R}_+^{|N|} \mid Bx = 1,\ Cx \leq c\}$$

are in one-to-one correspondence with the incidence vectors of the perfect matchings of G. Thus we consider (11.1) with respect to (11.35). This problem is solved in DERIGS [1979a] for coefficients with $a_{ij} \in H_+$ for all $(i,j) \in N$. An incidence vector x of a perfect matching and a dual feasible solution (u,v,s) in the sense of (11.6) are determined; i.e. (u,v,s) satisfies

$$u_i * u_j \leq a_{ij} * v_i * v_j * C_{ij}^T \square s$$

for all $(i,j) \in N$, $u_i, v_i \in H_+$ for all $i \in V$ and $s_T \in H$ for all $T \in FT$. Furthermore (u,v,s) is strongly dual feasible and satisfies

(11.36) $$(Cx)_T < d_T \Rightarrow s_T = e$$

$$1 = x_{ij} > 0 \Rightarrow u_i * u_j = a_{ij} * v_i * v_j * C_{ij}^T \square s .$$

(11.36) implies the complementarity conditions. Application of theorems (11.18) and (11.25) yields the following results on duality.

(11.37) Theorem (Perfect matching duality theorem)

Let H be a weakly cancellative d-monoid which is an extended semimodule over R_+. Then there exists an optimal feasible pair $(x;u,v,s)$ with (11.36) and

(1) $$1^T \square u = x^T \square a * 1^T \square v * d^T \square s ,$$

(2) $$1^T \square u * (-1)^T \square v(\tilde{\lambda}) * (-d)^T \square s(\tilde{\lambda}) = x^T \square a$$

$$\text{for } \tilde{\lambda} := \lambda(x^T \square a).$$

The solution method for such a combinatorial optimization problem is developed similar to the classical case (the vector-space $(\mathbb{R},+,\leq)$ over $\mathbb{R}$). Since feasibility and complementarity known from classical linear programming is used in such methods, the development of similar criteria was made at first in the form of a transformation concept. The duality concept described in this chapter shows a way to interpret these transformation concepts again by certain generalizations of the classical duality concept. It should be emphasized that the construction of optimal pairs as in theorem (11.37) fulfilling the generalized feasibility and complementarity criteria makes full use of the special combinatorial structure of the respective problems. In subsequent chapters we will discuss methods in network flow theory and matroid intersection theory. Matching theory can be investigated using the same concepts. The main difference lies in the combinatorial structure considered for the respective problems. Therefore for results in matching theory we refer to a series of papers by DERIGS ([1978a], [1978b], [1979a], [1979b]).

In the development of primal dual methods for the solution of combinatorial optimization problems it turned out to be very important to use lower bounds on the index of the optimal value for a reduction of the problem. Let $\mu \leq \lambda(\tilde{z})$ be a lower bound of the index of the optimal value of (11.26). Then

(11.26') $$\tilde{z} = \inf_{x \in P} x^T \Box a(\mu)$$

and $\tilde{x} \in P$ is an optimal solution of (11.26) if and only if it is an optimal solution of (11.26'). During the performance of the primal dual solution method a strongly dual feasible solution u,v is known. Then $u(\mu), v(\mu)$ is strongly dual feasible with respect to (11.26'). Further, if u,v is strongly dual feasible then

$$\mu := \max\{\lambda(u_i), \lambda(v_i) \mid i \in I\}$$

is a suitable lower bound of $\lambda(\tilde{z})$. Thus we may modify step 5 in the primal dual solution method by an additional revision of the dual variables.

(11.28') <u>Modification of step 5 in (11.28)</u>

Step 5' Let $d := \min\{(1/\pi^T A_j) \square \bar{a}_j \mid j \in \tilde{K}\}$;

let $\mu := \lambda(d)$;

let $u_i := u_i(\mu) * (\pi_+)_i \square d$,

$v_i := v_i(\mu) * (\pi_-)_i \square d$,

for all $i \in I$ and go to step 2.

Further the reduced cost coefficients $\bar{a}$ are determined from

(11.27') $$A_+^T \square u * A_-^T \square v * \bar{a} = a(\mu) * A_-^T \square u * A_+^T \square v.$$

In each iteration all dual variables have index μ or are equal to the neutral element e. As $\lambda(d) = \lambda(\bar{a}_j)$ for some $j \in \tilde{K}$ and as $\bar{a}_j$ is determined from (11.27') the index $\lambda(d)$ satisfies $\lambda(d) \geq \max\{\lambda(u_i, \lambda(v_i) \mid i \in I\}$ with respect to the current dual variables *before* the performance of step 5. Thus $\mu := \lambda(d)$ yields the same index as $\mu := \max\{\lambda(u_i), \lambda(v_i) \mid i \in I\}$ *after* the performance of step 5. The modification results in a possibly enlarged set K in step 2; therefore the number of iterations may be reduced. The fact that all dual variables which are not equal to e have the same index is very helpful in proving finiteness of special primal dual methods for combinatorial optimization problems.

Next we consider a solution of algebraic linear programs with primal methods. We develop primal optimality criteria for (11.26). In the discussion of the primal dual method we assumed that $u, v \in H_+^m$ were dual feasible. Now we assume only that the given vector x is primal feasible. The respective dual solutions are not necessarily dual feasible. Therefore optimality criteria as in theorems (11.12) and (11.18) cannot be applied. This is essentially due to the fact that a definition of reduced cost coefficients $\bar{a}$ is possible only if a dual feasible solution is known. In particular, a primal method will not be covered by the transformation approach. Primal methods for algebraic linear programs are considered in ZIMMERMANN, U. [1980a].

In the following discussion we assume that P (as defined for (11.26)) is nonempty and bounded. H is a weakly cancellative d-monoid which is an extended semimodule over R_+.

Now let $\bar{x} \in P$ and $\bar{\lambda} := \lambda(\bar{x}^T \square a)$. $w \in G_{\bar{\lambda}}^m$ is called *$\bar{\lambda}$-complementary* with respect to $\bar{x}$ if

$$(11.38) \qquad \bar{x}_j > 0 \quad \Rightarrow \quad A_j^T \square w = a(\bar{\lambda})_j \qquad \forall\, j \in J$$

(cf. 11.22). Further $w \in (G_{\bar{\lambda}})_+^m$ is called *$\bar{\lambda}$-dual feasible* if

$$(11.39) \qquad A^T \square w \le a(\bar{\lambda}).$$

If H is a module over R then $\bar{\lambda}$-complementarity and $\bar{\lambda}$-dual feasibility coincide with dual feasibility (cf. 11.5) and complementarity (cf. 11.10).

Let $D(\bar{\lambda})$ denote the set of all $\bar{\lambda}$-dual feasible vectors. Then composition of (11.39) with $\bar{x}$ yields $b^T \square w \leq \bar{x}^T \square a$ for all $w \in D(\bar{\lambda})$ and therefore with $\tilde{\lambda} := \lambda(\tilde{z})$ we find

$$(11.40) \qquad \tilde{h} = \sup_{w \in D(\tilde{\lambda})} b^T \square w \leq \tilde{z} \;.$$

This is another dual program for ALP. For a $\bar{\lambda}$-complementary vector w we find $b^T \square w = \bar{x}^T \square a$. This concept leads to the following optimality criterion.

(11.41) Theorem (Primal optimality criterion)

Let H be a weakly cancellative d-monoid which is an extended semimodule over R_+. Let $\bar{x} \in P$ and $\bar{\lambda} := \lambda(\bar{x}^T \square a)$. If w is $\bar{\lambda}$-complementary with respect to $\bar{x}$ and $\bar{\lambda}$-dual feasible, then $\bar{x}$ is an optimal solution of (11.26) and

$$(1) \qquad \tilde{h} = b^T \square w = \bar{x}^T \square a = \tilde{z} \;.$$

Proof. Due to the assumptions we find $\bar{x}^T \square a = b^T \square w \leq y^T \square a(\bar{\lambda})$ for all $y \in P$. If $\bar{\lambda} = \tilde{\lambda}$ then $y^T \square a(\bar{\lambda}) = y^T \square a$ for all $y \in P$. This implies optimality of $\bar{x}$. Otherwise let $\tilde{y} \in P$ denote an optimal solution. Then $\tilde{y}^T \square a(\bar{\lambda}) = e < \bar{x}^T \square a$ yields a contradiction. Now $\tilde{h} = \tilde{z}$ follows from (11.40).

■

Theorem (11.41) shows that a special type of complementarity and dual feasibility implies optimality without an additional index condition. Thus (11.38) can replace (11.18) if complementarity is used as optimality criterion. Furthermore it can be seen that if $\bar{x}$ is an optimal solution, w $\bar{\lambda}$-dual feasible and (11.41.1) holds then w is $\bar{\lambda}$-complementary. Otherwise if $\bar{x}$ is optimal, w is $\bar{\lambda}$-complementary and (11.41.1) holds then in general it is not guaranteed that w is $\bar{\lambda}$-dual feasible. In this sense (11.41) describes only a sufficient optimality criterion even if a result on strong duality holds.

A comparison of (11.40) with the other duals (11.20) and (11.24) shows that if the assumptions of (11.41) hold then all optimal objective values are equal. In this case all three approaches are equivalent.

If $\bar{\lambda}$-dual feasibility is used as optimality criterion and w is $\bar{\lambda}$-complementary then it suffices to check

(11.42) $$A_j^T \square w \leq a(\bar{\lambda})_j$$

for $\bar{x}_j = 0$ and $\lambda(a_j) \leq \bar{\lambda}$. Then (11.42) is equivalent to (11.39).

To use theorem (11.41) in a primal method we have to consider conditions for the existence of $\bar{\lambda}$-complementary solutions. In primal methods basic solutions play an important role. If B is a feasible basis for P then $x_B = A_B^{-1} b \in R_+^m$. If we assume

(11.43) $$\det A_B^{-1} \in R$$

for all feasible bases of P then the existence of a $\bar{\lambda}$-complementary solution with respect to a feasible basic solution $\bar{x}$ is guaranteed. The existence of an optimal basic solution for bounded, nonempty P follows from theorem (11.34) in the case that R is a subfield of $\mathbb{R}$, i.e. provided that H is a weakly cancellative d-monoid which is an extended semivectorspace over R_+. Clearly, (11.43) is valid in this case and we can w.l.o.g. assume that P is nondegenerate. Let $\tilde{P}$ denote the nonempty set of all feasible basic solutions. In the following we develop a primal simplex method in such semivectorspaces.

Let $\bar{x} \in \tilde{P}$ with basis $\bar{B}$. Then $x \in \tilde{P}$ with basis B is called a neighbor of $\bar{x}$ if $|B \cap \bar{B}| = m - 1$. Let $N(\bar{x})$ denote the set of all neighbors of $\bar{x}$. Then $\bar{x} \in \tilde{P}$ is called *locally optimal* if $\bar{x}^T \square a \leq x^T \square a$ for all $x \in N(\bar{x})$.

(11.44) Proposition

Let H be a weakly cancellative d-monoid which is an extended semivectorspace over R_+. Then $\bar{x} \in \tilde{P}$ is an optimal solution of the algebraic linear program (11.26) if and only if $\bar{x}$ is locally optimal.

Proof. We only show the nontrivial if-part. Let $\bar{x}$ be locally optimal and let $\tilde{x} \in P$ denote an optimal solution. Then we know from linear programming that

$$\tilde{x} = \bar{x} + \sum_{x \in N(\bar{x})} \mu_x (x - \bar{x})$$

with $\mu_x \in \mathbb{R}_+$. As $(R,+,\cdot)$ is a subfield of $(\mathbb{R},+,\cdot)$ we can show $\mu_x \in R_+$ for all $x \in N(\bar{x})$. Therefore

$$\tilde{x}^T \square a * \mu \square (\bar{x}^T \square a) \geq (1+\mu) \square (\bar{x}^T \square a)$$

with $\mu := \Sigma \mu_x$. Cancellation due to (6.17.7) yields $\tilde{x}^T \square a \geq \bar{x}^T \square a$.

■

From the proof we see that proposition (11.44) may be invalid if the semimodule is not extended. In particular this may happen for the linear bottleneck program. Proposition (11.44) again shows that the optimal value of (11.26) is attained at a feasible basic solution. We can state a necessary and sufficient optimality condition and another strong duality theorem.

(11.45) <u>Theorem</u> (Linear programming duality theorem)

Let H be a weakly cancellative d-monoid which is an extended semivectorspace over R_+. Let $\bar{x} \in \tilde{P}$ with basis $\bar{B}$ and $\bar{\lambda} := \lambda(\bar{x} \,\square\, a)$. Then the following statements hold:

(1) $\bar{x}$ is an optimal solution of (11.26) if and only if $w := (A_B^{-1})^T \,\square\, a(\bar{\lambda})_B$ is $\bar{\lambda}$-dual feasible;

(2) there exists an optimal basic solution;

(3) the optimal values of the considered dual programs (11.20), (11.24) and (11.40) coincide with $\tilde{z}$.

<u>Proof</u>. Due to (11.41) it suffices to show for (1) that $\bar{x}$ optimal implies that w is $\bar{\lambda}$-dual feasible. If w is not $\bar{\lambda}$-dual feasible then there exists $s \notin \bar{B}$ with $\lambda(a_s) \leq \bar{\lambda}$ and $A_s^T \,\square\, w = a_s(\bar{\lambda}) * \alpha$ with $\alpha \in H_{\bar{\lambda}}$, $\alpha > e$. Then $b^T \,\square\, w = x^T \,\square\, a(\bar{\lambda}) * (x_s \,\square\, \alpha)$ for $x \in \tilde{P}$ with basis $B = \bar{B} \setminus \{r\} \cup \{s\}$ (Existence of such a basic solution x follows from linear programming and (11.43)). Since $b^T \,\square\, w = \bar{x}^T \,\square\, a$ and $x_s > 0$ we find $\bar{x}^T \,\square\, a > x^T \,\square\, a(\bar{\lambda})$ which implies the contradiction $\bar{x}^T \,\square\, a > x^T \,\square\, a$. (2) follows from (11.44) and (3) follows from (11.25) if the equality constraints of (11.26) are replaced by inequality constraints in the usual manner.

∎

Similar to the classical case theorem (11.45) and proposition (11.44) imply the validity of the following primal simplex method for algebraic programming problems. We give an outline of the method, showing the differences to the classical case.

(11.46) <u>Primal simplex method</u>

Step 1 Find $\bar{x} \in \tilde{P}$ with basis $\bar{B}$; $\bar{\lambda} := \lambda(\bar{x}^T \square a)$.

Step 2 Calculate $w := (A_{\bar{B}}^{-1})^T \square a(\bar{\lambda})_{\bar{B}}$;
if $A_j^T \square w \leq a_j(\bar{\lambda})$ for all $j \notin \bar{B}$ with $\lambda(a_j) \leq \bar{\lambda}$ then stop.

Step 3 Choose $s \notin \bar{B}$ with $\lambda(a_s) \leq \bar{\lambda}$ and $A_s^T \square w > a_s(\bar{\lambda})$;
determine $x \in N(\bar{x})$ with basis B containing s;
redefine $\bar{x}$, $\bar{B}$ and $\bar{\lambda}$ and return to step 2.

If the method stops in step 2 then w is $\bar{\lambda}$-dual feasible and theorem (11.41) yields optimality. Finiteness of (11.46) can be seen from the proof of (11.45). A basis exchange step due to step 3 decreases strictly the value of the objective function. As $\tilde{P}$ is a finite set this implies finiteness. Primal degeneracy can be handled with some of the usual finiteness rules (cf. BLUM and OETTLI [1975] or BLAND [1977]).

If H is a weakly cancellative d-monoid which is an extended semimodule over R_+ then the primal simplex method can be applied provided that (11.43) is satisfied. In particular, (11.43) holds if the constraint matrix A is *totally unimodular*, i.e. A contains only submatrices of determinant 0 or ± 1.

The primal simplex method is applied to algebraic transportation problems in chapter 12. The optimality criterion (11.42) can in this special case be improved such that it suffices to check a less number of feasibility conditions.

We have developed two solution methods for algebraic linear programs which simultaneously solve the corresponding dual programs (11.20), (11.24) and (11.40). These duals have objective function *and* constraints with values in the semimodule H. Therefore, in general, such problems seem to have a structure differing from algebraic linear programs. We will now consider the solution of such problems. The dual (11.20) of (11.26) has an objective function $f\colon H_+^m \times H_+^m \to H$ defined by

$$b^T \square v * f(u,v) := b^T \square u$$

if $b^T \square v \leq b^T \square u$ or $\lambda(b^T \square v) = \lambda(b^T \square u) = \lambda_o$ and $f(u,v) := e$ otherwise. The set of all strongly dual feasible solutions $\tilde{D}$ consists in all $(u,v) \in H_+^m \times H_+^m$ with

(11.46) $$A_+^T \square u * A_-^T \square v \leq A_-^T \square u * A_+^T \square v * a,$$

(11.47) $$\lambda(u_i), \lambda(v_i) \leq \lambda(\tilde{z})$$

for all $i \in I$. If H is a vectorspace over R then

$$\sup_{(u,v) \in \tilde{D}} f(u,v)$$

is equivalent to

(11.48) $$\sup_{w \in \bar{D}} b^T \square w$$

with $\bar{D} := \{w \in H^m \mid A^T \square w \leq a\}$. This form looks much more familiar and does not contain an explicit index condition with respect to the optimal value of the original primal problem. Such a problem can be solved by the application of the generalized simplex methods to (11.26). Here we may interpret (11.26) as the dual of (11.48). In the general case of a semivectorspace difficulties arise from the fact that the external composition is only defined on R_+ (otherwise, a nontrivial extension to R implies that H is a modul over R) and that variables in H are more or less positive ($\lambda(a) > \lambda_o \Rightarrow a^{-1}$ does not exist). Therefore, splitting of variables and matrices leads to constraints of the form (11.46). If the strong index condition (11.47) can be assumed without changing the problem then (11.26) again may be interpreted as a suitable dual. If (11.47) is not added to the constraints (11.46) and is not implicitly implied by these constraints then even weak duality may fail for the pair of programs (11.26) and

$$(11.48) \qquad \sup_{(u,v) \in D'} f(u,v)$$

with $D' := \{(u,v) \in H_+^m \times H_+^m \mid A_+^T \square u * A_-^T \square v \leq A_-^T \square u * A_+^T \square v * a\}$.

A solution of (11.48) is not known in general.

Based on the max-matroid flow min-cocircuit theorem for algebraic flows in HAMACHER [1980a] we consider an example for a problem of the form (11.48). For fundamental definitions and properties of matroids we refer to chapter 1.

Let M denote a regular matroid defined on the finite set $I = \{0,1,2,...,m\}$. SC resp. SB denotes the set of all circuits resp. cocircuits. Let A denote the *oriented incidence matrix* of elements and cocircuits, i.e.

$$a_{iB} := \begin{cases} 1 & i \in B^+ \\ -1 & i \in B^- \\ 0 & i \notin B \end{cases}$$

for all $i \in I$ and all $B \in SB$. For a given capacity-vector $c \in H_+^m$ the elements of

$$D := \{(t,u) \mid A_+^T \square \binom{t}{u} = A_-^T \square \binom{t}{u},\ u \leq c,\ u \in H_+^m,\ t \in H_+\}$$

are called *algebraic matroid flows* in HAMACHER [1980a]. The *max-matroid flow problem* is

$$(11.49) \qquad \tilde{t} := \max_{(t,u) \in D} t .$$

Now we consider its formal dualization

$$(11.50) \qquad \tilde{z} := \min_{x \in P} x^T \square c$$

with the set of feasible solutions

$$P := \{x \in R_+^m \mid \exists\ y \in R^{SB} :\ Ay \geq \begin{bmatrix} 1 \\ -x \end{bmatrix}\} .$$

HAMACHER [1980a] develops an algorithm which determines an optimal solution $(\tilde{t},\tilde{u})$ of (11.49) with the property

$$(11.51) \qquad \lambda(\tilde{u}_i) \leq \lambda(\tilde{t})$$

for all $i = 1,2,\ldots,m$. Furthermore the algorithm yields a constructive proof for the existence of a cocircuit $\tilde{B}$ with $0 \in \tilde{B}^-$ such that

$$(11.52) \qquad \tilde{t} = (a_{1\tilde{B}}, a_{2\tilde{B}}, \ldots, a_{m\tilde{B}})_+^T \square c$$

and

$$\tilde{t} \leq (a_{1B}, a_{2B}, \ldots, a_{mB})_+^T \square c$$

for all $B \in SB$ with $0 \in B^-$. This implies directly the max-matroid flow - min cocircuit theorem in HAMACHER [1980a]. Furthermore it leads to the following strong duality result for (11.49) and (11.50).

(11.53) Theorem (Matroid flow duality theorem)
Let H be a weakly cancellative d-monoid which is an extended semimodule over R_+. Then there exists an optimal pair of a feasible primal solution $(\tilde{t},\tilde{u}) \in D$ and a feasible dual solution $\tilde{x} \in P$ such that $\tilde{t} = \tilde{x}^T \square c$.

Proof. At first we show weak duality, i.e. $t \leq x^T \square c$ for all $(t,u) \in D$, $x \in P$. Clearly $t \leq \tilde{t}$ for all optimal solutions $(\tilde{t},\tilde{u})$ of (11.49). W.l.o.g. we assume that (11.51) holds. $x \in P$ implies the existence of $y \in R^{SB}$ with $Ay \geq \begin{bmatrix} 1 \\ -x \end{bmatrix}$. Using distributivity and monotonicity in the semimodule we find

$$(A_-y_+ + A_+y_-)^T \square \begin{bmatrix} \tilde{t} \\ \tilde{u} \end{bmatrix} * \tilde{t} \leq (A_+y_+ + A_-y_-)^T \square \begin{bmatrix} \tilde{t} \\ \tilde{u} \end{bmatrix} * x^T \square \tilde{u} .$$

(11.51) shows that we can cancel the first terms on each side (cf. 6.17.7) which are equal due to the equality constraints in the definition of D. Now $\tilde{u} \leq c$ and monotonicity in the semimodule imply $\tilde{t} \leq x^T \square \tilde{u} \leq x^T \square c$.

Secondly cocircuits $\tilde{B} \in SB$ with $0 \in \tilde{B}^-$ correspond to certain elements of P defined by

$$\tilde{x}_i := \begin{cases} 1 & \text{if } i \in \tilde{B}^+ , \\ 0 & \text{otherwise}. \end{cases}$$

Indeed, $\tilde{y}_B := -1$ if $B = \tilde{B}$ and $\tilde{y}_B := 0$ otherwise, shows feasibility of $\tilde{x}$. Clearly $\tilde{x}^T \square c = (a_{1\tilde{B}}, a_{2\tilde{B}}, \ldots, a_{m\tilde{B}})_+^T \square c$. Therefore (11.52) implies $\tilde{t} = \tilde{x}^T \square c$.

■

In fact the proof shows that (11.50) has always a zero-one solution which is the incidence vector of the positively directed part of a cocircuit. As in particular we may consider the usual vectorspace $(\mathbb{R},+,\leq)$ over $\mathbb{R}$ a direct consequence is that all vertices of the polyhedron

$$P_{\mathbb{R}} = \{x \in \mathbb{R}_+^m \mid \exists\ y \in \mathbb{R}^{SB} : Ay \geq \begin{bmatrix} 1 \\ -x \end{bmatrix} \}$$

are in one-to-one correspondence with the set of all cocircuits $B \in SB$ with $0 \in B^-$ for a given regular matroid M.

A detailed study of the matroid flow problem is given in HAMACHER [1980b]. Related problems are discussed in BURKARD and HAMACHER [1979] and HAMACHER [1980c].

At the beginning of this section we gave some remarks on problem (11.3) which is also not equivalent to an algebraic linear program of the form (11.1), in general. (11.3) may be solved via the solution of an index problem of the form

(11.54) $$\tilde{\lambda} := \sup_{x \in P} \lambda(x^T \square a)$$

similar to (11.21). Its objective is

$$\lambda(x^T \square a) = \max(\{\lambda(a_j) \mid x_j > 0\} \cup \{\lambda_o\}) .$$

If $\tilde{\lambda}$ is known then (11.3) can be reduced to a problem over the module $G_{\tilde{\lambda}}$. Let $M := \{j \mid \lambda(a_j) \leq \tilde{\lambda}\}$. Then we consider the reduced problem

(11.55) $$\tilde{q}' := \sup_{y \in P_M} \quad y^T \square a(\tilde{\lambda})_M$$

with $P_M := \{y \in R_+^{|M|} \mid C_M y \geq c\}$. This is an optimization problem over the module $G_{\tilde{\lambda}}$ which is equivalent to the algebraic linear program

$$\inf_{y \in P_M} \; y^T \,\square\, b$$

with $b_j := a(\tilde{\lambda})_j^{-1}$ for all $j \in M$. The relationship between (11.3) and (11.55) is described by the following theorem.

(11.56) Theorem

Let H be a weakly cancellative d-monoid which is an extended semimodule over $\mathbb{R}_+$. Let $P \neq \emptyset$. Hence $\tilde{\lambda}$ exists. Then (11.3) and (11.55) are equivalent ($\tilde{q} = \tilde{q}'$). If $x \in P$ is an optimal solution for (11.3) then x_M is an optimal solution for (11.55). If $y \in P_M$ is an optimal solution of (11.55) then $(y,0)$ is an optimal solution of (11.3).

Proof. If $x \in P$ then $x_j = 0$ for all $j \notin M$. Thus $g\colon P \to P_M$ with $g(x) := x_M$ is a bijective mapping. Let $\tilde{x} \in P$ with $\lambda(\tilde{x}^T \,\square\, a) = \tilde{\lambda}$. Then $x^T \,\square\, a < \tilde{x}^T \,\square\, a$ and $x_M^T \,\square\, a(\tilde{\lambda})_M = e < \tilde{x}^T \,\square\, a$ for all $x \in P$ with $\lambda(x^T \,\square\, a) < \tilde{\lambda}$. On the other hand $x^T \,\square\, a = x_M^T \,\square\, a(\tilde{\lambda})_M$ for all $x \in P$ with $\lambda(x^T \,\square\, a) = \tilde{\lambda}$. Therefore $\tilde{q} = \tilde{q}'$.

■

We remark that a solution of the index problem (11.54) is trivial if there exists $x \in P$ such that $x_K > 0$ for

$$\lambda(a_K) = \max\{\lambda(a_j) \mid j \in J\} .$$

This condition is fulfilled if there is no trivial variable x_j, i.e. $x_j = 0$ for all $x \in P$. Such a reasonable assumption will often hold in the application of our duality theory to combinatorial optimization problems in subsequent chapters. In general, the determination of such trivial variables can be performed via suitable usual linear programs.

Finally we consider some examples for algebraic linear programs. As the difference to the usual linear program lies only in the choice of other objective functions which are linear with respect to the semimodule considered it suffices to discuss such objectives and the corresponding semimodule. All d-monoids at the end of chapter 4 lead to such examples as every d-monoid is a semimodule over $\mathbb{Z}_+$, i.e. a discrete semimodule.

(11.57) D-monoids in the real additive group

The linearly ordered modules $(\mathbb{R},+,\leq)$, $(\mathbb{Q},+,\leq)$ and $(\mathbb{Z},+,\leq)$ over $R \in \{\mathbb{R},\mathbb{Q},\mathbb{Z}\}$.
The objective function is the classical one, i.e.

$$x^T \Box a = \Sigma\, x_j\, a_j\,, \qquad \textit{(sum objective)}$$

We remark that the choice of R is restricted (cf. 6.6). Similar modules (cf. 4.20) lead to the objective functions

$$x^T \Box a = \Pi\, a_j^{x_j} \qquad \textit{(reliability objective)}$$

$$x^T \Box a = (\Sigma\, x_j\, a_j^p)^{1/p} \qquad \textit{(p-norm objective)}$$

with $0 < p < \infty$. The case $p = \infty$ may be interpreted as bottleneck objective (cf. 11.59).

(11.58) D-monoids in Hahn-groups

The linearly ordered modules $(\mathbb{R}^k,+,\preccurlyeq)$, $(\mathbb{Q}^k,+,\preccurlyeq)$, $(\mathbb{Z}^k,+,\preccurlyeq)$ over $R \in \{\mathbb{R},\mathbb{Q},\mathbb{Z}\}$. The objective function is

$$x^T \square a = \begin{bmatrix} \Sigma\ x_j a_{j1} \\ \vdots \\ \Sigma\ x_j a_{jk} \end{bmatrix} = x^T A \qquad \textit{(lexicographic objective)}$$

with n × k matrix $A = (a_{ji})$. A rather general objective is

$$x^T \square a = a_1^{x_1} * a_2^{x_2} * \dots * a_n^{x_n} \qquad \textit{(group-objective)}$$

for linearly ordered modules $(H,*,\leq)$ over $\mathbb{Z}$. Such a module can always be embedded in a Hahn-group (cf. 4.21), and thus in a linearly ordered vectorspace over $\mathbb{R}$ (cf. discussion at the end of chapter 6).

(11.59) Weakly cancellative d-monoids

The examples $([0,1],*,\leq)$ given in (4.22) with $a*b := a+b-ab$ and $a*b := (a+b)/(1+ab)$ lead to objective functions of the form

$$x^T \square a = 1 - \Pi_j (1-a_j)^{x_j} ,$$

$$x^T \square a = [1 - \Pi_j (\Delta a_j)^{x_j}]/[1 + \Pi_j (\Delta a_j)^{x_j}] ,$$

with $\Delta a_j := (1-a_j)/(1+a_j)$ for $j \in J$.

The naturally, linearly ordered bottleneck monoid $(\mathbb{R} \cup \{-\infty\}, \max, \leq)$ is a linearly ordered semimodule over R_+ with $R \in \{\mathbb{R}, \mathbb{Q}, \mathbb{Z}\}$. Its objective function reads

$$z(x) = x^T \square a = \max\{a_j \mid x_j > 0\}. \qquad \textit{(bottleneck objective)}$$

The corresponding extended semimodule is derived from the time-cost semigroup (cf. 4.22). Its objective function is

$$x^T \square (a,b) = (z(x), \sum_{a_j = z(x)} x_j b_j). \qquad \textit{(time-cost objective)}$$

Linear optimization problems with bottleneck objectives have been discussed by FRIEZE [1975] and GARFINKEL and RAO [1976]. Bottleneck- and time-cost objectives with one *nonlinear parameter* are considered in ZIMMERMANN, U. [1980b]. In general, $\bar{\lambda}$-dual feasibility characterizes the set of parameter values for which a given basis is optimal in a similar manner as in the classical case (cf. theorem 11.45.1).

A particular case of a homogeneous weakly cancellative d-monoid (cf. 4.22) is the set

$$\bar{H} := \{(0,a) \mid a \in \mathbb{R}^{n_o}\} \cup \bigcup_{\mu=1}^{k} \{(\mu,b) \mid b \in \mathbb{R}_+^{n_\mu} \setminus \{0\}\}.$$

An internal commutative composition on $\bar{H}$ is defined by

$$(\mu,a) * (\nu,b) := \begin{cases} (\mu,a) & \text{if } \mu > \nu \ , \\ (\mu,a+b) & \text{if } \mu = \nu \ . \end{cases}$$

Together with the lexicographical order relation this leads to an extended semivectorspace with objective function

$$x^T \square (\mu,a) = (\mu(x), \sum_{\mu_j=\mu(x)} x_j a_j)$$

with $\mu(x) = \max\{\mu_j \mid x_j > 0\}$ and $(\mu_j,a_j) \in \bar{H}$ for $j \in J$. Clearly, the time-cost objective is a special case of this objective.

(11.60) Conditionally complete d-monoids

The d-monoid $(\mathbb{R} \times [0,1], *, \preccurlyeq)$ as defined in (4.23) is an example for a discrete semimodule which is not weakly cancellative. The corresponding objective function is similar to the time-cost objective (cf. 11.59)

$$x^T \square (a,b) = (z(x), \min(\sum_{a_j = z(x)} x_j b_j , 1)) .$$

We remark that this semimodule cannot be extended to a semimodule over R_+ with $R_+ \supsetneqq \mathbb{Z}_+$ (cf. chapter 6). From the structure of the conditionally complete d-monoids described in (4.23) we see that the corresponding objectives have a similar structure.

(11.61) General d-monoids

CLIFFORD's example in (4.24) yields a discrete semimodule which cannot be subsumed under previously discussed classes.

An interpretation of the results in this chapter for the different types of objective functions mentioned is left to the reader. In particular, the strong duality theorems in this chapter as well as in the following chapters show the close relationship of all these optimization problems which are treated in a unifying way using an algebraic approach.

The following example contains an application of the results in this chapter to an eigenvalue problem discussed in chapter 9.

(11.62) A finite eigenvalue problem

Let $(R', \otimes, \leq)$ be a linearly ordered, divisible and commutative group. Then R' is a vectorspace over the field $\mathbb{Q}$ of all rational numbers with external composition defined by

$$(m/n) \square a := a^{m/n}$$

for all $m/n \in \mathbb{Q}$ and for all $a \in R'$. Theorem (6.5) shows that R'

is a linearly ordered vectorspace. If $|R'| > 1$ then R' is an extended vectorspace over $\mathbb{Q}$; this will be assumed w.l.o.g. in the following. The finite eigenvalue problem consists in the determination of $\lambda \in R'$, $y \in (R')^n$ such that

(11.63) $\max\{a_{ij} \otimes y_j | \ j = 1,2,\ldots,n\} = \lambda \otimes y_i$

for all $i \in N := \{1,2,\ldots,n\}$ with respect to a given $n \times n$ matrix $A = (a_{ij})$ over $R' \cup \{O\}$ (O denotes an adjoined zero). The solutions of such a problem are characterized in theorem (9.22) and are discussed subsequent to this theorem. G_A denotes the digraph with set of vertices N and arc set E defined by

$$(i,j) \in E \ :\Longleftrightarrow \quad a_{ij} \neq O$$

for $i,j \in N$. If a finite eigenvalue $\bar{\lambda} \in R'$ exists then G_A contains an elementary circuit p with length $l(p)$ and weight $w(p)$ such that $\bar{\lambda} = (1/l(p)) \ \square \ w(p)$. Then further

$$\bar{\lambda} = \lambda(A) := \max\left\{ (1/l(p)) \ \square \ w(p) \ \middle| \ \begin{array}{l} p \text{ elementary} \\ \text{circuit in } G_A \end{array} \right\}.$$

We assume that G_A contains at least one circuit. Otherwise the finite eigenvalue problem has no solution.

Let a denote the vector with components a_{ij}, $(i,j) \in E$. We consider the algebraic linear program

(11.64) $\tilde{w} := \max_{x \in P} \ x^T \ \square \ a$

with set P of feasible solutions described by the linear constraints

$$\begin{array}{ll} \sum_j x_{ij} - \sum_j x_{ji} = O & , \quad i \in N, \\[4pt] \sum_{ij} x_{ij} \quad\quad = 1 \end{array}$$

and $x \in \mathbb{Q}_{+}^{E}$. In matrix form we denote these equations by $Cx = c$. Then the dual of (11.64) is

$$(11.66) \qquad \tilde{\lambda} := \min_{s \in D} \; c^T \square s$$

with $D = \{s \in (R')^{n+1} \mid C^T \square s \geq a\}$. The inequality constraints $C^T \square s \geq a$ have the form

$$(11.67) \qquad s_i \otimes s_j^{-1} \otimes s_{n+1} \geq a_{ij}$$

for all $(i,j) \in E$. D is always nonempty as it contains s defined by $s_i := e$ for $i \in N$ and $s_{n+1} := \max\{a_{ij} \mid (i,j) \in E\}$ (e denotes the neutral element of R').

If $\bar{\lambda}$ is a finite eigenvalue and $y \in (R')^n$ a finite eigenvector then $s = (y,\lambda)^T$ is a dual feasible solution with

$$c^T \square s = \bar{\lambda} = \lambda(A) .$$

Let p denote an elementary circuit in G_A with $\bar{\lambda} = (1/l(p)) \square w(p)$. Then $x \in \mathbb{Q}_{+}^{E}$ defined by

$$x_{ij} = \begin{cases} 1/l(p) & \text{if } (i,j) \in p, \\ 0 & \text{otherwise} \end{cases} ,$$

is a primal feasible solution with

$$x^T \square a = (1/l(p)) \square w(p).$$

Due to the weak duality theorem $x^T \square a \leq c^T \square s$ for all feasible pairs $(x;s)$; thus $\tilde{\lambda} = \tilde{w} = \lambda(A) = \bar{\lambda}$.

If no finite eigenvalue exists then the existence of a circuit in G_A implies $\tilde{w} \geq \lambda(A)$. Further the strong duality theorem (11.34) shows $\tilde{\lambda} = \tilde{w} \geq \lambda(A)$.

In any case $\tilde{\lambda} = \tilde{w}$ can be determined using the methods proposed in this chapter for the solution of linear algebraic programs (cf. 11.28 and 11.46). As (11.64) is a maximization problem these methods which are formulated for minimization problems have to be modified in the usual manner (e.g. replace a_{ij} by a_{ij}^{-1} for 11.46, etc.). The optimal value $\tilde{\lambda}$ is used in chapter 9 either to compute the finite eigenvectors (if $\bar{\lambda}$ exists) or to verify that the finite eigenvalue problem has no solution.

For the special case of the additive group of real numbers this approach is proposed by DANTZIG, BLATTNER and RAO [1967] and is discussed in CUNINGHAME-GREEN [1979].

12. Algebraic Flow Problems

In this chapter we discuss the minimization of linear algebraic functions over the set of feasible flows in a network (algebraic flow problem). Although we assume that the reader is familiar with basic concepts from flow theory we begin with an introduction of the combinatorial structure which is completely the same as in the classical case. Then we develop the primal dual solution method which is a generalization of the min-cost max-flow algorithm in FORD and FULKERSON [1962]. We discuss three methods for the minimization of linear algebraic functions over the set of feasible flows in a transportation network (algebraic transportation problem). These methods are generalizations of the Hungarian method, the shortest augmenting path method and the stepping stone method.

The underlying d-monoid H is weakly cancellative and an extended semimodule over the positive cone R_+ of a subring of the linearly ordered field of real numbers (cf. chapter 6). Only in the discussion of the Hungarian method for algebraic transportation this assumption can be weakened. Then H is a d-monoid which is a linearly ordered semimodule over R_+.

We begin with an introduction of the combinatorial structure. $G = (N,E)$ denotes a simple, connected digraph with two special vertices σ and τ called *source* and *sink*. G contains no arc with terminal vertex σ or initial vertex τ. W.l.o.g. we assume that G is *antisymmetric*, i.e.

(12.1) $(i,j) \in E \quad \Rightarrow \quad (j,i) \notin E.$

B denotes the incidence matrix of vertices and arcs of G, i.e.

$$B_{ih} := \begin{cases} 1 & \text{if } h = (i,j)\ , \\ -1 & \text{if } h = (j,i)\ , \\ 0 & \text{otherwise} \end{cases}$$

for all $i \in N$ and for all $h \in E$. Then $x \in R_+^E$ is called a *flow* in G if $0 \leq B_\sigma x = -B_\tau x$ and

$$B_i x = 0$$

for all $i \in N \setminus \{\sigma,\tau\}$. Then $g := B_\sigma x$ is called the *flow value* of x. Let $c \in (R_+ \setminus \{0\})^E$ be a given *capacity vector*. Then a flow x is called *feasible* if $x \leq c$. A partition (S,T) of N is called a *cut* if $\sigma \in S$ and $\tau \in T$. Then the sum of all capacities c_{ij} with $(i,j) \in (I \times J) \cap E$ is called the *capacity of the cut* (S,T). The capacity is denoted by c(S,T). It is easy to establish the relation

(12.2) $$g = \sum_{(S\times T)\cap E} x_{ij} - \sum_{(T\times S)\cap E} x_{ij} \leq c(S,T)$$

for all feasible flows and for all cuts in G. A well-known result of flow theory (cf. FORD and FULKERSON [1962]) is the *max-flow min-cut* theorem which states that the *maximum value* f of a feasible flow is equal to the *minimum capacity* of a cut in G. A feasible flow of maximum value is called a *maximum flow*; similarly a *minimum cut* is defined.

In (11.49) we introduced algebraic flows in regular matroids. A generalization of the max-flow min-cut theorem for algebraic flows in regular matroids is developed in HAMACHER ([1980a],

[1980c]). A detailed study of such flows is given in HAMACHER [1980b].

There are many procedures for the determination of a flow with maximum flow value. Besides the flow augmenting path method in FORD and FULKERSON [1962] such methods have been investigated by DINIC [1970], EDMONDS and KARP [1972], ZADEH [1972], KARZANOV ([1974], [1975]), ČERKASSKIJ [1977], MALHOTRA, PRADMODH KUMAR and MAHESHWARI [1978], GALIL [1978], SHILOACH [1978] and GALIL and NAAMAD [1979]. EVEN [1980] has written an exposition of the method of KARZANOV which has a complexity bound of $O(|N|^3)$; numerical investigations are discussed by HAMACHER [1979].

With respect to a feasible flow x we introduce the *incremental* digraph $\Delta G = (N, E \cup \bar{E})$ with $\bar{E} := \{(j,i) \mid (i,j) \in E\}$ and with capacity vector $\Delta c \in R_+^{E \cup \bar{E}}$ defined by

$$\text{(12.3)} \qquad \Delta c_{ij} := \begin{cases} c_{ij} - x_{ij} & \text{if } (i,j) \in E , \\ x_{ji} & \text{if } (i,j) \in \bar{E} . \end{cases}$$

Δc is well-defined as R is a subring of $\mathbb{R}$. A feasible flow Δx in ΔG is called an *incremental flow*. Then $x \circ \Delta x$, defined by

$$\text{(12.4)} \qquad (x \circ \Delta x)_{ij} = x_{ij} + \Delta x_{ij} - \Delta x_{ji}$$

for all $(i,j) \in E$, is a feasible flow in G. For convenience we assume that an incremental flow has the property

$$\Delta x_{ij} > 0 \quad \Rightarrow \quad \Delta x_{ji} = 0$$

for all $(i,j) \in E \cup \bar{E}$. Conversely to (12.4), if $\tilde{x}$ is a feasible flow in G with $\tilde{g} \geq g$ then there exists an incremental flow Δx in ΔG such that $x \bullet \Delta x = x$. If Δg is the value of the incremental flow in Δx then $g + \Delta g = \tilde{g}$. Thus x is a maximum flow if and only if the value of a maximum incremental flow is 0.

Let p denote a path in G from σ to τ or a circuit in G. Then $p(\delta)$ with

$$(12.5)\qquad p_{ij}(\delta) := \begin{cases} \delta & (i,j) \in p\,, \\ 0 & \text{otherwise}\,, \end{cases}$$

for all $(i,j) \in E$ and $0 \leq \delta \in R$ is a flow in G. The *capacity* $\delta(p)$ of p is defined by

$$\delta(p) = \max\{c_{ij} \mid (i,j) \in p\}.$$

Then $p(\delta)$ is a feasible flow if $0 \leq \delta \leq \delta(p)$. Such a feasible flow is called a *path flow* or a *circuit flow*. The flow value of a path flow is δ and the flow value of a circuit flow is 0. Each flow x has a representation as the sum of certain path flows and circuit flows (cf. BERGE [1973]), i.e.

$$(12.6)\qquad x = \sum_{\rho=1}^{r} p_\rho(\delta_\rho) + \sum_{\rho=r+1}^{s} p_\rho(\delta_\rho)$$

where the first sum contains all path flows and the second sum contains all circuit flows. Similarly each incremental flow Δx in ΔG is the sum of *incremental path flows* and *incremental circuit flows*

$$(12.7)\qquad \Delta x = \sum_{\rho=1}^{r} \Delta p_\rho(\delta_\rho) + \sum_{\rho=r+1}^{s} \Delta p_\rho(\delta_\rho)\,.$$

The set P_g of all feasible flows in G with certain fixed flow value $g \in \mathbb{R}$ $(0 \le g \le f)$ can be characterized by linear constraints. Let $b \in \mathbb{R}^N_+$ be defined by $b_\sigma = -b_\tau = g$ and $b_i = 0$ for all $i \notin \{\sigma,\tau\}$. Then

$$(12.8) \qquad P_g = \{x \in \mathbb{R}^E_+ \mid Bx = b,\ x \le c\}.$$

For cost coefficients $a_{ij} \in H$, $(i,j) \in E$ we consider the *algebraic flow problem*

$$(12.9) \qquad \tilde{z} := \inf_{x \in P_g} x^T \square a.$$

H is a weakly cancellative d-monoid which is an extended semimodule over $\mathbb{R}_+$. A solution x of (12.9) is called a *minimum cost flow*. W.l.o.g. we assume that g = f, i.e. we solve a max-flow min-cost problem. The classical case is obtained from (12.9) for the special vectorspace $(\mathbb{R},+,\le)$ over the real field. A combinatorial generalization of the classical case is discussed in BURKARD and HAMACHER [1979]. They develop finite procedures for the determination of a maximum matroid flow (cf. theorem 11.53) with minimum cost in $\mathbb{R}$.
It is well-known that B is a totally unimodular matrix (e.g. LAWLER [1976]). Thus (11.43) is satisfied and the primal simplex method (11.46) can be applied for a solution of the algebraic flow problem (12.9). If the semimodule considered is not extended then the primal simplex method may terminate in a locally optimal solution. Later on in this chapter we will discuss the application of the primal simplex method to the special case of algebraic transportation problems (cf. 12.33); for classical transportation problems this method is known to be the most efficient solution procedure.

In the following we develop a primal dual method for the solution of the algebraic flow problem which generalizes the min-cost max-flow algorithm in FORD and FULKERSON [1962]. This method has previously been considered in BURKARD [1976], BURKARD, HAMACHER and ZIMMERMANN, U. ([1976], [1977a] and [1977b]). Based on involved combinatorial arguments BURKARD, HAMACHER and ZIMMERMANN, U. [1976] proved the validity of this method. Now we can provide a much simpler proof based on the duality theory in chapter 11. We assume that $a_{ij} \in H_+$ for all $(i,j) \in E$. This assumption allows an obvious initial solution and guarantees that the objective function is bounded from below by e. Self-negative cost coefficients ($a_{ij} < e$) will play a role in the solution of (12.9) only if

(12.10) $$\lambda(\tilde{z}) = \lambda_o .$$

(12.10) holds if and only if there exists a maximum flow (with value f) in the partial graph $G' = (N,E')$ with

$$E' := \{(i,j) \in E \mid \lambda(a_{ij}) = \lambda_o\}.$$

This can easily be checked using one of the usual maximum flow algorithms.

If $\lambda(\tilde{z}) = \lambda_o$ then it suffices to solve the algebraic flow problem in the graph G'. This is a problem in the module G_{λ_o} over R and can be solved similarly to the classical min-cost max-flow problem. For example, the necessary modification in the min-cost max-flow method in FORD and FULKERSON [1962] mainly consists in replacing usual additions by *-compositions and usual comparisons by $\leq$ - comparisons in G_{λ_o}.

Otherwise $\lambda(\tilde{z}) > \lambda_o$ and without changing the objective function value of any maximum flow we can replace $a_{ij} < e$ by $a_{ij} := e$.

In order to apply the duality theory of chapter 11 to the algebraic flow problem we assign dual variables u,v and t to the constraints $Bx \geq b$, $-Bx \geq -b$ and $-x \geq -c$. According to (11.6) and (11.17) (u,v,t) is strongly dual feasible if

$$(12.11)\qquad \begin{aligned} &u_i * v_j \leq a_{ij} * u_j * v_i * t_{ij} , \\ &t_{ij} \in H_+ , \ \lambda(t_{ij}) \leq \lambda(\tilde{z}) \end{aligned}$$

for all $(i,j) \in E$ and if

$$(12.12)\qquad u_i, v_i \in H_+ , \ \lambda(u_i) \leq \lambda(\tilde{z}), \ \lambda(v_i) \leq \lambda(\tilde{z})$$

for all $i \in N$. The complementarity condition (11.14.1) has the form

$$(12.13)\qquad (c - x)^T \square t = e = x^T \square \bar{a}$$

with reduced cost coefficients $\bar{a}$ defined by

$$(12.14)\qquad u_i * v_j * \bar{a}_{ij} = a_{ij} * u_j * v_i * t_{ij} .$$

As in the discussion of the modification (11.28') and (11.27') of the general primal dual method we consider only the case that all dual variables are elements of $H_\mu \cup \{e\}$ for a certain index $\mu \in \Lambda$ (cf. chapter 6 for the decomposition of H). Then a_{ij} in (12.13) and (12.14) is replaced by $a_{ij}(\mu)$.

For flow problems it suffices to consider only dual variables v_j , $j \in N$ with $v_\sigma = e$. A pair (x;v) of a feasible flow x and $v \in (H_\mu \cup \{e\})^N$ with $v_\sigma = e$ and $\mu \leq \lambda(\tilde{z})$ is called μ-*compatible* if

$$(12.15)\qquad \begin{aligned} x_{ij} < c_{ij} &\Rightarrow v_j \le a_{ij}(\mu) * v_i \\ x_{ij} > 0 &\Rightarrow v_j \ge a_{ij}(\mu) * v_i \end{aligned}$$

for all $(i,j) \in E$. If $(x;v)$ is μ-compatible then we can define a strongly dual feasible solution (u,v,t) by $u \equiv e$ and

$$(12.16)\qquad t_{ij} := \begin{cases} \varepsilon & \text{if } v_j > a_{ij}(\mu) * v_i \text{ and } v_j =: \varepsilon * a_{ij}(\mu) * v_i, \\ e & \text{if } v_j \le a_{ij}(\mu) * v_i \end{cases}$$

for all $(i,j) \in E$. All dual variables are elements of $H_\mu \cup \{e\}$ and $(x;u,v,t)$ satisfies the complementary condition (12.13) with reduced cost coefficients defined by

$$(12.17)\qquad u_i * v_j * \bar{a}_{ij} = a_{ij}(\mu) * u_j * v_i * t_{ij}$$

for all $(i,j) \in E$. (12.13) is a sufficient condition for (11.18.1). Therefore theorem (11.18) applied to

$$(12.18)\qquad \tilde{z} = \inf_{y \in P_g} y^T \square a(\mu)$$

shows that x is an optimal solution of (12.9) if x is a feasible solution with flow value g.

(12.18) <u>Proposition</u>

Let $(x;v)$ be μ-compatible and let g denote the flow value of x. Then x is a minimum cost flow of value g and

$$(1)\qquad g \square v_\tau = x^T \square a(\mu) * c^T \square t.$$

<u>Proof</u>: Due to our remarks it is obvious that x is a minimum cost flow. From theorem (11.18) we get

$$(-b)_+^T \square v = x^T \square a * (-b)_-^T \square v * c^T \square t.$$

The definition of b in (12.8) and $v_\sigma = e$ lead to (1). ∎

The method proceeds in the following manner. The initial compatible pair is $(x;v)$ with $x \equiv 0$ and $v \equiv e$ $(\mu = \lambda_o)$. Then we try to increase the value of x without violation of compatibility. If this is impossible then we try to increase the dual variables v without violation of compatibility. In this way the current flow is of minimum cost throughout the performance of the method. We will show that the procedure leads to a flow x of maximum value in a finite number of steps. Then due to proposition (12.18) x is an optimal solution of (12.9) with $g = f$.

At first we try to change x using an incremental flow Δx such that $(x, x \circ \Delta x)$ is μ-compatible, too. An arc $(i,j) \in E$ is called *admissible* if

(12.19) $$v_j = a_{ij}(\mu) * v_i \;;$$

an arc $(i,j) \in \bar{E}$ is called admissible if $(j,i) \in E$ is admissible. Let $\delta G = (N, \delta E)$ denote the partial graph of ΔG containing all admissible arcs. Then δG is called the *admissibility graph*. A feasible flow δx in δG is called an *admissible incremental flow*. Clearly the admissibility graph depends on the current flow x and the current variables v.

(12.20) Proposition

Let $(x;v)$ be μ-compatible, let g denote the flow value of x and let δx be an admissible incremental flow of value δg. Then

(1) $(x \circ \delta x; v)$ is μ-compatible,

(2) $(x \circ \delta x)^T \,\square\, a(\mu) = x^T \,\square\, a(\mu) * (\delta g \,\square\, v_\tau)$.

Proof: (1) Since δx is an incremental flow $x \circ \delta x$ is a feasible flow of value $g + \delta g$. Admissibility of δx shows that x is changed only in admissible arcs. Thus $(x \circ \delta x; v)$ is μ-compatible.

(2) Proposition (12.18) leads to

$$g \,\square\, v_\tau = x^T \,\square\, a(\mu) * c^T \,\square\, t$$

$$(g + \delta g) \,\square\, v_\tau = (x \circ \delta x)^T \,\square\, a(\mu) * c^T \,\square\, t \, .$$

These equations in the module G_μ over R imply (2).

■

Proposition (12.20) shows that augmentation of x with an admissible incremental flow δx does not violate compatibility. δx can be determined by a maximum flow procedure. In particular, this is a purely combinatorial step with complexity bound $O(|N|^3)$ if we use the method of KARZANOV [1974].

Secondly we discuss the case that the value of a maximum admissible incremental flow is 0. Let (S,T) be a minimum cut of δG. Then

(12.21) $$\Delta c_{ij} = 0$$

for all $(i,j) \in (S \times T) \cap \delta E$. (12.21) follows from the max-flow min-cut theorem. If $\Delta c_{ij} = 0$ for all $(i,j) \in (S \times T) \cap (E \cup \bar{E})$ then (S,T) is a minimum cut in ΔG of value 0 or, equivalently, (S,T) is a minimum cut in G of value f. Then x is a maximum flow in G and the procedure terminates. Otherwise we know that each flow x in G of flow value $\tilde{g} > g$ has either

(12.22) $$\tilde{x}_{rs} > x_{rs}$$

for some $(r,s) \in F := \{(i,j) \mid x_{ij} < c_{ij}\} \cap (S \times T)$ or

(12.23) $$\tilde{x}_{rs} < x_{rs}$$

for some $(r,s) \in B := \{(i,j) \mid 0 < x_{ij}\} \cap (T \times S)$. Since

$$v_s < a_{rs}(\mu) * v_r$$

for all $(r,s) \in F$ and

$$v_s > a_{rs}(\mu) * v_r$$

for all $(r,s) \in B$ we may define α_{rs}, $(r,s) \in F \cup B$ by

(12.24)
$$v_s * \alpha_{rs} := a_{rs}(\mu) * v_r \qquad \text{if } (r,s) \in F ,$$
$$v_s =: \alpha_{rs} * a_{rs}(\mu) * v_r \qquad \text{if } (r,s) \in B .$$

We remark that $\alpha_{rs} = \bar{a}_{rs}$ for $(r,s) \in F$ and $\alpha_{rs} = t_{rs}$ for $(r,s) \in B$. Therefore we may revise the dual variables in the following way.

(12.25) <u>Dual revision procedure</u>

Step 1 Determine the connected component S $(\sigma \in S \subseteq N)$ of the partial graph of δG which contains all arcs with $\Delta c_{ij} > 0$;
if $\tau \in S$ then return (exit 1).

Step 2 $T := N \setminus S$; $F := \{(i,j) \mid x_{ij} < c_{ij}\} \cap (S \times T)$;
$B := \{(i,j) \mid 0 < x_{ij}\} \cap (T \times S)$;
if $F \cup B = \emptyset$ then return (exit 2).

Step 3 Compute α_{rs} for $(r,s) \in F \cup B$ from (12.24);
$\alpha := \min\{\alpha_{rs} \mid (r,s) \in F \cup B\}$;
$\mu := \lambda(\alpha)$.

Step 4 Redefine all dual variables by $v_j := \begin{cases} v_j(\mu) & \text{if } j \in S, \\ v_j * \alpha & \text{if } j \in T, \end{cases}$
and go to step 1.

Due to our previous remarks it is clear that all steps are well-defined. In particular $\mu = \lambda_o$ in the first call of (12.25). If $F \cup B \neq \emptyset$ in step 2 then $|S|$ increases after a revision of the dual variables in step 4 at least by one. Thus the dual revision procedure terminates after at most $|N| - 1$ iterations either at exit 1 (δG contains an incremental flow of nonzero value) or at exit 2 (x is a maximum flow in G). Each iteration contains at most $O(|N|^2)$ usual operations, $O(|N|^2)$ $*$-compositions and $O(|N|^2)$ $\leq$ - comparisons.

(12.26) Proposition

Let $(x;v)$ be μ-compatible and let g denote the value of x ($g < f$). If the revision procedure leads to revised dual variables denoted by $\bar{v}$ with revised index $\bar{\mu}$ then

(1) $(x;\bar{v})$ is $\bar{\mu}$-compatible.

Proof: It suffices to prove (1) for one iteration in the dual revision procedure. The definition (12.24) of α_{rs} together with μ-compatibility implies $\lambda(\alpha) \geq \mu$ in step 3. If $\mu = \bar{\mu}$ then $\bar{\mu} \leq \lambda(\tilde{z})$. Otherwise $\alpha = a_{rs}(\mu) > e$ for some $(r,s) \in F$. Thus $\alpha = a_{rs}$. Since $\alpha_{ij} = t_{ij}$ for $(i,j) \in B$ we find $B = \emptyset$. Therefore each flow x in G with flow value $\tilde{g} > g$ has $\tilde{x}_{ij} > x_{ij}$ for some $(i,j) \in F$. Thus $\bar{\mu} \leq \lambda(\tilde{z})$. It remains to prove the inequalities (12.15) for $(x;\bar{v})$. It suffices to consider arcs $(i,j) \in [(S \times T) \cup (T \times S)] \cap E$. If $(i,j) \in F \cup B$ then $\alpha \leq \alpha_{ij}$ implies the corresponding inequality. If $(i,j) \in (S \times T) \setminus F$ then $x_{ij} = c_{ij}$ and $\bar{v}_j \geq a_{ij}(\bar{\mu}) * \bar{v}_i$. If $(i,j) \in (T \times S) \setminus B$ then $x_{ij} = 0$ and $\bar{v}_j \leq a_{ij}(\bar{\mu}) * \bar{v}_i$. ■

Proposition (12.26) shows that a dual revision step does not violate compatibility provided that $g < f$. In the last step it may happen that before actually finding a minimum cut μ is raised beyond $\lambda(\tilde{z})$. Therefore if a minimum cut is detected then an optimal solution of the possible dual programs (cf. chapter 11) has to be derived from the values of the dual variables before entering the dual revision procedure for the last time. Optimality of x is guaranteed in any case.
By alternately solving a max-flow problem in the admissibility graph and revising dual variables we construct a minimum cost flow $x \in P_f$. We summarize these steps in the following algorithm.

(12.27) Primal dual method for algebraic flows

Step 1 $x \equiv 0$; $v \equiv e$; $z := e$; $\mu := \lambda_o$.

Step 2 Determine a maximum flow δx of value δg in δG;

$x := x \circ \delta x$; $z := z * (\delta g \,\square\, v_\tau)$; $\bar{v} := v$.

Step 3 Revise dual variables by means of (12.25);

if (12.25) terminates at exit 2 then $v := \bar{v}$ stop;

otherwise go to step 2.

At termination x is an optimal solution of (12.9) for $g = f$ and v leads to an optimal solution (u,v,t) of the possible dual programs (cf. chapter 11). z is the optimal value; the recursion for z in step 2 follows from proposition (12.20). We remark that explicit knowledge of the ordinal decomposition (cf. chapter 6) of the semimodule is not necessary. The recursion for the index μ is deleted and $a_{rs}(\mu)$ in step 4 of the dual revision procedure is determined by

$$a_{rs}(\mu) := \begin{cases} a_{rs} & \text{if } v_\tau * a_{rs} > a_{rs} , \\ e & \text{if } v_\tau * a_{rs} = v_\tau . \end{cases}$$

In step 5 the dual variables $v_i(\mu)$ for $i \in S$ can be found by

$$v_i(\mu) := \begin{cases} v_i & \text{if } \alpha * v_i > \alpha , \\ e & \text{if } \alpha * v_i = \alpha . \end{cases}$$

This follows from proposition (6.17). Admissibility can be found similarly.

It remains to show that the primal dual method (12.27) is finite.

(12.28) Theorem

Let H be a weakly cancellative d-monoid which is an extended semimodule over $\mathbb{R}_+$. Then the primal dual method (12.27) is finite.

Proof: The performance of step 1, step 2 and step 3 is finite. Therefore it suffices to prove that step 3 in (12.27) is carried out only a finite number of times. At termination of step 2 the admissibility graph with respect to $(x;v)$ does not contain any path p from σ to τ of capacity $\bar{\delta}(p) > 0$. If the current flow x is not a maximum flow then at termination of step 3 the admissibility graph with respect to $(x;v)$ does contain such a path p. Then $v_\tau < \bar{v}_\tau$. With respect to x the arcs of p are partitioned in $F \cup B$ defined by

$$F := \{(i,j) \in p \mid x_{ij} < c_{ij}\} ,$$

$$B := \{(i,j) \in p \mid 0 < x_{ij}\} .$$

The number of partitions is finite as well as the number of different paths from σ to τ in G. Thus it suffices to prove that p did not appear in a previous iteration of the primal dual method with the same partition of its arcs. Admissibility and $\bar{v}_\sigma = e$ lead to

(12.29) $\qquad \bar{v}_\tau * \bar{\beta} = \bar{\gamma}$

where $\bar{\beta}$ $(\bar{\gamma})$ denotes the composition of all weights $a_{ij}(\bar{\mu})$ with $(i,j) \in B$ $((i,j) \in F)$.

Now assume that p previously appeared in an admissibility graph with respect to $(\tilde{x};\tilde{v})$, with capacity $\tilde{\delta}(p) > 0$ and with the same partition of its arcs. Similarly to (12.29) we get

(12.30) $\qquad \tilde{v}_\tau * \tilde{\beta} = \tilde{\gamma}$

with respect to weights $a_{ij}(\tilde{\mu})$. Further $\tilde{v}_\tau \leq v_\tau < \bar{v}_\tau$ and $\tilde{\mu} \leq \bar{\mu}$. Now $\bar{v}_\tau > e$ implies $\bar{\gamma} > e$ and from $\tilde{\mu} \leq \bar{\mu}$ we conclude $\tilde{\gamma} = \bar{\gamma}$. Thus

(12.31) $\qquad \tilde{v}_\tau * \tilde{\beta} = \bar{v}_\tau * \bar{\beta}$.

$\tilde{\mu}$-compatibility shows

$$a_{rs}(\tilde{\mu}) \leq \tilde{v}_r$$

for all $(r,s) \in B$. Thus $\lambda(\tilde{\beta}) \leq \tilde{\mu}$. Then $\bar{v}_\tau * \bar{\beta} \geq \bar{v}_\tau > e$ and (12.31) lead to $\bar{\mu} = \tilde{\mu}$ and $\bar{\beta} = \tilde{\beta}$. Cancellation in the module $G_{\tilde{\mu}}$ implies $\tilde{v}_\tau = \bar{v}_\tau$ contrary to $\tilde{v}_\tau < \bar{v}_\tau$.

■

Proposition (12.28) shows that the primal dual method is valid; therefore this method provides a constructive proof of the existence of an optimal solution of the algebraic flow problem. Further a strong duality theorem is implied.

(12.32) Theorem (Network flow duality theorem)

Let H be a weakly cancellative d-monoid which is an extended semimodule over R_+. For cost coefficient vector $a \in H_+^E$ an optimal feasible pair (x;u,v,t) for the algebraic flow problem exists. (x;u,v,t) is complementary and satisfies

$$(1) \qquad f \square v_\tau = x^T \square a * c^T \square t \ ,$$

$$(2) \qquad (-c)^T \square t(\mu) * f \square v_\tau = x^T \square a \qquad \text{for } \mu = \lambda(\tilde{z}).$$

In discrete semimodules ($R = \mathbb{Z}$) we get $\delta g \geq 1$ in step 2 of the primal dual method (12.27). Therefore at most O(f) iterations are performed until termination occurs. Each iteration consists in step 2 and step 3 and needs $O(|N|^3)$ usual operations, $O(|N|^2)$ $*$-compositions and $O(|N|^3)$ $\leq$ - comparisons (maximum flow with method of KARZANOV [1974]).
We remark that a solution of an equation $\alpha * \gamma := \beta$ for $\alpha \leq \beta$, $\alpha, \beta \in H$ is counted as one $*$-composition.

In the next part of this chapter we discuss three methods for the solution of an algebraic flow problem which generalizes the classical (balanced) transportation problem. We begin with a description of the underlying graph which shows the special combinatorial structure of this problem. The subgraph of $G = (N,E)$ induced by $N \setminus \{\sigma,\tau\}$ is bipartite.
Let $N = \{\sigma,\tau\} \cup I \cup J$ with $I := \{1,2,\ldots,m\}$ and $J := \{m+1,\ldots,m+n\}$. Then

$$E = \{\sigma\} \times I \cup I \times J \cup J \times \{\tau\} \ .$$

Capacities on arcs (σ,i) and (j,τ) are denoted by c_i , $i \in I$ and c_j , $j \in J$; there are no capacity constraints on all other arcs ($c_{ij} = \infty$). W.l.o.g. we assume that the capacities are *balanced*, i.e.

$$\Sigma_i c_i = \Sigma_j c_j \ .$$

Thus the value of a maximum flow is $f = \Sigma_i c_i = \Sigma_j c_j$. Cost coefficients on arcs (σ,i) and (j,τ) have value e. The other cost coefficients are elements of H. Such a graph is called a *transportation network* (Figure 6).

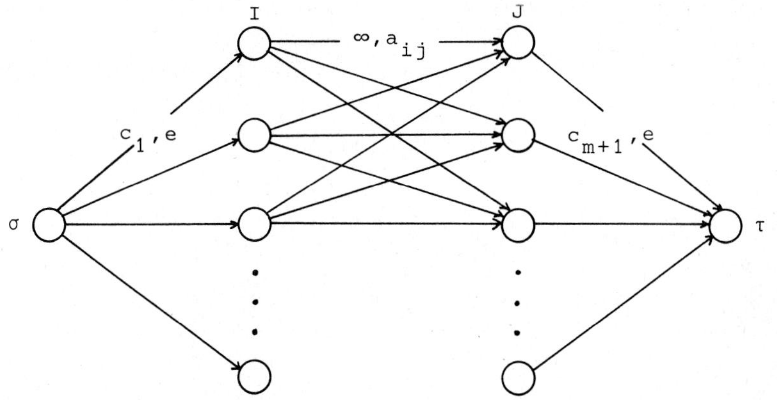

Figure 6 Transportation network with assigned capacities, cost coefficients

The corresponding algebraic flow problem is called *algebraic transportation problem*. The linear description of the set of feasible solutions can be simplified in this case. Let C denote the incidence matrix of vertices and arcs for the partial subgraph $(I \cup J, I \times J)$, i.e.

$$c_{kh} := \begin{cases} 1 & \text{if } h = (k,j) \text{ or } h = (i,k), \\ 0 & \text{otherwise.} \end{cases}$$

Now a flow x in G is uniquely defined by its values on arcs in $I \times J$; therefore it suffices to describe the set of all corresponding $y = x_{I\times J}$. The set P_T of all y corresponding to a maximum flow is

$$P_T = \{y \in \mathbb{R}_+^{mn} \mid \; Cy = c_{I\cup J}\} .$$

The *algebraic transportation problem* is

$$(12.33) \qquad \tilde{z} := \min_{y \in P_T} \; y^T \square a$$

for a cost coefficient vector $a \in H^{mn}$. If $a \notin H_+^{mn}$ then let (cf. proposition 4.7)

$$\alpha := (\min\{a_{ij} \mid \; a_{ij} < e\})^{-1} .$$

Then $\bar{a}$ defined by $\bar{a}_{ij} := \alpha * a_{ij}$ satisfies $\bar{a}_{ij} \in H_+$ for all $(i,j) \in I \times J$ and

$$(12.34) \qquad y^T \square a \le \tilde{y}^T \square a \iff y^T \square \bar{a} \le \tilde{y}^T \square \bar{a}$$

for all $\tilde{y}, y \in P_T$ provided that H is a d-monoid which is a linearly ordered semimodule over $\mathbb{R}_+$. (12.34) follows from

$$y^T \square a * f \square \alpha = y^T \square \bar{a}$$

for all $y \in P_T$. Therefore we can assume $a \in H_+^{mn}$ in the following w.l.o.g.

Due to the description of P_T it suffices to assign dual variables w and s to the constraints $Cy \ge c_{I\cup J}$ and $-Cy \ge -c_{I\cup J}$. Then (w,s) is strongly dual feasible if

$$(12.35) \qquad w_i * w_j \le a_{ij} * s_i * s_j$$

for all $(i,j) \in I \times J$ and if

$$w_k, s_k \in H_+, \quad \lambda(w_k) \leq \lambda(\tilde{z}), \quad \lambda(s_k) \leq \lambda(\tilde{z})$$

for all $k \in I \cup J$. Complementarity is satisfied if

$$y^T \boxdot \bar{a} = e$$

where $\bar{a} \in H_+^{mn}$ is defined by

$$(12.36) \qquad w_i * w_j * \bar{a}_{ij} := a_{ij} * s_i * s_j$$

for all $(i,j) \in I \times J$.

If H is a weakly cancellative d-monoid which is an extended semimodule over R_+ then we can apply the primal dual method (12.27). The constructed sequence of μ-compatible pairs $(x;v)$ defines a sequence of pairs $(y;w,s)$ by $y := x_{I \times J}$ and by

$$(12.37) \qquad w_k := \begin{cases} v_k & \text{if } k \in J, \\ e & \text{otherwise,} \end{cases} \qquad s_k := \begin{cases} v_k & \text{if } k \in I, \\ e & \text{otherwise.} \end{cases}$$

μ-compatibility implies $t_{ij} = e$ for all $(i,j) \in I \times J$ $(c_{ij} = \infty)$ and therefore (w,s) is strongly dual feasible with respect to (12.33). Further $(y;w,s)$ is complementary. Theorems (11.18) and (11.25) lead to the following strong duality theorem.

(12.38) Theorem (Transportation duality theorem I)

Let H be a weakly cancellative d-monoid which is an extended semimodule over R_+. For a cost coefficient vector $a \in H_+^{mn}$ there exists a feasible optimal pair $(y;w,s)$ for the algebraic transportation problem which is complementary and satisfies

(1) $c_J^T \square w_J = y^T \square a * c_I^T \square s_I$

(2) $(-c_I^T) \square s_I(\mu) * c_J^T \square w_J = y^T \square a \qquad \text{for } \mu = \lambda(\tilde{z})$.

The application of the primal dual method to the algebraic transportation problem is called the *Hungarian method*. In the following we will prove that this method is valid even if we only assume that H is a d-monoid which is a linearly ordered semimodule over R_+. We remark that all steps of the method are well-defined in such d-monoids. Nevertheless a proof of its validity in this case but for the general algebraic flow problem is not known. A more detailed analysis is necessary to develop new arguments on which we can base a proof of its validity for algebraic transportation problems. BURKARD [1978a] develops the Hungarian method based on the concept of 'admissible transformations'. We will derive a description of our method which shows that both methods proceed in almost the same manner. Many arguments are drawn from BURKARD [1978a].

Let H be a d-monoid which is a linearly ordered semimodule over R_+. Arguments with respect to μ-compatibility hold in this case, too. Thus the primal dual method constructs a sequence of μ-compatible pairs (x;v). On the other hand finiteness and optimality are proved using certain cancellation arguments. These do not hold in d-monoids, in general. The values $x_{\sigma i}$, $i \in I$ and $x_{j\tau}$, $j \in J$ are never decreased in the procedure. Therefore μ-compatibility shows that arcs (σ,i), $i \in I$ and (j,τ), $j \in J$ remain admissible until in a certain iteration such an arc is *saturated*, i.e.

$$x_{\sigma i} = c_i \quad \text{resp.} \quad x_{j\tau} = c_j .$$

We conclude that

$$(12.39) \qquad \begin{aligned} x_{\sigma i} < c_i \quad &\Rightarrow \quad v_i = v_\sigma = e , \\ x_{j\tau} < c_j \quad &\Rightarrow \quad v_j = v_\tau \end{aligned}$$

for all $i \in I$, $j \in J$ is satisfied. Further μ-compatibility implies

$$(12.40) \qquad v_j \leq a_{ij}(\mu) * v_i$$

for all $i \in I$, $j \in J$ which shows that

$$(12.41) \qquad x_{ij} > 0 \quad \Rightarrow \quad (i,j) \text{ admissible}$$

for all $i \in I$, $j \in J$. Let (S,T) be a cut determined in step 1 and step 2 of the dual revision procedure (12.25). Let

$$L := (S \cap I) \times (T \cap J) ,$$

$$K := (\{\sigma\} \times (T \cap I)) \cup ((S \cap J) \times \{\tau\}) ;$$

then $S \times T = L \cup K$. Further $Q = (T \times S) \cap E$ satisfies

$$Q := (T \cap I) \times (S \cap J)$$

(cf. Figure 7).

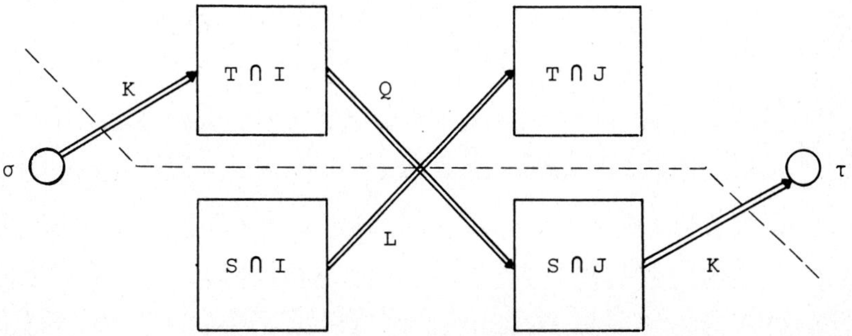

Figure 7 The cut (S,T) in a transportation network

Due to the construction of (S,T) and (12.41) we find

$$(12.42)\qquad \begin{aligned} x_{\sigma i} &= c_i && \text{for all } (\sigma,i) \in K \;, \\ x_{j\tau} &= c_j && \text{for all } (j,\tau) \in K \;, \\ x_{ij} &= 0 && \text{for all } (i,j) \in L \cup Q. \end{aligned}$$

This leads to a simple finiteness proof.

(12.43) Proposition

Let H be a d-monoid which is a linearly ordered semimodule over R_+. Then the Hungarian method terminates after finitely many steps with a maximum flow x.

Proof: Due to (12.42) the value g of the current flow x satisfies

$$(12.44)\qquad g = \sum_{T\cap I} c_i + \sum_{S\cap J} c_j$$

in step 3 of the Hungarian method. There are only finitely many flow values of such a form. As the flow value is strictly increased in an iteration of the Hungarian method after finitely many iterations a maximum flow x is constructed and termination occurs.

■

For a discussion of optimality of the final flow x a combinatorial property of maximum flows in transportation networks is of particular interest. A maximum flow $\bar{x}$ saturates all arcs (σ,i), $i \in I$ and all arcs (j,τ), $j \in J$. Therefore (12.44) implies $g = \Sigma_K \bar{x}_{ij}$. Then (12.2) applied to $\bar{x}$ leads to

$$(12.44)\qquad \Sigma_L \bar{x}_{ij} = f - g + \Sigma_Q \bar{x}_{ij} \;.$$

Another important combinatorial property is

$$\{(i,j) \in E \mid 0 < x_{ij}\} \cap (T \times S) = \emptyset.$$

Therefore in the dual revision procedure (12.25) the determination of α can be simplified. We get

(12.45) $\alpha = \min\{\bar{a}_{ij} \mid v_j * \bar{a}_{ij} := a_{ij}(\mu) * v_i,\ (i,j) \in L\}.$

Due to (12.40) reduced cost coefficients $\bar{a}_{ij}$ may be defined as solutions of

(12.46) $v_j * \bar{a}_{ij} := a_{ij}(\mu) * v_i$

for all $(i,j) \in I \times J$. Nevertheless these reduced cost coefficients are not uniquely determined by (12.46). If $v_j = a_{ij}(\mu) * v_i$ then we define $\bar{a}_{ij} = e$ as usual. Otherwise $v_j < a_{ij}(\mu) * v_i$ and $\bar{a}_{ij}$ is uniquely determined if H is weakly cancellative (cf. proposition 4.19.3). In general d-monoids, reduced cost coefficients are chosen as proposed in BURKARD [1978a]; initially $\bar{a} \equiv a$. Then after determination of α from

(12.47) $\alpha = \min\{\bar{a}_{ij} \mid (i,j) \in L\}$

in the dual revision procedure the reduced cost coefficients are redefined by

(12.48) $$\bar{a}_{ij} := \begin{cases} \bar{a}_{ij} * \alpha & \text{if } (i,j) \in Q, \\ \varepsilon & \text{if } (i,j) \in L \text{ and } \alpha * \varepsilon := \bar{a}_{ij}, \\ \bar{a}_{ij}(\mu) & \text{otherwise.} \end{cases}$$

In particular $\varepsilon := e$ if $\alpha = \bar{a}_{ij}$. It is easy to see that (12.46) is always satisfied. Even (12.48) does not uniquely determine all reduced cost coefficients. In any case $\mu = \lambda(\alpha)$ is uniquely determined. If we add $\bar{a} \equiv a$ in the initial step 1 of the primal dual method (12.27) then the following modification of the dual revision procedure is well-defined.

(12.49) <u>Modified dual revision procedure</u>

Step 1 Determine the connected component S ($\sigma \in S \subseteq N$) of the partial graph of ΔG which contains all arcs with $\Delta c_{ij} > 0$; if $\tau \in S$ then return (exit 1).

Step 2 $T := N \setminus S$; $L := (S \cap I) \times (T \cap J)$; $Q := (T \cap I) \times (S \cap J)$; if $L = \emptyset$ then return (exit 2).

Step 3 $\alpha := \min\{\bar{a}_{ij} \mid (i,j) \in L\}$;
$\mu := \lambda(\alpha)$.

Step 4 Redefine all dual variables by

$$v_j := \begin{cases} v_j(\mu) & \text{if } j \in S, \\ v_j * \alpha & \text{if } j \in T, \end{cases}$$

all reduced cost coefficients by (12.48) and go to step 3.

(12.49) specifies (12.25) for algebraic transportation problems in such a way that we can give a proof of the optimality of the final μ-compatible pair constructed by the following Hungarian method.

(12.50) <u>Hungarian method with dual variables</u>

Step 1 $x \equiv 0$; $v \equiv e$; $z := e$; $\mu = \lambda_0$; $\bar{a} := a$.

Step 2 Determine a maximum flow δx of value δg in δG;
$x := x \circ \delta x$; $z := z * (\delta g \,\square\, v_\tau)$; $\bar{v} := v$.

Step 3 Revise dual variables and reduced cost coefficients by means of (12.49);
if (12.49) terminates at exit 2 then $v := \bar{v}$ stop;
otherwise go to step 2.

The final μ-compatible pair $(x;v)$ is shown to be optimal in the proof of the following strong duality theorem.

(12.51) Theorem (Transportation duality theorem II)

Let H be a d-monoid which is a linearly ordered semimodule over R_+. For a cost coefficient vector $a \in H_+^{mn}$ there exists a feasible optimal pair $(y;w,s)$ for the algebraic transportation problem. $(y;w,s)$ is complementary and satisfies

(1) $$c_J^T \square w_J = y^T \square a * c_I^T \square s_I \; .$$

Further z defined recursively in step 2 of the Hungarian method (12.50) is the optimal value of the algebraic transportation problem.

Proof: We introduce an auxiliary variable β in the Hungarian method. Initially $\beta := e$. Then add

(12.52) $$\beta := \beta * [(f - g) \square \alpha]$$

in step 4 of the modified dual revision procedure (12.49). g denotes the flow value of the current flow x. We remark that $\lambda(\beta) = \mu$. We will prove that

(12.53) $$\beta * y^T \square \bar{a} = y^T \square a$$

is satisfied for all $y \in P_T$ throughout the performance of the Hungarian method (12.50). In the initial step (12.53) is obviously satisfied since $\bar{a} = a$. It suffices to consider an iteration in the modified dual revision procedure. Let (12.53) be satisfied before step 3 and denote the revised variables by $\tilde{v}$, $\tilde{a}$, $\tilde{\mu}$ and $\tilde{\beta}$. Let $y \in P_T$. Then $\tilde{\mu} \leq \lambda(\tilde{z}) \leq \lambda(y^T \square a)$ leads to

$$(12.54) \qquad y^T \square \bar{a} = y_Q^T \square \bar{a}_Q * y_L^T \square \bar{a}_L * y_H^T \square \tilde{a}_H$$

with $H := (I \times J) \setminus (Q \cup L)$. Further (12.48) yields

$$y_L^T \square \bar{a}_L = y_L^T \square \tilde{a}_L * [(\Sigma_L\, y_{ij}) \square \alpha] .$$

Due to (12.44) we get

$$y_L^T \square \bar{a}_L = y_L^T \square \tilde{a}_L * [(\Sigma_Q\, y_{ij}) \square \alpha] * [(f-g) \square \alpha] .$$

Together with (12.54) and (12.48) this leads to

$$(12.55) \qquad y^T \square \bar{a} = y^T \square \tilde{a} * [(f-g) \square \alpha] .$$

Then $\tilde{\beta} = \beta * [(f-g) \square \alpha]$ proves (12.53) with respect to the redefined variables. Let $(y;w,s)$ denote the final pair derived from the final compatible pair $(x;v)$ of the modified primal dual method. Then $\bar{y} \in P_T$ and $\bar{y}^T \square \bar{a} = e$. Therefore

$$(12.56) \qquad \begin{aligned} \bar{y}^T \square a &= \bar{\beta} * \bar{y}^T \square \bar{a} = \bar{\beta} \\ &\leq \bar{\beta} * y^T \square \bar{a} = y^T \square \bar{a} \end{aligned}$$

for all $y \in P_T$. Hence $\bar{y}$ is optimal.

The equation (1) is derived from the equations

$$w_j * \bar{a}_{ij} = a_{ij} * s_i$$

by collecting all these equations after composition with $\bar{y}$. The recursion for the optimal value is derived from (12.56) which shows $\bar{\beta} = \bar{y}^T \square a$ (another recursion!). Now let (α_μ, g_μ) be the constructed sequence of variables α in step 3 of (12.49) and corresponding current flow values g; then β satisfies (cf. 12.52)

$$\beta = \mathop{*}_{\mu=1,2,\ldots,r} [(f - g_\mu) \,\square\, \alpha_\mu] \;.$$

Let $\delta g_\mu := g_\mu - g_{\mu-1}$ for $\mu = 2,3,\ldots,r$ and let v_τ^μ, $\mu = 1,2,\ldots,r$ denote the corresponding sequence of values of the dual variable v_τ. Then $v_\tau^1 = e$ and

$$v_\tau^\mu = \alpha_1 * \alpha_2 * \ldots * \alpha_{\mu-1}$$

for $\mu = 2,3,\ldots,r$. Thus

$$\beta = \mathop{*}_{\mu=2,3,\ldots,r} \delta g_\mu \,\square\, v_\tau^\mu = z.$$

■

If we are not interested in the optimal dual variables then we may delete dual variables from (12.50). In this case it is necessary to identify admissible arcs without explicit use of dual variables. Due to the remarks on admissibility before (12.39) we may consider all arcs (σ,i), $i \in I$ and (j,τ), $j \in J$ as admissible throughout the performance of the Hungarian method without changing the generated sequence of compatible pairs. Now $(i,j) \in I \times J$ is admissible (cf. 12.19 and 12.46) if and only if

(12.57) $\qquad v_j * \bar{a}_{ij} = v_j \;.$

Then (j,i) is admissible, too. If H is weakly cancellative then this is equivalent to $\bar{a}_{ij} = e$ as well as to

(12.58) $\qquad \beta * \bar{a}_{ij} = \beta$

where β is recursively defined by (12.52). BURKARD [1978a] uses (12.58) for a definition of admissibility. In general d-monoids H it is possible that these definitions lead to differing admissibility graphs. Let $(i,j) \in I \times J$ be β-*admissible* if (12.58) is satisfied. Then (j,i) is called β-*admissible*, too. Let $\delta G(\beta)$ denote the partial subgraph of ΔG containing all arcs from $(E \cup \bar{E}) \setminus ((I \times J) \cup (J \times I))$ and all β-admissible arcs. This leads to the following variant of the Hungarian method (12.50).

(12.59) Hungarian method without dual variables

Step 1 $x \equiv 0$; $g := 0$; $\beta := e$; $\bar{a} := a$; $f := \Sigma_i c_i$.

Step 2 Determine a maximum flow δx of value δg in $\delta G(\beta)$;
$x := x \circ \delta x$; $g := g + \delta g$.

Step 3 Revise reduced cost coefficients and β by means of (12.60);
if (12.60) terminates at exit 2 then stop;
otherwise go to step 2.

(12.60) Revision of reduced cost coefficients

Step 1 Determine the connected component S ($\sigma \in S \subseteq N$) of the partial graph of $\delta G(\beta)$ which contains all arcs with $\Delta c_{ij} > 0$;
if $\tau \in S$ then return (exit 1).

Step 2 $T := N \setminus S$; $L := (S \cap I) \times (T \cap J)$; $Q := (T \cap I) \times (S \cap J)$;
if $L = \emptyset$ then return (exit 2).

Step 3 $\alpha := \min\{\bar{a}_{ij} \mid (i,j) \in L\}$;
$\beta := \beta * (f - g) \square \alpha$.

Step 4 Redefine all reduced cost coefficients by

$$\bar{a}_{ij} := \begin{cases} \bar{a}_{ij} * \alpha & \text{if } (i,j) \in Q \,, \\ \varepsilon & \text{if } (i,j) \in L \text{ and } \alpha * \varepsilon := \bar{a}_{ij} \,, \\ \bar{a}_{ij} & \text{otherwise} \,, \end{cases}$$

and go to step 1.

Obviously all steps in the Hungarian method are well-defined. The reduced cost coefficients $\bar{a}_{ij}$ satisfy $\bar{a}_{ij} \geq e$ for all $(i,j) \in I \times J$ throughout the performance of the Hungarian method. Further

(12.61) $x_{ij} > 0 \quad \Rightarrow \quad (i,j)$ is β-admissible

for all $(i,j) \in I \times J$ is always satisfied. This follows from the transformation of the reduced cost coefficients in step 4 of (12.60). $\bar{a}_{ij}$ is increased only on arcs $(i,j) \in Q$; therefore $x_{ij} = 0$ for $(i,j) \in Q$ and

$$\operatorname{dom} \beta \subseteq \operatorname{dom}(\beta * (f-g) \,\square\, \alpha)$$

(cf. 4.18.2) lead to (12.61). The cardinality of S is increased at least by one after such a transformation. Thus (12.60) terminates after at most $(n+m-1)$ transformations. The same argument as used in the proof of proposition (12.43) shows that the Hungarian method (12.59) is finite and leads to a maximum flow $\bar{x}$. Then due to (12.61) $\bar{y}^T \,\square\, \bar{a} = e$ for $\bar{y} := \bar{x}_{I \times J}$. (12.53) can be proved in a similar manner as before. Thus

$$\bar{y}^T \,\square\, a = \beta \leq \beta * y^T \,\square\, \bar{a} = y^T \,\square\, a$$

for all $y \in P_T$ since $\bar{a} \in H_+^{mn}$. Hence $\bar{y}$ is optimal.

(12.59) is developed in BURKARD [1978a]. The particular case of *algebraic assignment problems* ($c_k = 1$, $k \in I$ and $R = \mathbb{Z}$) is solved in the same manner in BURKARD, HAHN and ZIMMERMANN, U. [1977]. It should be noted that (12.59) and (12.50) lead to the same sequence of flows x if H is weakly cancellative. In the form (12.59) the Hungarian method can be interpreted as a sequence of (admissible) transformations of the reduced cost coefficients $\bar{a} \to \bar{\bar{a}}$ such that

$$y^T \square \bar{a} = \gamma * y^T \square \bar{\bar{a}}$$

for all $y \in P_T$. The composition of a finite number of (admissible) transformations is again an (admissible) transformation. The transformations are constructed in a systematic manner such that a final admissible transformation $a \to \bar{a} \in H_+^{mn}$ is generated satisfying

$$y^T \square a = \beta * y^T \square \bar{a}$$

for all $y \in P_T$ and

$$\beta = \beta * \bar{y}^T \square \bar{a}$$

for some $\bar{y}^T \in \bar{a}$. As we have shown before such $\bar{y}$ are optimal.

In discrete semimodules ($R = \mathbb{Z}$) we get similar complexity bounds as for the algebraic network flow problem. In particular for algebraic assignment problems an optimal solution is found in $O(n^4)$ usual operations, $O(n^4)$ $*$-compositions and $O(n^4)$ $\leq$ -comparisons.
In the classical case the Hungarian method [(12.50) as well as (12.59)] reduce to the Hungarian method for transportation

problems (cf. MURTY [1976]) which is developed for assignment problems in KUHN [1955]. For bottleneck objectives it reduces to the second (threshold) algorithm of GARFINKEL and RAO [1971]. In this case the algorithm remains valid even in the case of finite capacities ($c_{ij} < \infty$) on arcs $(i,j) \in I \times J$. This follows from its threshold character (cf. BURKARD [1978a], EDMONDS and FULKERSON [1970]).

We will discuss two further methods for a solution of algebraic transportation problems. Both methods are well-defined in d-monoids H which are linearly ordered semimodules over R_+; on the other hand a proof of validity and finiteness is only known in the case that H is a weakly cancellative d-monoid which is an extended semimodule over R_+.

Now we develop an algorithm proposed in BURKARD and ZIMMERMANN, U. [1980]. The method proceeds similar to the Hungarian method by alternately flow augmentations and dual revision steps. Different from the Hungarian method flow augmentation is performed along certain shortest paths. The method consists of m stages ($k = 1,2,\dots,m$). In the k-th stage the flow x satisfies

$$x_{\sigma i} = \begin{cases} c_i & \text{if } i < k \\ 0 & \text{if } i > k \end{cases}$$

for all $i \in I$ and $0 \le x_k < c_k$. In order to augment such a flow x we consider paths in the *reduced incremental graph* ΔG_k with set of arcs

$$\{(\sigma,k)\} \cup I \times J \cup \{(j,i) \mid (i,j) \in I \times J,\ x_{ij} > 0\} \cup \{(j,\tau) \mid x_{j\tau} < c_j\}.$$

If $x_k < c_k$ then ΔG_k contains at least one elementary path p from σ to τ. Then $\delta(p) > 0$ (capacity of p), $p(\delta(p))$ is an incremental path flow, and we may augment x to $x \circ p(\delta(p))$. At the k-th stage the current flow x and the current dual variables v_j , $j \in N$ will be *almost μ-compatible*, i.e. all conditions (cf. 12.15) are satisfied with the possible exception of $v_\sigma \geq v_i$ which is necessary while $x_{\sigma i} < c_i$ for $i \geq k$. At the end of the k-stage $x_{\sigma k} = c_k$. If we augment x by $p(\delta(p))$ then almost μ-compatibility is violated, in general. Due to the choice of p it will be possible to revise the dual variables in such a manner that the new current x and v are almost μ-compatible, again.

For a determination of p we assign weights a to the arcs of ΔG_k. Since x and v are almost μ-compatible the reduced cost coefficients $\bar{a}_{ij}$ can be defined by

$$v_j * \bar{a}_{ij} := a_{ij}(\mu) * v_i$$

for all $(i,j) \in I \times J$. Let $\tilde{a}_{ij} := \bar{a}_{ij}$ for $(i,j) \in I \times J$ and $\tilde{a}_{ij} := e$ otherwise. We remind that by definition

$$x_{ij} > 0 \;\Rightarrow\; \bar{a}_{ij} = e$$

for all $(i,j) \in I \times J$ (cf. 12.46). Let $w(p)$ denote the weight of a path p in ΔG_k with respect to weights $\tilde{a}_{ij}$ (cf. chapter 8). Let $N' := \{1,2,\dots,k\} \cup J \cup \{\tau\}$. Shortest paths from σ to $i \in N'$ can be determined using the algebraic version of (8.4) together with (8.10) since $\tilde{a}_{ij} \in H_+$ for all arcs (i,j) of ΔG_k.

The following shortest path method is a suitable version of (8.4) applied to ΔG_k. During its performance a label $[\pi_i; p(i)]$ is attached to vertices $i \in N'$. π_i is the weight of the current detected shortest path from σ to i and p(i) is the predecessor of i on this path.

(12.61) Shortest paths in ΔG_k

Step 1 Label vertex k with $[e;\sigma]$ and label all vertices $j \in J$ with $[\tilde{a}_{kj};k]$; $I' := \{1,2,\dots,k-1\}$; $J' := J$.

Step 2 Determine $h \in J'$ with

$$\pi_h = \min\{\pi_j \mid j \in J'\}$$

and redefine $J' := J' \setminus \{h\}$.

Step 3 If $x_{h\tau} < c_h$ then go to step 6;
let $S(h) := \{i \in I' \mid x_{ih} > 0\}$;
if $S(h) = \emptyset$ then go to step 2;
otherwise label all vertices $i \in S(h)$ with $[\pi_h;h]$ and redefine $I' := I' \setminus S(h)$.

Step 4 For all $(i,j) \in S(h) \times J'$ do
if $\pi_i * \tilde{a}_{ij} < \pi_j$ then label vertex j with $[\pi_i * \tilde{a}_{ij};i]$.

Step 5 Go to step 2.

Step 6 Label τ with $[\pi_h;h]$ and define

$$\pi_i := \pi_h$$

for all vertices $i \in I \cup J$ which satisfy $\pi_i > \pi_h$ or which are not labeled.

At termination the sink τ is labeled and π_τ is the weight of a shortest path $\tilde{p}$ from σ to τ which can easily be found by back-tracing using the predecessor labels. If the weight of a shortest path from σ to $i \in N'$ is not greater than π_τ then it is equal to π_i. Otherwise $\pi_i = \pi_\tau$. The algorithm terminates after $O(|N|^2)$ usual operations, $O(|N|^2)$ $*$-compositions and $O(|N|^2)$ $\leq$ - comparisons.

(12.62) Augmenting path method

Step 1 $k := 1;\ x \equiv 0;\ v \equiv e;\ \mu := \lambda_o$.

Step 2 Determine weights π_i, $i \in N \setminus \{\sigma\}$ and the shortest path $\tilde{p}$ with capacity $\tilde{\delta}$ using the shortest path method (12.61).

Step 3 $x := x \circ \tilde{p}(\tilde{\delta})$, $\mu := \lambda(v_\tau * \pi_\tau)$.

Step 4 Redefine the dual variables by

$v_i := (v_i * \pi_i)(\mu)$

for all $i \in N \setminus \{\sigma\}$.

Step 5 If $x_{\sigma k} < c_k$ go to step 2;

if $k = m$ stop;

$k := k+1$ and go to step 2.

All steps in (12.62) are well-defined in a d-monoid H provided that $(x;v)$ is almost μ-compatible throughout the performance of (12.62). The necessary determination of the reduced cost coefficients is not explicitly mentioned. As in the Hungarian method it is possible to derive a recursion for the reduced cost coefficients; then the dual variables can be deleted from the algorithm with the exception of v_τ. The determination of π_i in step 2 is also possible using cuts and dual revisions as in the

Hungarian method; but the shortest path method leads to a better complexity bound: $O(|N|^2)$ instead of $O(|N|^3)$ with respect to all types of operations considered (cf. 12.25). This is due to the fact that we assign nonnegative weights $\tilde{a}_{ij}$ to the arcs of ΔG_k which admits the application of a fast shortest path procedure. It should be noted that in the case $\pi_\tau = e$ the subsequent determination of shortest paths of zero-weight can be replaced by the determination of a maximum incremental flow in ΔG_k using only arcs with weight $\tilde{a}_{ij} = e$. For a proof of finiteness and validity of the augmenting path method we assume that H is a weakly cancellative d-monoid which is an extended semimodule over R_+. A proof for general d-monoids which are linearly ordered semimodules is not known.

(12.63) Proposition

The augmenting path method (12.62) generates a sequence of almost μ-compatible $(x;v)$. Further x is of minimum cost among all flows $\tilde{x}$ with $x_{\sigma i} = \tilde{x}_{\sigma i}$ for all $i \in I$.

Proof: In step 1 $x \equiv 0$ and $v \equiv e$ are λ_o-compatible. Thus it suffices to prove that an iteration does not violate almost μ-compatibility. If at the end of the k-th stage $(x;v)$ is almost μ-compatible then the same holds at the beginning of the k+1th stage since $x_{\sigma k} = c_k$. Thus it suffices to discuss an iteration at the k-th stage. Let $(x;v)$ denote the current pair in step 2 and let $(x';v')$ denote the redefined variables. We assume that $(x;v)$ is almost μ-compatible and we will prove that $(x';v')$ is almost μ'-compatible $(\mu' = \lambda(v_\tau * \pi_\tau))$.

At first we consider the validity of the inequalities in (12.15). Obviously, these inequalities are satisfied on arcs (σ,i), $1 \leq i < k$ and (j,τ), $j \in J$ in the same way as in the Hungarian method (cf. 12.39). Let $(i,j) \in I \times J$. Then

(12.64) $$v_j * \bar{a}_{ij} = a_{ij}(\mu) * v_i$$

due to our assumption. If $\pi_j \leq \pi_i$ then $v_j' \leq a_{ij}(\mu') * v_i'$. Otherwise $\pi_j > \pi_i$ implies $\pi_\tau > \pi_i$. Therefore π_i is the value of a shortest path in ΔG_k and $\pi_i * \bar{a}_{ij} \geq \pi_j$. Hence $v_j * \pi_j \leq v_j * \bar{a}_{ij} * \pi_i = a_{ij}(\mu) * v_i * \pi_i$. $\mu \leq \mu'$ implies $\bar{v}_j \leq a_{ij}(\mu') * \bar{v}_i$.

Now let $x_{ij}' > 0$. Then either $x_{ij} > 0$ or $(i,j) \in \tilde{p}$, the shortest path from σ to τ. If $x_{ij} > 0$ then $v_j = a_{ij}(\mu) * v_i$ implies $\bar{a}_{ij} = e$. Therefore $\pi_i = \pi_j$ which leads to $v_j' = a_{ij}(\mu') * v_i'$. We remark that

(12.65) $$x^T \square \tilde{a} = e.$$

Otherwise $(i,j) \in \tilde{p}$. Then $\pi_j = \bar{a}_{ij} * \pi_i$ which implies $v_j * \pi_j = a_{ij}(\mu) * v_i * \pi_i$. $\mu \leq \mu'$ leads to $v_j' = a_{ij}(\mu') * v_i'$.

Secondly we remark $v_\sigma' = e$ and $v' \in (H_{\mu'} \cup \{e\})^N$. Now we will show $\mu' \leq \lambda(\tilde{z})$. (12.65) shows that x is a minimum cost flow with respect to weights $\tilde{a}$. Let $\tilde{x}$ be a flow in G with $\tilde{x}_{\sigma i} = x_{\sigma i}'$ for all $i \in I$. Then $\tilde{x} = x \circ \Delta x$ where Δx is an incremental flow in ΔG_k. Due to (12.7) we know that Δx has a representation as sum of incremental path flows and circuit flows

$$\Delta x = \sum_{\rho=1}^{r} p_\rho(\delta_\rho) + \sum_{\rho=r+1}^{s} p_\rho(\delta_\rho)$$

and weights with respect to $\tilde{a}$ satisfy

$$\tilde{x}^T \square \tilde{a} = \mathop{*}_{\rho=1}^{r} (\delta_\rho \square \tilde{w}(p_\rho)) * \mathop{*}_{\rho=r+1}^{s} (\delta_\rho \square \tilde{w}(p_\rho)).$$

If $\tilde{x}$ is of minimum cost among all such flows then

$$\bar{x} := x \circ (\sum_{\rho=1}^{r} p_\rho(\delta_\rho)) ,$$

too. Since $\tilde{w}(\tilde{p}) \leq w(p_\rho)$ for all $\rho = 1,2,\ldots,r$ we find that x' is of minimum cost with respect to $\tilde{a}$, too.

Now let $\tilde{x}$ be a maximum flow in G of minimum cost with respect to $\tilde{a}$. $\tilde{x}$ contains a partial flow $\bar{x} \leq \tilde{x}$ with $\bar{x}_{\sigma i} = x'_{\sigma i}$ for all $i \in I$. Then

$$(x')^T \square \tilde{a} \leq \bar{x}^T \square \tilde{a} \leq \tilde{x}^T \square \tilde{a}.$$

By theorem (12.38) the algebraic transportation problem has an optimal solution $y \in P_T$. (12.64) leads to

$$c_J^T \square v_J * y^T \square \tilde{a} \leq y^T \square a * c_I^T \square v_I = \tilde{z} * c_I^T \square v_I .$$

Therefore $\tilde{x}^T \square \tilde{a} \leq y^T \square \tilde{a}$ implies

$$c_J^T \square v_J * (x')^T \square \tilde{a} \leq \tilde{z} * c_I^T \square v_I .$$

Then $\lambda(v_i) \leq \mu \leq \lambda(\tilde{z})$ for all $i \in I \cup J$ and $(x')^T \square \tilde{a} = \delta(\tilde{p}) \square \tilde{w}(\tilde{p})$ lead to

$$\lambda(\pi_\tau) \leq \lambda(\tilde{z}) .$$

Since $\mu' = \lambda(v_\tau * \pi_\tau)$ we conclude that x' and v' are almost μ'-compatible.

Finally we consider a network G' which differs from G in the capacities $c_i' := x_{\sigma i}$ for $i \in I$ and contains an additional arc (τ,τ') to an auxiliary sink τ' with capacity $c_{\tau\tau'} := \Sigma_i c_i'$ and $a_{\tau\tau'} := e$. Let $x_{\tau\tau'} := c_{\tau\tau'}$ and $v_{\tau'} := v_\tau$. Then the extended pair $(x;v)$ is μ-compatible and therefore due to proposition (12.18) of minimum cost. This implies that x in G is of minimum cost. ∎

Proposition (12.63) shows that all steps in the augmenting path method are well-defined. Further, if the method is finite, then at the end of the k-th stage a maximum flow x of minimum cost is determined.

(12.66) Theorem

Let H be a weakly cancellative d-monoid which is an extended semimodule over R_+. Then

(1) the augmenting path method (12.62) is finite,

(2) the final pair $(x;v)$ is μ-compatible,

(3) $(y;w,s)$ defined by (12.37) satisfies

$$c_J^T \square w_J = y^T \square a * c_I^T \square s_I \ ,$$

$$(-c_I)^T \square s_I(\mu) * c_J^T \square w_J = y^T \square a \qquad \text{for } \mu = \lambda(\tilde{z}) \ .$$

Proof: At first we assume finiteness. Then the final pair is μ-compatible. Hence $(y;w,s)$ is complementary and satisfies the claimed equations (cf. theorem 12.38).
Secondly we prove finiteness in a similar manner as in the proof of (12.28). It is sufficient to prove that the number of performances of step 4 is finite for fixed parameter k. Assume that we find a shortest augmenting path of weight e in step 2. Then in step 4 the dual variables are not changed. Subsequent iterations of this type are equivalent to the determination of a maximum incremental flow in the partial graph of ΔG_k con-

taining only arcs with $\tilde{a}_{ij} = e$. The number of subsequent iterations is finite if we choose augmenting paths subject to the finiteness rule of PONSTEIN [1972] or if we directly apply a maximum flow algorithm (e.g. KARZANOV [1974]). Otherwise $\pi_\tau > e$ and $\lambda(\pi_\tau) \geq \mu$ since $\tilde{a}_{ij} \in H_\mu \cup \{e\}$. Let $(h,\tau) \in \tilde{p}$, the shortest path determined in step 2. Then $\pi_\tau = \pi_h$ and

$$\bar{v}_h = v_h(\bar{\mu}) * \pi_h > v_h$$

where $\bar{v}$ denotes the redefined dual variables with redefined index $\bar{\mu}$. The arcs of $\tilde{p}$ are partitioned into $\tilde{F} \subseteq E$ and $\tilde{B} \subseteq \bar{E}$. Suppose that $\tilde{p}$ appeared in a previous iteration with the same partition of its arcs. Let $\tilde{v}$ denote the corresponding redefined dual variables with index $\tilde{\mu}$ after the previous determination of $\tilde{p}$. Then

(12.67) $$\bar{v}_h > v_h \geq \tilde{v}_h \ .$$

Further we remind that $\bar{v}_k = \tilde{v}_k$ since $\pi_k \equiv e$ at the k-th stage. Then $\tilde{\mu} \leq \bar{\mu}$ and

(12.68) $$\begin{aligned} \bar{v}_j &= a_{ij}(\bar{\mu}) * \bar{v}_i \ , \\ \tilde{v}_j &= a_{ij}(\tilde{\mu}) * \tilde{v}_i \end{aligned}$$

for all $(i,j) \in \tilde{F} \cup \{(s,r) \mid (r,s) \in \tilde{B}\}$ which leads to

(12.69) $$\begin{aligned} \bar{v}_h * \beta(\bar{\mu}) &= \alpha(\bar{\mu}) * \bar{v}_k \\ \tilde{v}_h * \beta(\tilde{\mu}) &= \alpha(\tilde{\mu}) * \tilde{v}_k \end{aligned}$$

for constants α, β defined by

$$\beta := \mathop{*}_{(j,i) \in \tilde{B}} a_{ij}$$

$$\alpha := \mathop{*}_{(i,j) \in \tilde{F}} a_{ij} \ .$$

(12.68) implies $\lambda(a_{ij}(\tilde{\mu})) \leq \tilde{\mu}$ for all $(j,i) \in \tilde{B}$. Therefore $\lambda(\alpha) \leq \tilde{\mu}$ and $\lambda(\beta) \leq \tilde{\mu}$. If $\tilde{\mu} < \bar{\mu}$ then

$$\bar{v}_h = \bar{v}_k = \tilde{v}_k \leq \tilde{v}_h$$

contrary to (12.67). Otherwise cancellation of $\beta(\bar{\mu})$ in

$$\bar{v}_h * \beta(\bar{\mu}) = \tilde{v}_h * \beta(\bar{\mu})$$

leads to $\bar{v}_h = \tilde{v}_h$ contrary to (12.67).

■

Again we may derive complexity bounds in the discrete case ($R = \mathbb{Z}$) from the observation that $\delta(\tilde{p}) \geq 1$ in step 2. Then any iteration mainly consists in a determination of a shortest path and the redefinition of flow values and dual variables. These steps need $O((m+n)^2)$ usual operations, $O((m+n)^2)$ $*$-compositions, and $O((m+n)^2)$ $\leq$ - comparisons. There are at most $f = \Sigma_i c_i$ such iterations. In the particular case of assignment problems this leads to a complexity bound of $O(n^3)$ usual operations, $O(n^3)$ $*$-compositions and $O(n^3)$ $\leq$ - comparisons. These bounds are better than corresponding bounds for the Hungarian method (by a factor of $O(n+m)$).
We remark that in the classical case the augmentation method reduces to a method proposed by TOMIZAWA [1972]; computational experience is reported in DORHOUT [1973] in the case of assignment problems. For bottleneck objectives computational experience is discussed in DERIGS and ZIMMERMANN, U. ([1978b],[1979]).

All methods that we previously described in this chapter can be interpreted as primal dual methods. Nevertheless, we shortly mentioned the possible application of the primal simplex method (cf. 11.46) to the algebraic flow problem. In the following we discuss this method in the case of algebraic transportation problems.

We assume that H is a weakly cancellative d-monoid which is an extended semimodule over R_+. W.l.o.g. let $a \in H_+^{mn}$. Since it is well-known (cf. LAWLER [1976]) that the constraint matrix describing the polyhedron P_T in the algebraic transportation problem (12.33) is totally unimodular all steps of the primal simplex method are well-defined. In particular, for transportation problems it is easy to find an initial feasible basic solution. Only slight modifications are necessary (if any) in order to find such a solution for algebraic transportation problems. Then the primal simplex method generates a sequence of feasible basic solutions. We assume that cycling does not occur - the same assumption is usually made in the classical transportation problem. Thus we may forgo an explicit discussion of finiteness. Due to the analysis of the simplex method for linear algebraic programs we know that the final feasible basic solution x is $\bar{\lambda}$-dual feasible with $\bar{\lambda} := \lambda(x^T \square a)$. Theorem (11.41) implies that x is optimal. In the following we discuss the primal simplex method in more detail and derive a particular form of the optimality criterion (11.42).

Let $x \in R_+^{mn}$ be a basic feasible solution of P_T (cf. 12.33) and let B denote the set of all arcs (i,j) for basic variables x_{ij}. Let $\mu := \lambda(x^T \square a)$. Then a μ-complementary solution (cf. 11.38)

$s \in (G_\mu)^{m+n}$ can be determined solving the system of equations

(12.70) $\qquad s_j = a_{ij}(\mu) * s_i$

for all $(i,j) \in B$. We remark that G_μ is the group containing H_μ (cf. discussion of extended semimodules in chapter 6). The neutral element of G_μ is identified with e. It is well-known that B is the arcset of a connected partial graph $T = (I \cup J, B)$ of the bipartite graph $(I \cup J, I \times J)$. T contains no cycle and is called a *spanning tree*.

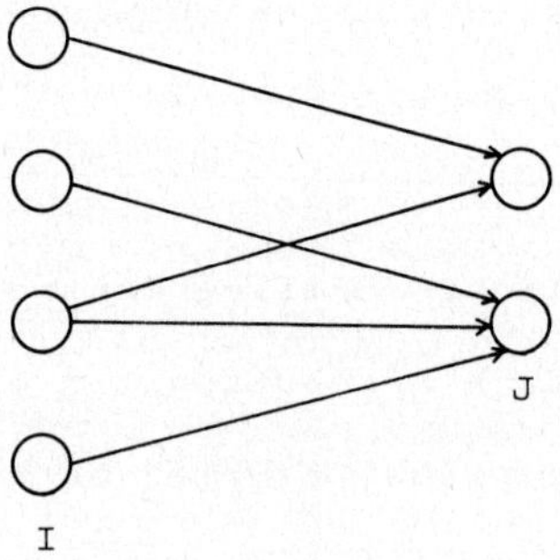

Figure 8. Spanning tree T in $(I \cup J, I \times J)$

If $a_{ij} = e$ for all $(i,j) \in B$ then x is an optimal feasible solution. Otherwise a solution of (12.70) can be determined in the following way. Let $(r,t) \in B$ with $e < a_{rt}(\mu) \in G_\mu$. Let $s_r := e$, $s_t := a_{rt}$ and solve (12.70) along the tree using (12.70) recursively. In general, this leads to a solution $s \in G_\mu^{m+n}$ which contains some negative components. Then let

$$\delta := [\min\{s_k \mid k \in I \cup J\}]^{-1}$$

which exists in the group G_μ. We define a solution $v \in (H_\mu \cup \{e\})_+^{m+n}$ of (12.70) by

$$v_k := s_k * \delta$$

for all $k \in I \cup J$. Then v contains some components of value e; v is the *minimal positive solution* of (12.70) in G_μ. Since $|B| = m+n-1$ such a solution can be determined with $O(m+n-1)$ $*$-compositions and $O(m+n-1)$ $\leq$-comparisons. We remark that

(12.71) $\quad v_i < v_j = a_{ij} * v_i$

for all $(i,j) \in B$ with $a_{ij} \in H_\mu \setminus \{e\}$.

(12.72) Theorem

Let H be a weakly cancellative d-monoid which is an extended semimodule over R_+. Let $a \in H_+^{mn}$. Let $x \in P_T$ be a basic solution with basis B and $\mu := \lambda(x^T \square a)$. Let v denote the minimal positive solution of (12.70). If

(1) $\quad v_j \leq a_{ij}(\mu) * v_i$

for all $(i,j) \notin B$ with $v_j > e$ and $\lambda(a_{ij}) \leq \mu$ then x is an optimal basic feasible solution of the algebraic transportation problem (12.33).

Proof: We define dual variables $w_k \in G_\mu$, $k \in I \cup J$ by

$$w_k := \begin{cases} v_j & j \in J \,, \\ v_j^{-1} & j \in I \,. \end{cases}$$

Then (1) implies that w is μ-dual feasible and μ-complementary (cf. 11.38 and 11.39). Theorem (11.41) shows that x is an optimal basic solution of the algebraic transportation problem (12.33) and w is an optimal solution of the corresponding dual problems.

■

An optimality criterion similar to (12.72) is discussed in ZIMMERMANN, U. [1979d]. Here the involved combinatorial arguments in that paper are not necessary since we can apply theorem (11.41) based on the duality principles from ZIMMERMANN, U. [1980a]. The main difference lies in the fact that instead of a 'minimal' solution $v \in H_+^{m+n}$ of the system of equations

$$v_j = a_{ij} * v_i$$

for all $(i,j) \in B$ with $\lambda(v_j) \leq \mu$ a minimal positive solution of (12.70) is considered.

The advantage of using (1) and not (cf. 11.42)

$$w_i * w_j \leq a_{ij}(\mu)$$

as optimality criterion lies in the fact that due to the choice of a minimal solution several v_j will have value O. Thus we expect that (1) may lead to computational savings.

If the current basic solution x does not satisfy (1) then we may use a 'most negative reduced cost coefficient rule' similar to the classical case. Let

$$K := \{(i,j) \notin B \mid v_j > e,\ \lambda(a_{ij}) \leq \mu,\ v_j > a_{ij}(\mu) * v_i\}$$

and let α_{ij} be defined by

(12.73) $$v_j =: \alpha_{ij} * a_{ij}(\mu) * v_i$$

for $(i,j) \in K$. Then choose x_{rs} with $(r,s) \in K$ and with

(12.74) $$\alpha_{rs} = \max\{\alpha_{ij} \mid (i,j) \in K\}$$

as new basic variable in the new basis.

It is easy to derive a generalization of the primal simplex method applied to transportation problems. An efficient implementation using the tree-structure of the basic solution is described in MURTY [1976]. We give only an outline of the method showing the differences to the classical case.

(12.75) Primal transportation algorithm

Step 1 Apply any of the methods for the determination of an initial feasible basic solution $\bar{x}$ with basis $\bar{B}$; let $\mu := \lambda(\bar{x}^T \square a)$.

Step 2 Determine a minimal positive solution $v \in (H_\mu \cup \{e\})_+^{m+n}$ of the equations (12.70);
if $v_j \leq a_{ij}(\mu) * v_i$ for $(i,j) \notin \bar{B}$ with $v_j > e$ and $\lambda(a_{ij}) \leq \mu$ then stop ($\bar{x}$ is optimal).

Step 3 Choose $(r,s) \in K$ with
$\alpha_{rs} := \max\{\alpha_{ij} \mid (i,j) \in K,\ v_j := \alpha_{ij} * a_{ij}(\mu) * v_i\}$
and determine the new basis by introducing x_{rs} into the basis;
redefine $\bar{x}$, $\bar{B}$ and μ and go to step 2.

We remark that with these modifications of the general primal simplex method (11.46) it is possible to avoid an explicit use of the associated negative part of the groups G_μ. All variables in (12.75) are elements of the given semimodule H.

The choice of the nonbasic variable in step 3 which is introduced into the basis leads in the classical case to the classical stepping stone method; for bottleneck objectives the method is equivalent to those of BARSOV [1964], SWARCZ ([1966], [1971]) and HAMMER ([1969], [1971]) in the sense that it generates the same sequence of basic solutions provided that the starting solutions are the same. In all these methods the entering nonbasic variable is determined investigating certain cycle sets defined with respect to the bottleneck values in the current basic solution. The above proposed rule using dual variables looks more promising as well from a theoretical as from a computational point of view.

Finally we give some remarks on particular algebraic transportation problems which have been considered in more detail. We forgo an explicit discussion of the classical problem and refer for this purpose to standard textbooks.
Algebraic assignment problems were solved first in BURKARD, HAHN and ZIMMERMANN, U. [1977] using a generalized Hungarian method. The augmenting path method was directly developed for algebraic transportation problems by BURKARD and ZIMMERMANN, U. [1980]; that paper contains a short discussion of the specific case of algebraic assignment problems, too. The primal simplex method for algebraic transportation problems is given in ZIMMERMANN, U. [1979d] without explicit discussion of algebraic assignments. FRIEZE [1979] develops a method for the algebraic assignment problem similar to the classical method of DINIC and KRONROD [1969]. This method is of order $O(n^3)$ similar to the augmenting path method.
The solution of algebraic transportation problems is used for

deriving bounds in a branch and bound method for the solution of certain algebraic scheduling problems in BURKARD [1979c].

Particular objective functions differing from the classical case which have found considerable interest in the literature are bottleneck objectives and lexicographical objectives (cf. 11.58 and 11.59).

For lexicographical objectives a FORTRAN program is developed from the algebraic approach in ZIMMERMANN, U. [1976]. Although the program is structured in order to show the impact of the algebraic approach it is easy to derive an efficient version for lexicographic objectives, in particular. Computational experience is discussed, too.

Bottleneck assignment and bottleneck (or time) transportation problems have been considered by many authors. The following list will hardly be complete but hopefully provides an overview which seems to be necessary since even today similar approaches are separately published from time to time. Such problems were posed and solved by BARSOV [1964] (1959 in russian). He proposed a primal solution method. Independently GROSS [1959] stimulated by FULKERSON, GLICKSBERG and GROSS [1953] investigated the same objective function for assignment problems. In the following years GRABOWSKI ([1964], [1976]), SWARCZ ([1966], [1971]) and HAMMER ([1969], [1971]) developed solution methods. In particular HAMMER ([1969], [1971]), SWARCZ [1971], SRINIVASAN and THOMPSON [1976], ZIMMERMANN, U. ([1978a], [1979d]) consider primal methods equivalent to BARSOV's method. They only differ in the method for the determination of the pivot element. GARFINKEL [1971] and

GARFINKEL and RAO [1971] discuss primal dual methods based on the threshold algorithm in EDMONDS and FULKERSON [1970]. The augmenting path method is finally developed within the scope of an algebraic approach by BURKARD and ZIMMERMANN, U. [1980]. Further discussions can be found in SLOMINSKI ([1976], [1977], [1978]).

Computational experience is discussed by PAPE and SCHÖN [1970], BURKARD [1975a], SRINIVASAN and THOMPSON [1976], DERIGS and ZIMMERMANN, U. ([1978b], [1979]), FINKE and SMITH [1979], and DERIGS [1979d]. FORTRAN programs of the augmenting path method are contained in DERIGS and ZIMMERMANN, U. ([1978b], [1979]). Computational efficiency has shown to be highly dependent on the choice of initial heuristics; by now, augmenting path methods and primal methods have been developed to nearly the same computational efficiency. Both seem to be faster than versions of the Hungarian method (also called primal dual or threshold method).

13. Algebraic Independent Set Problems

In this chapter we discuss the solution of linear algebraic optimization problems where the set of feasible solutions corresponds to special independence systems (cf. chapter 1). In particular, we consider matroid problems and 2-matroid intersection problems; similar results for matching problems can be obtained. We assume that the reader is familiar with solution methods for the corresponding classical problems (cf. LAWLER [1976]).

Due to the combinatorial structure of these problems it suffices to consider discrete semimodules ($R = \mathbb{Z}$); in fact, the appearing primal variables are elements of $\{0,1\}$.

We begin with a discussion of linear algebraic optimization problems in rather a general combinatorial structure. In chapter 1 we introduce independence systems F which are subsets of the set P(N) of all subsets of $N := \{1,2,\ldots,n\}$. W.l.o.g. we assume that F is *normal*, i.e. $\{i\} \in F$ for all $i \in N$. By definition (1.23) F contains all subsets J of each of its elements $I \in F$. In the following we identify an element I of F with its *incidence vector* $x \in \{0,1\}^n$ defined by

$$(13.1) \qquad x_j = 1 \iff j \in I$$

for all $j \in N$. Vice versa $I = I(x)$ is the *support* of x defined by $I = \{j \in N \mid x_j = 1\}$.

A linear description of an independence system F can be derived in the following way. Let H denote the set of all flats (or closed sets) with respect to F and let $r: P(N) \to \mathbb{Z}_+$ denote the rank function of F. A denotes the matrix the rows of which are the incidence vectors of the closed sets $C \in H$; b denotes a vector with components $b_C := r(C)$, $C \in H$. Then the set P of all incidence vectors of independent sets is

(13.2) $\qquad P = \{x \in \mathbb{Z}_+^n \mid Ax \le b\}.$

We remark that such a simple description contains many redundant constraints; for irredundant linear descriptions of particular independence systems we refer to GRÖTSCHEL [1977].

Let H be a linearly ordered, commutative monoid and let $a \in H^n$. We state the following linear algebraic optimization problems:

(13.3) $\qquad \min\{x^T \square a \mid x \in P\}$,

(13.4) $\qquad \max\{x^T \square a \mid x \in P\}$,

(13.5) $\qquad \min\{x^T \square a \mid x \in P_k\}$,

(13.6) $\qquad \max\{x^T \square a \mid x \in P_k\}$,

with $P_k := \{x \in P \mid \Sigma x_j = k\}$ for some $k \in \mathbb{N}$. (13.3) and (13.5) can trivially be transformed into a problem of the form (13.4) and (13.6) in the dual ordered monoid. Since P and P_k are finite sets all these problems have optimal solutions for which the optimal value is attained. Theoretically a solution can be determined by enumeration; in fact, we are only interested in more promising solution methods.

An element in P of particular interest is the *lexicographically maximum vector* $\bar{x}(P)$. A related partial ordering $\leqq$ on $\mathbb{Z}_+^n$ is defined by

(13.7) $$x \leqq y \iff \forall\, k \in N: \sum_{j=1}^{k} x_j \le \sum_{j=1}^{k} y_j$$

for $x, y \in \mathbb{Z}_+^n$.

(13.8) <u>Proposition</u>

If $x \in P$ is a maximum of P with respect to the partial ordering (13.7) then $x = \bar{x}(P)$.

<u>Proof</u>: It suffices to show that $x \leqq y$ implies $x \preccurlyeq y$ for all $x, y \in P$. Let $x \succ y$. Then $x_i = y_i$ for all $1 \le i < k$ but $x_k > y_k$ for some $k \in N$. Therefore

$$\sum_{j=1}^{k} x_j > \sum_{j=1}^{k} y_j$$

which shows that $x \not\leqq y$.

■

We introduce two vectors x^+ and x^k derived from $x \in \mathbb{Z}_+^n$ by

(13.9) $$x_j^+ := \begin{cases} x_j & \text{if } a_j \ge e \,, \\ 0 & \text{otherwise} \,, \end{cases}$$

(13.10) $$x_j^k := \begin{cases} x_j & \text{if } j \le j(k), \\ 0 & \text{otherwise} \end{cases}$$

for all $j \in N$ and for all $k \in \mathbb{Z}_+$ with $j(k) := \max\{j \in N \mid \sum_{i=1}^{j} x_j \le k\}$. The relationship of these vectors and the partial ordering (13.7) is described in the following proposition.

(13.11) Proposition

Let H be a linearly ordered, commutative monoid. Hence H is a linearly ordered semimodule over $\mathbb{Z}_+$. Let $x, y \in \mathbb{Z}_+^n$ with $\mu = \Sigma x_j^+$ and $\nu = \Sigma y_j$. Then

(1) $\quad e \leq (x^+)^T \square a$.

If $a_1 \geq a_2 \geq \ldots \geq a_n$ then

(2) $\quad y \leqq x \;\Rightarrow\; y^T \square a \leq (x^\nu)^T \square a$,

(3) $\quad (x^r)^T \square a \leq (x^s)^T \square a \qquad \forall\, r,s \in \mathbb{Z}_+ : r \leq s \leq \mu \vee \mu \leq s \leq r$,

(4) $\quad y \leqq x \;\Rightarrow\; y^T \square a \leq (x^+)^T \square a$.

Proof: (1) is an immediate implication of (13.9).

(2) $y \leqq x$ implies $\Sigma x_j^\nu = \nu$. Assume $y \neq x' := x^\nu$ and let $k(x') := \min\{j \mid y_j \neq x_j'\}$. Then $\delta := x_k' - y_k > 0$. Define x'' by

$$x'' := \begin{cases} x_j' - \delta & \text{if } j = k \;, \\ x_j' + \delta & \text{if } j = k+1, \\ x_j' & \text{otherwise} \;. \end{cases}$$

Then $y \leqq x''$ and due to $\delta \square a_{k+1} \leq \delta \square a_k$ we get $(x'')^T \square a \leq (x')^T \square a$. If $y = x''$ then (2) is proved. Otherwise we repeat the procedure for x''. Since $k(x'') > k(x')$ after a finite number of steps we find $y = x''$.

(3) If $r \leq s \leq \mu$ then $x_j^r = x_j^s = 0$ for all j with $a_j < e$. Therefore $x_j^r \leq x_j^s$ for all $j \in N$ implies the claimed inequality. If $\mu \leq s \leq r$ then $x_j^r = x_j^s$ for all j with $a_j \geq e$. Further $x_j^r \geq x_j^s$ for all $j \in N$ implies $x_j^r \square a_j \leq x_j^s \square a_j$ for all j with $a_j < e$. This proves the claimed inequality.

(4) From (2) we get $y^T \square a \leq (x^\nu)^T \square a$. From (3) we know

$(x^k)^T \,\Box\, a \leq (x^+)^T \,\Box\, a$ for all $k \in \mathbb{Z}_+$. For $k = \nu$ we find the claimed inequality.

■

The importance of proposition (13.11) for the solution of the linear algebraic optimization problems (13.4) and (13.6) can be seen from the following theorem which is directly implied by (13.11) and (13.8).

(13.12) Theorem

Let H be a linearly ordered, commutative monoid. Hence H is a linearly ordered semimodule over $\mathbb{Z}_+$. Let $\bar{x}$ denote the lexicographically maximum solution in P and let $k \in N$ with $1 \leq k \leq$ $\leq \max\{\Sigma x_j \mid x \in P\}$. If

(1) $a_1 \geq a_2 \geq \dots \geq a_n$,

(2) P has a maximum with respect to the partial ordering (13.7)

then

(3) $(\bar{x}^+)^T \,\Box\, a = \max\{x^T \,\Box\, a \mid x \in P\}$,

(4) $(\bar{x}^k)^T \,\Box\, a = \max\{x^T \,\Box\, a \mid x \in P_k\}$.

Theorem (13.12) gives sufficient conditions which guarantee that a solution of (13.4) and (13.6) can easily be derived from the lexicographically maximum solution $\bar{x}$ in P. It is well-known (cf. LAWLER [1976]) that the following algorithm determines the lexicographic maximum $\bar{x}(P)$; e_j, $j \in N$ denotes the j-th unit vector of $\mathbb{Z}^n$.

(13.13) Greedy algorithm

Step 1 $x \equiv 0$; $j := 1$.

Step 2 If $x + e_j \in P$ then $x := x + e_j$.

Step 3 If $j = n$ then stop;

$j := j + 1$ and go to step 2.

If we assume that an efficient procedure is known for checking $x + e_j \in P$ in step 2 then the greedy algorithm is an efficient procedure for the determination of $\bar{x}(P)$. It is easy to modify the greedy algorithm such that the final x is equal to $\bar{x}^{+}$ or $\bar{x}^{k}$.

Condition (13.12.1) can obviously be achieved by rearranging the components of vectors in $\mathbb{Z}^n$. Let Π denote the set of all permutations $\pi: N \to N$. For $\pi \in \Pi$ we define a mapping $\bar{\pi}: \mathbb{Z}^n \to \mathbb{Z}^n$ by

$$\bar{\pi}(x_1, x_2, \ldots, x_n) := (x_{\pi^{-1}(1)}, x_{\pi^{-1}(2)}, \ldots, x_{\pi^{-1}(n)})$$

which permutes the components of $x \in \mathbb{Z}$ accordingly (π^{-1} is the inverse permutation of π). If π is the necessary rearrangement to achieve (13.12.1) then $\bar{\pi}[P] := \{\bar{\pi}(x) \mid x \in P\}$ is the corresponding new independence system. We remark that $\bar{\pi}[P]$ does not necessarily have a maximum with respect to the partial ordering (13.7) if P satisfies (13.12.2). If $\bar{\pi}[P]$ satisfies (13.12.2) for all $\pi \in \Pi$ then the lexicographically maximum solution $\bar{x}(\bar{\pi}[P])$ leads to an optimal solution if

$$a_{\pi^{-1}(1)} \geq a_{\pi^{-1}(2)} \geq \ldots \geq a_{\pi^{-1}(n)} .$$

Such independent systems are matroids as the following theorem shows which can be found in ZIMMERMANN, U. [1977]. Mainly the same result is given in GALE [1968]; the difference lies in the fact that GALE explicitly uses assigned real weights.

(13.14) Theorem

Let T(P) denote the independence system corresponding to P. Then the following statements are equivalent:

(1) T(P) is a matroid,

(2) for all $\pi \in \Pi$ there exists a maximum of $\bar{\pi}[P]$ with respect to the partial order (13.7).

Proof: (1) ⇒ (2). If T(P) is a matroid then $T(\bar{\pi}[P])$ is a matroid since the definition of a matroid does not depend on a permutation of the coordinates in P. Thus it suffices to show that P contains a maximum with respect to the partial ordering (13.7). Let $\bar{x} = \bar{x}(P)$ be the lexicographical maximum of P and let $y \in P$. Suppose $y \not\leq \bar{x}$. Then there exists a smallest $k \in N$ such that

$$\sum_{j=1}^{k} y_j > \sum_{j=1}^{k} \bar{x}_j \; .$$

Let $J := \{j \mid y_j = 1,\; j \leq k\}$; $I := \{j \mid \bar{x}_j = 1,\; j < k\}$. Then $|J| > |I|$ and due to (1.24) there exists an independent set $I \cup \{\mu\}$ for some $\mu \in J \setminus I$. The incidence vector x' of I satisfies $x' \succ \bar{x}$ contrary to the choice of $\bar{x}$.

(2) ⇒ (1). Assume that T(P) is not a matroid. Then there exist two independent sets I,J with $|I| < |J|$ such that for all $j \in J \setminus I$ the set $I \cup \{j\}$ is dependent. We choose $\pi \in \Pi$ such that

$$\pi[I] < \pi[J \setminus I] < \pi[N \setminus (I \cup J)] \; .$$

Let u denote the incidence vector of J. If $\bar{\pi}(P)$ contains a maximum $\bar{\pi}(y)$ with respect to the partial ordering (13.7) then proposition (13.8) shows that the lexicographical maximum $\bar{x} = \bar{x}(\bar{\pi}(P))$ satisfies $\bar{x} = \bar{\pi}(y)$. Then $y_i = 1$ for all $i \in I$ and $y_i = 0$ for all $i \in J \setminus I$. Therefore $\bar{\pi}(u) \nleq \bar{\pi}(y)$ which implies that $\bar{\pi}[P]$ does not contain a maximum with respect to the particular ordering (13.7).

■

Due to the importance of the lexicographical maximum $\bar{x}(P)$ we describe a second procedure for its determination.
Let $\lesssim$ be a partial ordering on $\mathbb{Z}^n$. Then we define $\lesssim'$ on $\mathbb{Z}^n$ by

(13.15) $\qquad x \lesssim' y \iff \bar{\sigma}(y) \lesssim \bar{\sigma}(x)$

for $\sigma \in \Pi$ with $\sigma(i) = n - i + 1$ for all $i \in N$. In particular we may consider the partial orderings $\leqq'$ and $\preccurlyeq'$. We remark that $x \leqq' y$ implies $x \preccurlyeq' y$.

(13.16) <u>Modified greedy algorithm</u>

Step 1 $\quad x \equiv 0$; $j := n$.

Step 2 $\quad$ If $x + e_j \in P$ then $x := x + e_j$.

Step 3 $\quad$ If $j = 1$ then stop;

$j := j - 1$ and go to step 2.

The final vector x in this algorithm is denoted by $x'(P)$. Since the application of this algorithm to P is equivalent to the application of the greedy algorithm to $\bar{\sigma}[P]$ we get $x'(P) = \bar{x}(\bar{\sigma}[P])$. (13.15) shows that x' is the minimum of P with respect to $\preccurlyeq'$.

Let $r := \max\{\Sigma x_j \mid x \in P\}$ and define

$$P^* := \{y \in \{0,1\}^n \mid \exists\, x \in P_r : y \leq 1 - x\}.$$

Then P^* corresponds to a certain independence system $T(P^*)$. In particular, if $T(P)$ is a matroid then $T(P^*)$ is the dual matroid (cf. chapter 1). Therefore we call P^* the *dual of P*. Let $s := \max\{\Sigma y_i \mid y \in P^*\}$. Then $r + s = n$.

(13.17) Proposition

If P has a maximum with respect to the partial ordering (13.7) then $\bar{x}(P) = 1 - x'(P^*)$.

Proof: Since $\bar{x}(P)$ is the maximum of P with respect to the partial ordering (13.7), $\bar{x} \in P_r$. Thus $1 - \bar{x} \in P_s^*$. Let $x \in P_r$. Then $x \leqq \bar{x}$ means by definition

$$\sum_{j=1}^{k} x_j \leq \sum_{j=1}^{k} \bar{x}_j$$

for all $k \in N$. Further $\Sigma x_j = \Sigma \bar{x}_j = r$. Thus

$$\sum_{j=k+1}^{n} x_j \geq \sum_{j=k+1}^{n} \bar{x}_j$$

for all $k = 0,1,\ldots,n-1$ which shows $\bar{\sigma}(\bar{x}) \leqq \bar{\sigma}(x)$. Therefore $\bar{x}$ is the maximum of P_r with respect to $\leqq'$. Hence $\bar{x}$ is the maximum of P_r with respect to $\preccurlyeq'$. Thus

$$1 - \bar{x} \preccurlyeq' 1 - x$$

for all $x \in P_r$ and $1 - \bar{x}$ is the minimum of P_s^* with respect to $\preccurlyeq'$. Let $y \in P^* \setminus P_s$. Then there exists $\tilde{y} \in P_s^*$ such that $y \leq \tilde{y}$ which implies $\tilde{y} \preccurlyeq' y$. Therefore $1 - \bar{x}$ is the minimum of P^* with respect to $\preccurlyeq'$. Hence $1 - \bar{x} = x'(P^*)$.

■

Proposition (13.17) shows that the lexicographical maximum $\bar{x}(P)$ can be derived as the complement of $x'(P^*)$ which is determined by (13.16) applied to P^*. The application of the modified greedy algorithm to P^* is called the *dual greedy algorithm*. This is quite a different method for the solution of the optimization problems (13.4) and (13.6). In particular, checking $x + e_j \in P$ in step 2 of the greedy algorithm is replaced by checking $x + e_j \in P^*$ in the dual greedy algorithm. Therefore the choice of the applied method will depend on the computational complexity of the respective checking procedures.

For a more detailed discussion on properties of related partial orders we refer to ZIMMERMANN, U. ([1976], [1977]). The greedy algorithm has been treated before by many authors. KRUSKAL [1956] applied it to the shortest spanning tree problem, RADO [1957] realized that an extension to matroids is possible. WELSH [1968] and EDMONDS [1971] discussed it in the context of matroids. The name 'greedy algorithm' is due to EDMONDS. Related problems are discussed in EULER [1977]; a combinatorial generalization is considered in FAIGLE [1979]. For further literature, in particular for the analysis of computational complexity and for related heuristics we refer to the bibliographies of KASTNING [1976] and HAUSMANN [1978]. For examples of matroids we refer to LAWLER [1976] and WELSH [1976].

The following allocation problem is solved in ALLINEY, BARNABEI and PEZZOLI [1980] for nonnegative weights in the group $(\mathbb{R}, +, \leq)$. Its solution can be determined using a special (simpler) form of the greedy algorithm (13.13).

Example (Allocation Problem)

A system with m units (denoted as set T) is requested to serve n users (denoted as set S). It is assumed that each user can engage at most one unit. Each user i requires a given amount a_i of resources and each unit j can satisfy at most a given required amount b_j of resources. Thus user i can be assigned to unit j iff $a_i \leq b_j$. Let $G = (S \cup T, L)$ denote the bipartite graph with vertex sets S and T and edge set L where $(i,j) \in L$ iff $a_i \leq b_j$. A matching I is a subset of L such that no two different edges in I have a common endpoint. A set $A \subseteq S$ is called assignable if there exists a matching I such that $A = \{i \mid (i,j) \in I\}$. The set of all assignable sets is the independence system of a matroid (cf. EDMONDS and FULKERSON [1965] or WELSH [1976]). In particular, all maximal assignable sets have the same cardinality. We attach a weight $c_i \in H_+$ to each user i where $(H,*,\leq)$ is a linearly ordered commutative monoid. (13.12) and (13.14) show that we can find an optimal solution of the problem $\max\{x^T \square c \mid x \in P\}$ by means of the greedy algorithm (13.13) (P denotes the set of all incidence vectors of assignable sets). Due to the special structure of the matroid it is possible to simplify the independence test $(x + e_j \in P)$ in step 2 of the greedy algorithm. In fact, we will give a separate proof of the validity of the greedy algorithm for the independence system $T(P)$. Together with theorem (13.14) we find that $T(P)$ is a matroid.

Let $c_i = \max\{c_k \mid k \in S\}$. If $T_i := \{j \in T \mid a_i \leq b_j\} = \emptyset$ then there exists no unit which can be assigned to user i. Then G' denotes the subgraph generated by $(S \setminus \{i\}) \cup T$. If we denote the optimal value of an assignable set of G by z(G) then

(1) $z(G') = z(G)$.

If $T_i \neq \emptyset$ then let $b_j = \min\{b_k \mid k \in T_i\}$. Let I denote a matching corresponding to an optimal assignable set $\{\nu \mid (\mu,\nu) \in I\}$. Every such matching is called an "optimal matching". We claim that we can assume w.l.o.g. $(i,j) \in I$. Let $V := \{\nu,\mu \mid (\mu,\nu) \in I\}$. If $i,j \notin V$ then $I \cup \{(i,j)\}$ is an optimal matching, too. If $i \in V$, $j \notin V$ then we replace $(i,k) \in I$ by (i,j). If $i \notin V$, $j \in V$ then we replace $(k,j) \in I$ by (i,j); since c_i is of maximum value the new matching is optimal, too. If $i,j \in V$ but $(i,j) \notin I$ then we replace $(i,\nu),(\mu,j) \in I$ by (i,j) and (μ,ν). We remark that $(\mu,\nu) \in L$ since $a_\mu \leq b_j \leq b_\nu$. Thus we can assume $(i,j) \in I$. Let G' denote the subgraph generated by $(S \setminus \{i\}) \cup (T \setminus \{j\})$. Then $I \setminus \{(i,j)\}$ is a matching in G'. Therefore $z(G) \leq z(G') * c_i$. On the other hand, if I' is an optimal matching in G', then $I' \cup \{(i,j)\}$ is a matching in G and thus $z(G') * c_i \leq z(G)$. Therefore in the case $T_i \neq \emptyset$ we find

(2) $z(G') * c_i = z(G)$.

(1) and (2) show the validity of the following variant of the greedy algorithm. We assume $c_1 \geq c_2 \geq \ldots \geq c_n$.

Step 1 $I = \emptyset$; $i := 1$; $T := \{1,2,\ldots,m\}$.

Step 2 If $T_i := \{b_k \mid a_i \leq b_k, k \in T\} = \emptyset$ then go to step 3;
find $j \in T_i$ such that $b_j = \min T_i$;
$I := I \cup \{(i,j)\}$;
$T := T \setminus \{j\}$.

Step 3 If $i = n$ then stop;

$i := i + 1$ and go to step 2.

At termination I is an optimal matching which describes the assignment of the users i of an optimal assignable set $A^* = \{i \mid (i,j) \in I\}$ to units j. It should be clear (cf. theorem (13.12)) that $\{i \mid (i,j) \in I\}$ is optimal among all assignable sets of the same cardinality at any stage of the algorithm and for arbitrary $c_i \in H$ $(i \in S)$ satisfying $c_1 \geq c_2 \geq \ldots$ $\ldots \geq c_n$. In particular, we may consider $(\mathbb{R}, +, \leq)$ and $(\mathbb{R} \cup \{\infty\}, \min, \leq)$. Then for $c_i \in \mathbb{R}$, $i \in S$ we find

$$(3) \qquad \Sigma_{i \in A^*}\, c_i \geq \Sigma_{i \in A}\, c_i \,,$$

$$\min\{c_i \mid i \in A^*\} \geq \min\{c_i \mid i \in A\}$$

for all assignable sets A with $|A| = |A^*|$; i.e. the sum as well as the minimum of all weights is maximized simultaneously.

In the remaining part of this chapter we discuss linear algebraic optimization problems for combinatorial structures which do not contain a maximum with respect to the partial ordering (13.7), in general. We assume that H is a weakly cancellative d-monoid. Hence w.l.o.g. H is an extended semimodule over $\mathbb{Z}_+$ (cf. 5.16 and chapter 6). From the discussion in chapter 11 we know that max- and min-problems are not equivalent in d-monoids due to the asymmetry in the divisor rule (4.1). In chapter 11 we developed reduction methods for both types of algebraic optimization problems which reduce the original problem to certain equivalent problems in irreducible sub-d-monoids of H; since H is weakly cancellative these sub-d-monoids are parts of groups.

Let $(H_\lambda;\ \lambda \in \Lambda)$ denote the ordinal decomposition of H. Then

$$\lambda(x^T \square a) = \max(\{\lambda(a_j) \mid x_j = 1\} \cup \{\lambda_o\})$$

for an incidence vector x of an independent set ($\lambda_o := \min \Lambda$). According to (11.21) and (11.54) the optimization problems (13.3)-- (13.6) are reduced to optimization problems in H_μ with

(13.18) $$\mu_1 = \min\{\lambda(x^T \square a) \mid x \in P\} ,$$

(13.19) $$\mu_2 = \max\{\lambda(x^T \square a) \mid x \in P\} ,$$

(13.20) $$\mu_3 = \min\{\lambda(x^T \square a) \mid x \in P_k\},$$

(13.21) $$\mu_4 = \max\{\lambda(x^T \square a) \mid x \in P_k\}.$$

Since $o \in P$ the reduction (13.18) of (13.3) is trivial: $\mu_1 = \lambda_o$. The reduction (13.19) of (13.4) is also simple; since the corresponding independence system is normal the unit vectors e_j , $j \in N$ are elements of P. Therefore

$$\mu_2 = \max\{\lambda(a_j) \mid j \in N\}.$$

For the determination of μ_3 we assume that it is possible to compute the value of the rank function $r\colon P(N) \to \mathbb{Z}_+$ for a given subset of N by means of an efficient (polynomial time) procedure. This is a reasonable assumption since we will consider only such independence systems for which the classical combinatorial optimization problems can efficiently be solved; then an efficient procedure for the determination of an independent set of maximum cardinality in a given subset of N is known, too. In chapter 1 the closure operator σ: $P(N) \rightarrow P(N)$ is defined by

$$\sigma(I) = \{i \in N \mid r(I \cup \{i\}) = r(I)\}.$$

A subset I of N is called *closed* if $\sigma(I) = I$. Clearly, if the rank function can efficiently be computed then we may efficiently construct a closed set J containing a given set I. J is called a *minimal closed cover of I*. Every set $I \subseteq N$ has at least one minimal closed cover J, but in general J is not unique. Let $s := k - r(J) \geq 0$ (cf. 13.5 and 13.6). Then we define the *best remainder set* ΔJ by

$$(13.22) \qquad \Delta J := \begin{cases} \emptyset & \text{if } s = 0, \\ \{i_1, i_2, \ldots, i_s\} & \text{otherwise,} \end{cases}$$

where $N \setminus J = \{i_1, i_2, \ldots, i_t\}$ and $a_{i_1} \leq a_{i_2} \leq \ldots \leq a_{i_t}$. The following method for the determination of μ_3 is a refinement of the threshold method in EDMONDS and FULKERSON [1970].

(13.23) Reduction method for (13.5)

Step 1 $\mu := \max\{\lambda(a_j) \mid j \in \Delta\emptyset\}$; $K := \{j \in N \mid \lambda(a_j) \leq \mu\}$.

Step 2 If $r(K) \geq k$ then stop;

determine a minimal closed cover J of K.

Step 3 $\mu := \max(\{\lambda(a_j) \mid j \in \Delta J\} \cup \{\mu\}$;

$K := \{j \in N \mid \lambda(a_j) \leq \mu\}$;

go to step 2.

In the performance of (13.23) we assume $k \leq r(N)$. Then the method terminates in $O(n)$ steps. A similar method for the derivation of thresholds for Boolean optimization problems in linearly ordered commutative monoids is proposed in ZIMMERMANN, U. [1978c].

(13.24) Proposition

The final parameter μ in (13.23) is equal to μ_3 in (13.20).

Proof: Denote the value of μ before the last revision in step 3 by $\bar{\mu}$. The corresponding sets are denoted by $\bar{K}$, $\bar{J}$ and $\Delta\bar{J}$. Then $r(\bar{K}) < k$. Since $P_k \neq \emptyset$ we conclude $\mu_3 > \bar{\mu}$. Further $r(\bar{J}) = r(\bar{K})$ shows that an independent set with k elements has at least $s = k - r(\bar{K})$ elements in $N \setminus \bar{J}$. Thus

$$\mu_3 \geq \max\{\lambda(a_j) \mid j \in \Delta\bar{J}\}$$

which implies $\mu_3 \geq \mu$. Now $r(K) \geq k$ shows $\mu_3 \leq \mu$.

■

We remark that for bottleneck objectives (cf. 11.58) the reduction method (13.21) leads to the optimal value of (13.5) since $\lambda(x^T \square a) = x^T \square a$ in this case.

In the determination of μ_4 we consider certain systems F^j, $j \in N$ derived from the underlying independence system F. We define

(13.25) $$F^j := \{I \in F \mid I \cup \{j\} \in F,\ j \notin I\} \qquad \text{for } j \in N.$$

Obviously F^j is an independence system for each $j \in N$. Its rank function is denoted by r^j. W.l.o.g. we assume

$$a_1 \geq a_2 \geq \dots \geq a_n$$

in the following method.

(13.26) Reduction method for (13.6)

Step 1 $\nu := 1$.

Step 2 If $r^\nu(N) \geq k-1$ then stop ($\mu = \lambda(a_\nu)$).

Step 3 $\nu := \nu + 1$;

go to step 2.

Finiteness of this method is obvious.

(13.27) Proposition

The final parameter ν in (13.26) satisfies $\mu_4 = \lambda(a_\nu)$ for μ_4 in (13.21).

Proof: Let ν denote the final parameter in (13.26). Then there exists an independent set $I \in F$ of cardinality k such that $j \in I$. Therefore $\lambda(a_\nu) \leq \mu_4$. If $\nu = 1$ then equality holds. Otherwise $r^{\nu-1}(N) < k-1$ shows that there exists no $I \in F$ of cardinality k such that $j \in I$. Thus $\mu_4 \leq \lambda(a_\nu)$.

■

Again, we remark that for bottleneck objectives the reduction method (13.27) leads to an optimal value of (13.6).

After determination of the corresponding index μ the algebraic optimization problems (13.3) - (13.6) are reduced according to

(11.23) and (11.55). The sets of feasible solutions P and P_k are replaced by

$$P_M := \{x_M \mid x \in P \text{ and } x_j = 0 \text{ for all } j \in M\},$$

$$(P_k)_M := \{x_M \mid x \in P_k \text{ and } x_j = 0 \text{ for all } j \in M\}$$

with $M = M(\mu) = \{j \in N \mid \lambda(a_j) \leq \mu\}$. If F denotes the corresponding independence system then (13.28) shows that the reduced sets of feasible solutions correspond to the restriction F_M of F to M, i.e.

(13.28) $\qquad F_M := \{I \in F \mid I \subseteq M\}$

which is again a normal independence system. Theorems (11.25) and (11.56) show that (13.3) - (13.6) are equivalent to the following *reduced linear algebraic optimization problems*

(13.29) $\qquad \min\{x^T \square a(\mu)_M \mid x \in P_M\}$,

(13.30) $\qquad \max\{x^T \square a(\mu)_M \mid x \in P_M\}$,

(13.31) $\qquad \min\{x^T \square a(\mu)_M \mid x \in (P_k)_M\}$,

(13.32) $\qquad \max\{x^T \square a(\mu)_M \mid x \in (P_k)_M\}$

with respect to $M = M(\mu)$ and $\mu = \mu_i$, $i = 1,2,3,4$ defined by (13.18) - (13.21). Since H is weakly cancellative (13.29) - (13.32) are problems in the extended module G_μ over $\mathbb{Z}$ for $\mu = \mu_i$, $i = 1,2,3,4$.

Next we consider two particular independence systems. Let F denote the intersection $F_1 \cap F_2$ of two matroids F_1 and F_2 where F_1 and F_2 are the corresponding independence systems. An element $I \in F$ is called an *intersection*. Then (13.3) - (13.6) are called *algebraic matroid intersection problems*. In particular, (13.5) and (13.6) are called *algebraic matroid k-intersection problems*. Let (V,N) denote a graph. Then $I \subseteq N$ is called a *matching* if no two different edges in I have a common endpoint. We remark that in a graph (i,j) and (j,i) denote the same edge. The set of all matchings is an independence system F. Then (13.3) - (13.6) are called *algebraic matching problems*. In particular, (13.5) and (13.6) are called *algebraic k-matching problems*.

The classical matroid intersection problem is well-solved; efficient solution methods are developed in LAWLER ([1973], [1976]), IRI and TOMIZAWA [1976], FUJISHIGE [1977] and EDMONDS [1979]. An augmenting path method and the primal dual method are described in the textbook of LAWLER [1976]; FUJISHIGE [1977] considers a primal method. The classical matching problem is well-solved, too; an efficient primal dual method for its solution is given in EDMONDS [1965] and is described in the textbook of LAWLER [1976]. A primal method is considered in CUNNINGHAM and MARSH [1976]. The necessary modification of the primal dual method for the classical versions of (13.5) and (13.6) is discussed in WHITE [1967]. For the classical matroid intersection problem such a modification is not necessary since in the shortest augmenting path method optimal intersections of cardinality 0,1,2,... are generated subsequently (cf. LAWLER [1976]).

In particular, efficient procedures for the determination of a matching (matroid intersection) of maximum cardinality in a given subset $N' \subseteq N$ are described in LAWLER [1976]. Thus the evaluation of the rank function in step 2 of the reduction method (13.23) is possible in polynomial time for both problems.

Let F be the set of all matchings in the graph (V,N). Then F^j (cf. 13.25) is the set of all matchings in the graph (V,N^j) where N^j is the set of all edges which have no endpoint in common with j ($j \in N$). Let $F = F_1 \cap F_2$ be the set of all intersections. Then $F^j = F_1^j \cap F_2^j$. Further F_1^j (and F_2^j) is the set of all independent sets of a certain matroid (called contraction of the original matroid to $N \setminus \{j\}$, cf. WELSH [1976]). Thus the evaluation of the rank function r^ν, $\nu \in N$ in step 2 of the reduction method (13.26) is possible in polynomial time for both problems.

The restriction F_M (cf. 13.28) leads to the set of all matchings in the graph (V,M) and to the intersection of the two restricted matroids $(F_1)_M \cap (F_2)_M$. Therefore the reduced problems (13.29) - (13.32) are algebraic matching (matroid intersection) problems provided that (13.3) - (13.6) are algebraic matching (matroid intersection) problems.

It remains to solve the reduced problems in the respective modules. All methods for the solution of the classical matroid intersection problem are valid and finite in modules. A reformulation of these methods in groups consists only in replacing the usual addition of real numbers by the internal composition in the underlying group and in replacing the usual linear ordering of the real numbers by the linear ordering in the underlying group. Optimality of the augmenting path method is proved by KROGDAHL (cf. LAWLER [1976]) using only combinatorial arguments and the group structure (i.e. mainly cancellation arguments) of the real additive group. Thus his proof remains valid in modules. Optimality of the primal method in FUJISHIGE [1977] is based on similar arguments which remain valid in modules, too. Optimality of the primal dual method is based on the classical duality principles. From chapter 11 we know how to apply similar arguments in modules. In the following we develop these arguments explicitly; for a detailed description of the primal dual method we refer to LAWLER [1976].

Let $Ax \leq b$ and $\tilde{A}x \leq \tilde{b}$ be the constraint systems (cf. 13.2) with respect to the restricted matroids. Then

$$(13.33) \qquad P_M = \{x \in \mathbb{Z}_+^M \mid Ax \leq b,\ \tilde{A}x \leq \tilde{b}\}\ .$$

In modules it suffices to consider max-problems since a min-problem can be transformed into a max-problem replacing the cost coefficients by inverse cost coefficients. We assign dual variables u_i and v_k to the closed sets i and k of the restricted matroids. Then (u,v) is dual feasible if

$$(13.34) \qquad A^T \square u * \tilde{A}^T \square v \geq \tilde{a},\ u_i \geq e,\ v_k \geq e$$

where $\tilde{a} = a(\mu)_M$ and where μ denotes the respective index of the reduction. The set of all dual feasible (u,v) is denoted by D_M. Then the algebraic dual of (13.30) is

$$(13.35) \qquad \min\{b^T \square u * \tilde{b}^T \square v \mid (u,v) \in D_M\}.$$

In the usual manner we find weak duality, i.e.

$$x^T \square \tilde{a} \leq b^T \square u * \tilde{b}^T \square v.$$

The corresponding complementarity conditions are

(13.36) $\quad x_j > 0 \Rightarrow (A^T \square u * \tilde{A}^T \square v)_j = \tilde{a}_j$

(13.37) $\quad u_i > e \Rightarrow (Ax)_i = b_i$

(13.38) $\quad v_k > e \Rightarrow (\tilde{A}x)_k = \tilde{b}_k$

for all $j \in M$ and all closed sets i and k. If $\tilde{a}_j \leq e$ for all $j \in M$ then $x \equiv 0$, $u \equiv e$, $v \equiv e$ is an optimal pair of primal and dual feasible solutions. Otherwise the initial solution $x \equiv 0$, $v \equiv e$, $u_i = e$ for all $i \neq M$ and

$$u_M := \max\{\tilde{a}_j \mid j \in M\}$$

is primal and dual feasible. Further all complementarity conditions with the possible exception of

(13.39) $\quad u_M > e \Rightarrow \sum_{j \in M} x_j = b_M$

are satisfied. We remark that b_M is the maximum cardinality of an independent set in one of the restricted matroids. Such a pair (x;u,v) is called *compatible*, similarly to the primal dual method for network flows. Collecting all equations (13.36) after external composition with x_j we find

(13.40) $\quad x^T \square \tilde{a} * [(b_M - \sum_{j \in M} x_j) \square u_M] = b^T \square u * \tilde{b}^T \square v$

for compatible pairs. The primal dual method proceeds in stages. At each stage either the primal solution is augmented or the values of the dual variables are revised. At each stage no more than $2|M|$ dual variables are permitted to be non-zero. Throughout the performance of the method the current pair (x;u,v) is compatible. This is achieved by an alternate solution of

(13.41) $\max\{x^T \Box \tilde{a} \mid x \in P_M, (x;u,v) \text{ compatible}\}$

for fixed dual feasible solution (u,v) and of

(13.42) $\min\{b^T \Box u * \tilde{b}^T \Box v \mid (u,v) \in D_M, (x;u,v) \text{ compatible}\}$

for fixed primal feasible solution x. From (13.40) we conclude that (13.41) is equivalent to

(13.43) $\max\{\sum_{j \in M} x_j \mid x \in P_M, (x;u,v) \text{ compatible}\}$

and that (13.42) is equivalent to

(13.44) $\min\{u_M \mid (u,v) \in D_M, (x;u,v) \text{ compatible}\}.$

An alternate solution of these problems is constructed in the same way as described in LAWLER [1976] for the classical case. In particular, the method is of polynomial time in the number of usual operations, $*$-compositions and $\leq$-comparisons. The final compatible pair (x;u,v) is complementary. Thus it satisfies $x^T \Box \tilde{a} = b^T \Box u * \tilde{b}^T \Box v$; then weak duality shows that (x;u,v) is an optimal pair. Therefore this method provides a constructive proof of a duality theorem for algebraic matroid intersection problems.

Let $Cx \leq c$ and $\tilde{C}x \leq \tilde{c}$ be the constraint systems (cf. 13.2) with respect to the two matroids considered. Then

(13.45) $P = \{x \in \mathbb{Z}_+^n \mid Cx \leq c, \tilde{C}x \leq \tilde{c}\}.$

A closed set I of one of the matroids with $I \subseteq M$ is closed, with respect to the restricted matroid, too. We remark that $c_j = b_j$ ($\tilde{c}_j = \tilde{b}_j$) for all $j \in M$.

We assign dual variables u_i, v_k to the closed sets i and k of the matroids. Then (u,v) is dual feasible with respect to

(13.4') $\max\{x^T \square a(\mu_2) \mid x \in P\}$

if

(13.46) $c^T \square u * \tilde{c}^T \square v \geq a(\mu_2),\ u_i \geq e,\ v_k \geq e$.

The set of all dual feasible (u,v) is denoted by D_2. Then the algebraic dual of (13.4) is

$$\min\{c^T \square u * \tilde{c}^T \square v \mid (u,v) \in D_2\}.$$

Algebraic dual programs with respect to

(13.3') $\min\{x^T \square a(\mu_1) \mid x \in P\}$

are considered in chapter 11. (u,v) is strongly dual feasible with respect to (13.3') if

$$e \leq a(\mu_1) * c^T \square u * \tilde{c}^T \square v ,$$

$$\lambda(u_i) \leq \mu_1 ,\ \lambda(v_k) \leq \mu_1 ,\ e \leq u_i ,\ e \leq v_k .$$

An objective function is defined according to (11.19).

(13.47) Theorem (Matroid intersection duality theorem)

Let H be a weakly cancellative d-monoid. Hence H is an extended semimodule over $\mathbb{Z}_+$. There exist optimal feasible pairs (x;u,v) for the algebraic matroid intersection problems (13.3') and (13.4') which are complementary and satisfy

(1) $e = x^T \square a * c^T \square u * \tilde{c}^T \square v$ (for 13.3')

(2) $(-c)^T \square u(\mu_1) * (-\tilde{c})^T \square v(\mu_1) = x^T \square a$ (for 13.3'),

(3) $c^T \square u * \tilde{c}^T \square v = x^T \square a$ (for 13.4').

Proof: (1). The min-problem (13.29) is transformed into a max-problem of the form (13.30) replacing $a_j(\mu_1)$ by its inverse in the group G_{μ_1} for all $j \in M = M(\mu_1)$. Let $(\bar{x};\bar{u},\bar{v})$ denote the final pair generated in the application of the primal dual method to this max-problem. Then for $j \in M$:

$$a_j(\mu_1)^{-1} \leq (A^T \square \bar{u} * \tilde{A}^T \square \bar{v})_j$$

when A and $\tilde{A}$ are submatrices of C and $\tilde{C}$ (columns $j \in M$ and rows of closed sets $I \subsetneqq M$ and row $\sigma(M)$ with the closure function of the respective matroid). Let

$$x_j := \begin{cases} \bar{x}_j & \text{if } j \in M\,, \\ 0 & \text{otherwise} \end{cases}$$

then $x \in P$. Further let

$$u_i := \begin{cases} \bar{u}_i & \text{if } i \subsetneqq M\,, \\ \bar{u}_M & \text{if } i = \sigma_1(M)\,, \\ e & \text{otherwise} \end{cases}$$

where σ_1 is the closure function of the matroid F_1 and let

$$v_k := \begin{cases} \bar{v}_k & \text{if } k \subsetneqq M\,, \\ \bar{v}_M & \text{if } k = \sigma_2(M)\,, \\ e & \text{otherwise} \end{cases}$$

where σ_2 is the closure function of the matroid F_2. Then

$$e \leq a_j(\mu_1) * (C^T \square u * \tilde{C}^T \square v)_j$$

for all $j \in M$. Now $\lambda(a_j) > \mu_1$ for $j \notin M$ shows

$$e \leq a(\mu_1) * C^T \square u * \tilde{C}^T \square v\,.$$

Therefore (u,v) is strongly dual feasible, (x;u,v) is complementary and (1) and (2) are satisfied.

(3) Let (x;u,v) denote the final pair generated in the application of the primal dual method to (13.30). Since $N = M(\mu_2)$

we find

$$C^T \square u * \tilde{C}^T \square v \geq a(\mu_2) .$$

Therefore (u,v) is dual feasible, (x;u,v) is complementary and (3) is satisfied.

■

A valuable property of compatible pairs leads to the solution of algebraic matroid k-intersection problems. Again we consider a solution in the respective group. Then

$$(P_k)_M = \{x \in \mathbb{Z}^n_+ \mid Ax \leq b, \tilde{A}x \leq \tilde{b}, \sum_{j \in M} x_j = k\}.$$

We assign a further dual variable λ to $\Sigma x_j = k$. Then (u,v,λ) is dual feasible if

(13.48) $$A^T \square u * \tilde{A}^T \square v * [\lambda] \geq \tilde{a}, \quad u_i \geq e, \; v_k \geq e.$$

The dual variable λ is unrestricted in sign in the respective group. The set of all dual feasible (u,v) is denoted by $(D_k)_M$. Then the algebraic dual of (13.32) is

(13.49) $$\min\{b^T \square u * b^T \square v * (k \square \lambda) \mid (u,v) \in (D_k)_M\}.$$

Again we find weak duality

$$x^T \square \tilde{a} \leq b^T \square u * \tilde{b}^T \square v * (k \square \lambda)$$

for all primal and dual feasible pairs $(x;u,v,\lambda)$. We apply the primal dual method to (13.30), but now for $\mu := \mu_4$ and $M := M(\mu_4)$. A sequence of compatible pairs $(x^\nu;u^\nu,v^\nu)$ is generated with $\Sigma x_j^\nu = \nu$ for $\nu = 0,1,\ldots$.

We modify the dual revision procedure slightly in order to admit negative u_M (cf. LAWLER [1976], p. 347: let $\delta = \min\{\delta_u, \delta_v, \delta_w\}$). The stop-condition $u_M = 0$ is replaced by $\Sigma x_j = k$.

Then it may happen that u_M becomes negative during the performance of the primal dual method. All other compatibility conditions remain valid. Now replace u by u' defined by

$$u_i' := \begin{cases} u_i & i \neq M, \\ e & i = M \end{cases}$$

and let $\lambda := u_M$. Then $(x;u',v,\lambda)$ satisfies (13.48) and

$$x^T \square \tilde{a} = b^T \square u * \tilde{b}^T \square v * (k \square \lambda).$$

Hence $(x;u',v,\lambda)$ is optimal for (13.32) and (13.49) if x is primal feasible, i.e. if $\Sigma x_j = k$. Similarly to theorem (13.47) we find a strong duality theorem. From (13.45) we get

$$P_k = \{x \in \mathbb{Z}_+^n \mid Cx \leq c,\ \tilde{C}x \leq \tilde{c},\ \Sigma x_j = k\}.$$

We assign dual variables u_i and v_k to the closed sets of the matroids and two further dual variables λ_- to $\Sigma x_j \geq k$ and λ_+ to $\Sigma x_j \leq k$. In the case of a max-problem with $\tilde{M} = M \setminus N(\mu_4) \neq \emptyset$ we have to adjoin the inequality

$$\Sigma_{\tilde{M}}\, x_j \leq 0$$

with assigned dual variable γ. Then

$$P_k' := \{x \in P_k \mid \Sigma_{\tilde{M}}\, x_j \leq 0\}.$$

$(u,v,\lambda_+,\lambda_-,\gamma)$ is dual feasible with respect to

(13.6') $$\max\{x^T \square a(\mu_4) \mid x \in P_k'\}$$

if

(13.50) $$C^T \square u * C^T \square v * [^e_\gamma] * [\lambda_+] \geq a(\mu_4) * [\lambda_-]\ ,$$

$$u_i \geq e,\ v_k \geq e,\ \lambda_- \geq e,\ \lambda_+ \geq e,\ \gamma \geq e\ ,$$

where $[^e_\gamma]$ denotes the vector with j-th component γ for $j \in \tilde{M}$ and j-th component e otherwise. Without this additional dual variable γ which does not appear in the objective function it is necessary that at least one of the other dual variables has the index $\lambda(\max\{a_j \mid j \in \widetilde{M}\}) > \mu_4$. This leads to a duality gap which can be avoided by the introduction of $\gamma := \max\{a_j \mid j \in \widetilde{M}\}$. The set of all dual feasible solutions according to (13.50) is denoted by D_k'. Then a dual objective function $f: D_k' \to H$ can be defined with respect to (13.6') similarly as in (11.19). Let $\alpha := k \,\square\, \lambda_-$ and

$$\beta := c^T \,\square\, u * \tilde{c}^T \,\square\, v * (k \,\square\, \lambda_+) \;.$$

Then f is defined by

$$\alpha * f(u,v,\lambda_+,\lambda_-,\gamma) := \beta$$

if $\alpha \leq \beta$ or $\lambda(\alpha) = \lambda(\beta) = \lambda_o$. Otherwise let $f(u,v,\lambda_+,\lambda_-,\gamma) := \infty$ where ∞ denotes a possibly adjoint maximum of H. The dual of (13.6') is

$$\min\{f(u,v,\lambda_+,\lambda_-) \mid (u,v,\lambda_+,\lambda_-) \in D_k'\}.$$

Dual programs with respect to

(13.5') $$\min\{x^T \,\square\, a(\mu_3) \mid x \in P_k\}$$

are defined as in chapter 11. (u,v) is strongly dual feasible if

$$[\lambda_-] \leq a(\mu_3) * C^T \,\square\, u * \tilde{C}^T \,\square\, v * [\lambda_+]$$

$$\lambda(\lambda_-) \leq \mu_3 \;,\; \lambda(\lambda_+) \leq \mu_3,\; \lambda(u_i) \leq \mu_3,\; \lambda(v_k) \leq \mu_3,$$

$$e \leq \lambda_- \;,\; e \leq \lambda_+ \;,\; e \leq u_i \;,\; e \leq v_k \;.$$

An objective function is defined according to (11.19).

(13.51) Theorem (Matroid k-intersection duality theorem)

Let H be a weakly cancellative d-monoid. Hence H is an extended semimodule over $\mathbb{Z}_+$. There exist optimal feasible pairs $(x;u,v,\lambda_+,\lambda_-)$ and $(x;u,v,\lambda_+,\lambda_-,\gamma)$ for the algebraic matroid k-intersection problems (13.5') and (13.6') which are complementary and satisfy

(1) $k \square \lambda_- = x^T \square a * c^T \square u * \tilde{c}^T \square v * (k \square \lambda_+)$ (for 13.5')

(2) $k \square \delta(\mu_3) * (-c)^T \square u(\mu_3) * (-\tilde{c})^T \square v(\mu_3) = x^T \square a$ (for 13.5'),

(3) $x^T \square a * (k \square \lambda_-) = c^T \square u * \tilde{c}^T \square v * (k \square \lambda_+)$ (for 13.6'),

(4) $x^T \square a = c^T \square u * \tilde{c}^T \square v * k \square \varepsilon(\mu_4)$ (for 13.6')

with $\delta(\mu_3) := \lambda_- * (\lambda_+)^{-1}$ and $\varepsilon(\mu_4) := \lambda_+ * (\lambda_-)^{-1}$.

Proof: The proof of (1) and (2) is quite similar to the proof of (1) and (2) in theorem (13.47). The value of the variable in the final pair generated by the primal dual method is assigned to λ_+ or λ_- in an obvious manner.
The proof of (3) and (4) follows in the same manner as the proof of (3) in theorem (13.47) if we choose

$$\gamma = \max\{a_j \mid j \in N\} ;$$

the variable γ does not appear in the objective function.

■

If we are not interested in the determination of the solutions of the corresponding duals then we propose to use the augmenting path method instead of the primal dual method for a solution of the respective max-problem in a group. This method generates subsequently primal feasible solutions x_ν, $\nu = 1,2,\ldots$

of cardinality ν which are optimal among all intersections of the same cardinality. It should be noted that such solutions are not necessarily optimal among all intersections of the same cardinality with respect to the original (not reduced) problem; optimality with respect to the original problem is implied only for those ν which satisfy

$$\mu = \max\{\lambda(x^T \square a) \mid x \in P_\nu\}$$

where μ denotes the index used in the reduction considered. The solution of algebraic matroid intersection (k-intersection) problems is previously discussed in ZIMMERMANN, U. ([1976], [1978b], [1978c], [1979a]) and DERIGS and ZIMMERMANN, U. [1978a].

A solution of the reduced problems in the case of matching problems can be determined in the same manner. Again it suffices to consider max-problems. For a detailed description of the primal dual method we refer to LAWLER [1976]. We may w.l.o.g. assume that the respective group G is divisible (cf. proposition 3.2). Hence G is a module over $\mathbb{Q}$. The primal dual method remains valid and finite in such modules. Again a reformulation of the classical primal dual method for such modules consists only in replacing usual additions and usual comparisons in the group of real numbers by the internal compositions and comparisons in the group considered. All arguments used for a proof of the validity and finiteness in the classical case carry over to the case of such modules. Optimality follows from similar arguments as in the case of matroid intersections.

Let A denote the incidence matrix of vertices and edges in the underlying graph (V,N). Let S_k be any subset of V of odd cardinality $2s_k+1$. Then

$$\sum_{i\in S_k} \sum_{j\in S_k} x_{ij} \leq s_k$$

is satisfied by the incidence vector x of any matching. We represent all these constraints in matrix form by $Sx \leq s$. Then

(13.52) $$P = \{x \in \mathbb{Z}_+^n \mid Ax \leq 1,\ Sx \leq s\}.$$

For the reduced problem with $M = M(\mu)$ we find

(13.53) $$P_M = \{x \in \mathbb{Z}_+^n \mid A_M x \leq 1,\ S_M x \leq s\}.$$

We assign dual variables u_i to the vertices $i \in V$ and dual variables v_k to the odd sets S_k. Then (u,v) is dual feasible if

(13.54) $$A_M^T \square u * S_M^T \square v \geq \tilde{a}$$

where $\tilde{a}_j := a(\mu)_j$ for $j \in M = M(\mu)$. The set of all dual feasible solutions is denoted by D_M. In the usual manner we find weak duality, i.e.

$$x^T \square \tilde{a} \leq 1^T \square u * s^T \square v$$

for primal feasible x and dual feasible (u,v). The corresponding complementarity conditions are

(13.55) $$x_{ij} > 0 \Rightarrow u_i * u_j * \mathop{\ast}_{T_k \supseteq \{i,j\}} v_k = \tilde{a}_{ij}\ ,$$

(13.56) $$u_i > e \Rightarrow \Sigma_j x_{ij} = 1,$$

(13.57) $$v_k > e \Rightarrow (S_M x)_k = s_k$$

for all $(i,j) \in N$, for all vertices V and for all odd sets $T_k \subseteq N$.

Let

(13.58) $$V(x) := \{i \in V \mid \Sigma_j x_{ij} < 1\},$$

i.e. $V(x)$ is the set of all vertices which are not endpoint of some edge in the matching corresponding to x. If $\tilde{a}_{ij} \leq e$ for all $(i,j) \in M$ then $x \equiv 0$, $u \equiv e$, $v \equiv e$ is an optimal pair of primal and dual feasible solutions. Otherwise let

$$\delta := (1/2) \square (\max\{\tilde{a}_{ij} \mid (i,j) \in M\})$$

which is well-defined in the module considered. The initial solution $x \equiv 0$, $v \equiv e$ and $u \equiv \delta$ is primal and dual feasible, satisfies (13.55) and (13.57) and all dual variables v_i for $i \in V(x)$ have an identical value η. Such a pair is called *compatible*, similarly to previously discussed primal dual methods. From (13.55) we conclude

$$x^T \square \tilde{a} * (1 - A_M x)^T \square u = 1^T \square u * s^T \square v$$

for a compatible pair $(x;u,v)$. Since all variables u_i with $(1 - A_M x)_i \neq 0$ have identical value η we get

$$(13.59) \qquad x^T \square \tilde{a} * (|V| - 2 \Sigma_M x_{ij}) \square \eta = 1^T \square u * s^T \square v.$$

Similar to the primal dual method for matroid intersections the primal dual method for matchings alternately proceeds by revisions of the current matching and the current dual solution without violation of compatibility. This is achieved by an alternate solution of

$$(13.60) \qquad \max\{x^T \square \tilde{a} \mid x \in P_M,\ (x;u,v) \text{ compatible}\}$$

for fixed dual solution (u,v) and of

$$(13.61) \qquad \min\{1^T \square u * s^T \square v \mid (u,v) \in D_M,\ (x;u,v) \text{ compatible}\}$$

for fixed primal solution x. From (13.59) we conclude that

(13.60) is equivalent to

(13.62) $\max\{\Sigma_M x_{ij} \mid x \in P_M\ ,\ (x;u,v) \text{ compatible}\}$

and that (13.61) is equivalent to

(13.63) $\min\{\eta \mid (u,v) \in D_M\ ;\ (x;u,v) \text{ compatible}\}$

where η is the common value of all variables u_i with $i \in V(x)$ (cf. 13.58). Finiteness and validity of the method can be shown in the same manner as described in LAWLER [1976] for the classical case. In particular, the method is of polynomial time in the number of usual operations, $*$-compositions and $\leq$ - comparisons. The final compatible pair is complementary. Thus it is optimal. Therefore the method provides a constructive proof of a duality theorem for algebraic matching problems.

We consider the algebraic matching problems

(13.3") $\min\{x^T \square a(\mu_1) \mid x \in P\}$,

(13.4") $\max\{x^T \square a(\mu_2) \mid x \in P\}$.

The dual of (13.4") is

(13.64) $\min\{1^T \square u * s^T \square v \mid A^T \square u * S^T \square v \geq a(\mu_2),\ u_i \geq e,\ v_k \geq e\}$.

Strong dual feasibility with respect to (13.3") is defined by

(13.65)
$$e \leq a(\mu_1) * A^T \square u * S^T \square v$$
$$\lambda(u_i) \leq \mu_1\ ,\ \lambda(v_k) \leq \mu_1\ ,\ e \leq u_i\ ,\ e \leq v_k$$

and an objective function is defined according to (11.19).

(13.66) Theorem (Matching duality theorem)

Let H be a weakly cancellative d-monoid. Hence w.l.o.g. H is an extended semimodule over $\mathbb{Q}_+$. There exist optimal feasible pairs (x;u,v) for the algebraic matching problems (13.3") and (13.4") which are complementary and satisfy

(1) $e = x^T \square a * 1^T \square u * s^T \square v$ (for 13.3"),

(2) $(-1)^T \square u(\mu_1) * (-s)^T \square v(\mu_1) = x^T \square a$ (for 13.3"),

(3) $1^T \square u * s^T \square v = x^T \square a$ (for 13.4").

Proof: Similar to the proof of theorem (13.47).

■

WHITE [1967] develops a parametric approach for k-matching problems (the classical case of (3.6)) which remains valid for modules over $\mathbb{Q}$. DERIGS [1978a] proposes a modification of the primal dual method which remains valid and finite in such modules. Compatibility is slightly modified, too. Then similar results as for the corresponding algebraic k-intersection problems (cf. theorem 13.51) can be derived. Algebraic matching problems have previously been discussed by DERIGS ([1978a], [1978b]). In particular, solution methods for the perfect matching problem (cf. 11.37) are proposed in DERIGS ([1978b], [1979a], [1979b]). A short joint discussion of algebraic problems is given in DERIGS and ZIMMERMANN, U. [1978a].
Matroid intersection problems for particular matroids are discussed in further papers. ZIMMERMANN, U. [1976] gives an extension of the Hungarian method for assignment problems using the concept of admissible transformations (cf. 12.59). If the

set of feasible solutions consists in the intersections of two matroids then a method based on admissible transformations is valid if and only if the two matroids are partition matroids (cf. theorem 12.15 and 12.16 in ZIMMERMANN, U. [1976]). A further extension to three partition matroids is considered in BURKARD and FRÖHLICH [1980] within a branch and bound scheme. Algebraic matroid intersection problems in the particular case that one of the matroids is a partition matroid are solved by DETERING [1978]. Algebraic branching problems are solved by HAAG [1978] with generalizations of the various known methods. In particular, the derivation of EDMONDS' method (cf. EDMONDS [1967]) in KARP [1971] (cf. also CHU and LIU [1965]) remains valid in arbitrary d-monoids.

CONCLUSIONS

Although the material in this book covers several classes of optimization problems it should be emphasized that there are many areas of optimization problems which have not been analyzed in algebraic terms, by now.

At first, we mention nonlinear optimization. For example, a formulation of a function which is quadratic over an ordered algebraic structure can easily be given. Some remarks on quadratic assignment problems can be found in BURKARD [1975b]. In particular, the duality theory in chapter eleven has consequences for such problems similarly to the classical case. Combinatorial optimization problems with such objectives have linear relaxations which can be treated using methods described in this book. Piecewise linear algebraic functions and quotients of such functions are discussed in CUNINGHAME-GREEN and MEIJER [1978], and CUNINGHAME-GREEN ([1978], [1979a]).

Secondly, the greater part of the results in part two of this book applies only to problems over linearly ordered algebraic structures. With the exception of path problems it seems that linear and combinatorial optimization problems become much more difficult if linearity of the order relation is not assumed. For example, one of the simpler combinatorial problems, the determination of a maximum spanning tree in a weighted graph, is efficiently solved by means of the greedy method provided that the weights are linearly ordered. The algebraic structure of the underlying semigroup is quite general. But if linearity is not assumed then no efficient solution method is known for this problem. A related bottleneck problem is proposed in ZIMMERMANN, U. [1980].

Thirdly, hard combinatorial optimization problems are usually solved using simpler relaxations within a branch and bound method. Again the methods developed in this book may be applied for a solution of algebraic versions of such relaxations. For example SCHOLZ [1978] and BURKARD [1979c] discuss certain algebraic scheduling problems which have algebraic transportation problems as relaxation, and BURKARD and FRÖHLICH [1980] consider 3-dimensional assignment problems. Since the algebraic nature of the objective function is only of interest for deriving bounds, i.e. for solving the relaxed problems, the enumeration scheme remains unchanged. In this way a branch and bound method for the algebraic case of the original combinatorial optimization problem is derived.

BIBLIOGRAPHY

The following bibliography contains all papers and books referenced in this monograph. Some related papers, books and bibliographies are added.
Journal titles are abbreviated according to Mathematical Reviews.

AHO, A.V.; HOPCROFT, J.E.; ULLMAN, J.D.: *The Design and Analysis of Computer Algorithms*, Addison-Wesley, Amsterdam (1974).

ALLINEY, S.; BARNABEI, M.; PEZZOLI, L.: W-optimal assignable sets and allocation problems, *Boll. U. Mat. Ital. A (5)* 17 (1980) 131 - 136.

ARLAZAROV, V.L.; DINIC, E.A.; KRONROD, M.A.; FARADŽEV, I.A.: On economical construction of the transitive closure of an oriented graph, *Soviet Math. Dokl.* 11 (1970) 1209 - 1210.

BACKHOUSE, R.C.; CARRÉ, B.A.: Regular algebra applied to path-finding problems, *J. Inst. Math. Appl.* 15 (1975) 161 - 186.

BARSOV, A.: *What is linear programming*, Heath, Boston (1964).

BELLMAN, R.: On a routing problem, *Quart. Appl. Math.* 16 (1958) 88 - 90.

BELLMAN, R.; KALABA, R.: On k-th best policies, *SIAM J. Appl. Math.* 8 (1960) 582 - 588.

BENZAKEN, C.: Structures algêbraiques des cheminements: pseudo-treillis, gerbiers de carrê nul; in *Network and Switching Theory* (Biorci, G.; Ed.), Academic Press, New York (1968) 40 - 47.

BERGE, C.: *Graphs and Hypergraphs*, North Holland, Amsterdam (1973).

BERGE, C.; GHOUILA-HOURI, A.: *Programme, Spiele, Transportnetze*, Teubner, Leipzig (1969).

BIRKHOFF, G.: *Lattice Theory*, American Mathematical Society, Providence (1967).

BLAND, R.G.: New finite pivoting rules for the simplex method, *Math. Oper. Res.* 2 (1977) 103 - 107.

BLONIARZ, P.: A shortest path algorithm with expected time $O(n^2 \log n \log^* n)$, *Proceedings of the 12th ACM Symposium on the Theory of Computing*, Los Angeles (1980).

BLONIARZ, P.; FISCHER, M.J.; MEYER, A.R.: A note on the average time to compute transitive closures, in *Automata Languages and Programming*, (Michaelson; Milner; Eds.), Edinburgh University Press (1976).

BLUM, E.; OETTLI, W.: *Mathematische Optimierung: Grundlagen und Verfahren*, Springer, Berlin (1975).

BLYTH, T.S.: Matrices over ordered algebraic structures, *J. London Math. Soc.* 39 (1964) 427 - 432.

BLYTH, T.S.: *Module Theory, An Approach to Linear Algebra*, Clarendon Press, Oxford (1977).

BLYTH, T.S.; JANOWITZ, M.F.: *Residuation Theory*, Pergamon Press, Oxford (1972).

BORŮVKA, O.: O jistêm problêmu minimâlnim. *Prace Moravské Přirodovědecké Společnosti* 3 (1926) 37 - 58 (partially translated to German).

BRUCKER, P.: Verbände stetiger Funktionen und kettenwertige Homomorphismen, *Math. Ann.* 187 (1970) 77 - 84.

BRUCKER, P.: R-Netzwerke und Matrixalgorithmen, *Computing* 10 (1972) 271 - 283.

BRUCKER, P.: *Theory of Matrix Algorithms*, Mathematical systems in economics 13, Hain, Meisenheim am Glan (1974).

BURKARD, R.E.: Kombinatorische Optimierung in Halbgruppen, in *Optimization and Optimal Control* (Bulirsch, R.; Oettli, W.; Stoer, J.; Eds.), Lecture notes in mathematics, Springer, Berlin, 477 (1975a) 1 - 17.

BURKARD, R.E.: Numerische Erfahrungen mit Summen- und Bottleneck-Zuordnungsproblemen, in *Numerische Methoden bei graphentheoretischen und kombinatorischen Problemen* (Collatz, L.; Meinardus, G.; Werner, H.; Eds.), Birkhäuser, Basel, (1975b) 9 - 25.

BURKARD, R.E.: Flüsse in Netzwerken mit allgemeinen Kosten, in *Graphen, Algorithmen, Datenstrukturen* (Noltemeier, H.; Ed.), Hauser, München (1976) 123 - 134.

BURKARD, R.E.: A general Hungarian method for the algebraic transportation problem, *Discrete Math.* 22 (1978a) 219 - 232.

BURKARD, R.E.: An application of algebraic transportation problems to scheduling problems, in *Proceedings of the Polish-Danish Mathematical Programming Seminar* (Krarup, J.; Walukiewicz, S.; Eds.), Instytut Badan Systemowych, Polska Akademia Nauk, Warsaw (1978b) 124 - 138.

BURKARD, R.E.: Über eine Anwendung algebraischer Transportprobleme bei Reihenfolgeproblemen, in *Numerische Methoden bei graphentheoretischen und kombinatorischen Problemen* (Collatz, L.; Meinardus, G.; Wetterling, W.; Eds.), Birkhäuser, Basel, 2 (1979a) 22 - 36.

BURKARD, R.E.: On the use of algebraic objective functions in combinatorial programming, Report 79-3, Mathematisches Institut der Universität zu Köln (1979b).

BURKARD, R.E.: Remarks on some scheduling problems with algebraic objective functions, *Operations Research Verfahren* 32 (1979c) 63 - 77.

BURKARD, R.E.; FRÖHLICH, K.: Some remarks on 3-dimensional assignment problems (extended abstract), *Operations Research Verfahren* 36 (1980) 31 - 35.

BURKARD, R.E.; HAHN, W.; ZIMMERMANN, U.: An algebraic approach to assignment problems, *Math. Programming* 12 (1977) 318 - 327.

BURKARD, R.E.; HAMACHER, H.: Minimal cost flows in regular matroids, to appear in *Math. Programming Stud.* (1981).

BURKARD, R.E.; HAMACHER, H.; ZIMMERMANN, U.: The algebraic network flow problem, Report 76-7, Mathematisches Institut der Universität zu Köln, Köln (1976).

BURKARD, R.E.; HAMACHER, H.; ZIMMERMANN, U.: Einige Bemerkungen zum algebraischen Flußproblem, *Operations Research Verfahren* 23 (1977a) 252 - 254.

BURKARD, R.E.; HAMACHER, H.; ZIMMERMANN, U.: Flußprobleme mit allgemeinen Kosten, in *Numerische Methoden bei graphentheoretischen Problemen* (Collatz, L.; Meinardus, G.; Wetterling, W.; Eds.), Birkhäuser, Basel 3 (1977b) 9 - 22.

BURKARD, R.E.; ZIMMERMANN, U.: The solution of algebraic assignment and transportation problems, in *Optimization and Operations Research* (Henn, R.; Korte, B.; Oettli, W.; Eds.), Lecture notes in economics and mathematical systems, Springer, Berlin, 157 (1978) 55 - 65.

BURKARD, R.E.; ZIMMERMANN, U.: Weakly admissible transformations for solving algebraic assignment and transportation problems, *Math. Programming Stud.* 12 (1980) 1 - 18.

CARRÉ, B.A.: An algebra for network routing problems, *J. Inst. Math. Appl.* 7 (1971) 273 - 294.

CARRÉ, B.A.: *Graphs and Networks*, Oxford University Press (1979).

ČERKASSKIJ, B.V.: An algorithm for the construction of the maximal flow in a network with a labor expenditure of $O(|V|^2 \sqrt{|E|})$ actions, *Mathematical methods for the solution of economic problems (Suppl. to Ekonom. i Mat. Metody)* 7 (1977) 117 - 126.

CHAČIJAN, L.G.: A polynomial algorithm in linear programming, *Soviet Math. Dokl.* 20 (1979) 191 - 194.

CHRISTOFIDES, N.: *Graph Theory, An Algorithmic Approach*, Academic Press, New York (1975).

CHU, Y.J.; LIU, T.H.: On the shortest arborescence of a directed graph, *Sci. Sinica* 14 (1965) 1396 - 1400.

CLIFFORD, A.H.: Naturally totally ordered commutative semigroups, *Amer. J. Math.* 76 (1954) 631 - 646.

CLIFFORD, A.H.: Totally ordered commutative semigroups, *Bull. Amer. Math. Soc.* 64 (1958) 305 - 306.

CLIFFORD, A.H.: Completion of semi-continuous ordered commutative semigroups, *Duke Math. J.* 26 (1959) 44 - 59.

CONRAD, P.: Ordered semigroups, *Nagoya Math. J.* 16 (1960) 51 - 64.

CONRAD, P.: *Lattice Ordered Groups*, Preprints and Lecture Notes in Mathematics, Tulane University, New Orleans (1970).

CONRAD, P.; HARVEY, J.; HOLLAND, C.: The Hahn embedding theorem for lattice-ordered groups, *Trans. Amer. Math. Soc.* 108 (1963) 143 - 169.

CONWAY, J.H.: *Regular Algebra and Finite Machines*, Chapman and Hall, London (1971).

CRUON, R.; HERVÉ, P.: Quelques résultats rélatifs à une structure algêbraique et à son application au problème central de l'ordonnancement, *Revue française de recherche opérationelle* 34 (1965) 3 - 29.

CUNNINGHAM, W.H.; MARSH, A.B.: A primal algorithm for optimum matching, Technical report 262, Department of Mathematical Sciences, John Hopkins University, Baltimore (1976).

CUNINGHAME-GREEN, R.A.: Process synchronisation in a steelwork-a problem of feasibility, in *Proceedings of the 2nd International Conference on Operational Research*, English University Press (1960) 323 - 328.

CUNINGHAME-GREEN, R.A.: Describing industrial processes with inference and approximating their steady-state behaviour, *Operational Res. Quart.* 13 (1962) 95 - 100.

CUNINGHAME-GREEN, R.A.: Projections in minimax algebra. *Math. Programming* 10 (1976) 111 - 123.

CUNINGHAME-GREEN, R.A.: An algebra for the absolute centre of a graph, Report, Department of Mathematical Statistics, University of Birmingham (1978).

CUNINGHAME-GREEN, R.A.: *Minimax Algebra*, Lecture notes in economics and mathematical systems 166, Springer, Berlin (1979).

CUNINGHAME-GREEN, R.A.: An algorithms for the absolute centre of a graph, Report, Department of Mathematical Statistics, University of Birmingham (1979a).

CUNINGHAME-GREEN, R.A.; BORAWITZ, W.C.: Some algebraic ideas relevant to operations research I, the structures J,K,L,M, Report, Department of Mathematical Statistics, University of Birmingham (1978a).

CUNINGHAME-GREEN, R.A.; BORAWITZ, W.C.: Some algebraic ideas relevant to operations research II, non-commuting z-transforms, Report, Department of Mathematical Statistics, University of Birmingham (1978b).

CUNINGHAME-GREEN, R.A.; MEIJER, P.F.J.: An algebra for piece-wise-linear minimax problems, Report, Department of Mathematical Statistics, University of Birmingham (1978).

DANTZIG, G.B.: On the shortest route through a network, *Management Sci.* 6 (1960) 187 - 190.

DANTZIG, G.B.: All shortest routes in a graph, in *Theory of Graphs* (Rosenstiehl, P.; Ed.), Gordon and Breach, New York (1967) 91 - 92.

DANTZIG, G.B.; BLATTNER, W.O.; RAO, M.R.: Finding a cycle in a graph with minimum cost to time ratio with application to a ship routing problem, in *Theory of Graphs* (Rosenstiehl, P.; Ed.), Gordon and Breach, New York (1967a) 77 - 83.

DANTZIG, G.B.; BLATTNER, W.O.; RAO, M.R.: All shortest routes from a fixed origin in a graph, in *Theory of Graphs* (Rosenstiehl, P.; Ed.), Gordon and Breach, New York (1967b) 85 - 90.

DEO, N.; PANG, C.: Shortest path algorithms: taxonomy and annotation, Report CS-80-057, Washington State University (1980).

DERIGS, U.: Algebraische Matchingprobleme, Doctoral thesis, Mathematisches Institut der Universität zu Köln, Köln (1978a).

DERIGS, U.: On solving symmetric assignment and perfect matching problems with algebraic objectives, in *Optimization and Operations Research* (Henn, R.; Korte, B.; Oettli, W.; Eds.), Lecture notes in economics and mathematical systems, Springer, Berlin, 157 (1978b) 79 - 86.

DERIGS, U.: Duality and the algebraic matching problem, *Operations Research Verfahren* 28 (1978c) 253 - 264.

DERIGS, U.: A generalized Hungarian method for solving minimum weighted perfect matching problems with algebraic objective, *Discrete Appl. Math.* 1 (1979a) 167 - 180.

DERIGS, U.: A shortest augmenting path method for solving minimal perfect matching problems, Report 79-3, Mathematisches Institut der Universität zu Köln (1979b).

DERIGS, U.: Duality and admissible transformations in combinatorial optimization, *Z. Operations Res. Ser. A* 23 (1979c) 251 - 267.

DERIGS, U.: On three basic methods for solving bottleneck transportation problems, Report, Industrieseminar der Universität zu Köln (1979d).

DERIGS, U.: On two methods for solving the bottleneck matching problem, in *Optimization Techniques* (Iracki, K.; Malanowski, K.; Walukiewicz, S.; Eds.), Lecture notes in control and information sciences, Springer, Berlin, 23 (1980a) 176 - 184.

DERIGS, U.; ZIMMERMANN, U.: Duality principles in algebraic linear programming, Report 78-6, Mathematisches Institut der Universität zu Köln, Köln (1978a).

DERIGS, U.; ZIMMERMANN, U.: An augmenting path method for solving linear bottleneck assignment problems, *Computing* 19 (1978b) 285 - 295.

DERIGS, U.; ZIMMERMANN, U.: An augmenting path method for solving linear bottleneck transportation problems, *Computing* 22 (1979) 1 - 15.

DERNIAME, J.C.; PAIR, C.: *Problèmes de Cheminement dans les Graphes*, Dunod, Paris (1971).

DETERING, R.: Ein Lösungsverfahren für eine spezielle Klasse von algebraischen Matroid-Intersektionen Problemen mit Anwendung auf das algebraische verallgemeinerte Zuordnungsproblem, Diploma thesis, Mathematisches Institut der Universität zu Köln (1978).

DIJKSTRA, E.W.: A note on two problems in connexion with graphs, *Numerische Mathematik* 1 (1959) 269 - 271.

DINIC, E.A.: Algorithm for solution of a problem of maximum flow in a network with power estimation, *Soviet Math. Dokl.* 11 (1970) 1277 - 1280.

DINIC, E.A.; KRONROD, M.A.: An algorithm for the solution of the assignment problem, *Soviet Math. Dokl.* 10 (1969) 1324 - 1326.

DOMSCHKE, W.: *Kürzeste Wege in Graphen: Algorithmen, Verfahrensvergleiche*, Mathematical systems in economics 2, Hain, Meisenheim am Glan (1972).

DORHOUT, B.: Het Lineaire Toewijzungsproblem, Vergelijken von Algorithmen, Report BU 21/73, Stichting Mathematisch Centrum, Amsterdam (1973).

DRǍGAN, I.: Un algorithme pour la résolution de certains problèmes paramétriques, avec un seul paramètre contenu dans la fonction économique, *Rev. Roumaine Math. Pures Appl.* 11 (1966) 447 - 451.

EDMONDS, J.: Maximum matching and a polyhedron with 0 - 1 vertices, *J. Res. Nat. Bur. Standards Sect. B* 69 (1965) 125 - 130.

EDMONDS, J.: Optimum branchings, in *Mathematics of the Decision Sciences* (Dantzig, G.B.; Veinott, A.F.; Eds.), American Mathematical Society, Providence, 1 (1968) 346 - 361.

EDMONDS, J.: Submodular functions, matroids and certain polyhedra, in *Combinatorial structures and their applications* (Guy, R.; Hanam, H.; Schonheim, J.; Eds.), Gordon and Breach, New York (1970) 69 - 87.

EDMONDS, J.: Matroids and the greedy algorithm, *Math. Programming* 1 (1971) 127 - 136.

EDMONDS, J.: Matroid intersection, *Ann. Discrete Math.* 4 (1979) 39 - 51.

EDMONDS, J.; FULKERSON, D.R.: Transversals and matroid partition, *J. Res. Nat. Bur. Standards Sect. B* 69 (1965) 147 - 153.

EDMONDS, J.; FULKERSON, D.R.: Bottleneck extrema, *J. Combinatorial Theory* 8 (1970) 229 - 306.

EDMONDS, J.; KARP, R.M.: Theoretical improvement in algorithmic efficiency for network flow problems, *J. Assoc. Comput. Mach.* 19 (1972) 248 - 264.

EDMONDS, J.: PULLEYBLANK, W.: Faces of 1-matching polyhedra, in *Hypergraph Seminar* (Berge, C.; Ray-Chaudhuri, D.; Eds.), Lecture notes in mathematics, Springer, Berlin, 411 (1974) 214 - 242.

EILENBERG, S.: *Automata, Languages and Machines*, Volume A, Academic Press, New York (1974).

EULER, R.: Die Bestimmung maximaler Elemente in monotonen Mengensystemen, Report 7761, Institut für Ökonometrie und Operations Research, Universität Bonn (1977).

EULER, R.: On rank functions of general independence systems, Report 7893, Institut für Ökonometrie und Operations Research, Universität Bonn (1978a).

EULER, R.: Rank-axiomatic characterizations of independence systems, Report 78126, Institut für Ökonometrie und Operations Research, Universität Bonn (1978b).

EVEN, SH.: The maximal flow algorithm of Dinic and Karzanov, to appear in *Theor. Comput. Sci.* (1980).

FAIGLE, U.: The greedy algorithm for partially ordered sets, *Discrete Math.* 28 (1979) 153 - 159.

FAUCITANO, G.; NICAUD, J.F.: Algorithme de Gauss pour la quasi-inversion de matrices, rapport de DEA sur la complexité des algorithmes (J.Berstel), Université de Paris VI (1975).

FINKE, G.; AHRENS, J.H.: A variant of the primal transportation algorithm, *INFOR* 16 (1978) 35 - 46.

FINKE, G.; SMITH, P.A.: Primal equivalents to the threshold algorithm, *Operations Research Verfahren* 31 (1979) 185 - 198.

FISCHER, M.J. ; MEYER, A.R.: Boolean matrix multiplication and transitive closure, *Conference Record IEEE 12th Annual Symposium on Switching and Automata Theory* (1971) 129 - 131.

FLETCHER, J.G.: A more general algorithm for computing closed semiring costs between vertices of a directed graph, *Comm. ACM* 23 (1980) 350 - 351.

FLOYD, R.W.: Algorithm 97, shortest path, *Comm. ACM* 5 (1962) 345.

FORD, L.R.; FULKERSON, D.R.: *Flows in Networks*, Princeton University Press (1962).

FORTET, R.: Application de l'algèbre de Boole en recherche opérationelle, *Revue Française Recherche Opérationelle* 4 (1960) 17 - 26.

FOX, B.: Calculating k-th shortest paths, *INFOR* 11 (1973) 66 - 70.

FRATTA, L.; MONTANARI, U.: A vertex elimination algorithm for enumerating all simple paths in a graph, *Networks* 5 (1975) 151 - 177.

FREDMAN, M.L.: New bounds on the complexity of the shortest path problem, *SIAM J. Comput.* 5 (1976) 83 - 89.

FRIEZE, A.M.: Bottleneck linear programming, *Operational Res. Quart.* 26 (1975) 871 - 874.

FRIEZE, A.M.: Algebraic linear programming, Report, Department of Computer Science and Statistics, University of London (1978).

FRIEZE, A.M.: An algorithm for algebraic assignment problems, *Discrete Appl. Math.* 1 (1979) 253 - 259.

FUCHS, L.: *Teilweise geordnete algebraische Strukturen*, Vandenhoeck and Ruprecht, Göttingen (1966).

FUJISHIGE, S.: A primal approach to the independent assignment problem, *J. Operations Res. Soc. Japan* 20 (1977) 1 - 14.

FULKERSON, R.; GLICKSBERG, I.; GROSS, O.: A production line assignment problem, RAND Research Memorandum RM-1102 (1953).

FURMAN, M.E.: Application of a method of fast multiplication of matrices in the problem of finding the transitive closure of a graph, *Soviet Math. Dokl.* 11 (1970) 1252.

GABOVIČ, E.J.: An equivalency problem in discrete programming over ordered semigroups, *Semigroup Forum* 8 (1974) 69 - 73.

GABOVIČ, E.J.: Constant problems of discrete optimization over permutation sets, *Kybernetika* 5 (1976a) 128 - 134 (Russian).

GABOVIČ, E.J.: Fully ordered semigroups and their applications, *Russian Math. Surveys* 31 (1976b) 147 - 216.

GALE, D.: Optimal assignments in an ordered set. An application of matroid theory, *J. Combinatorial Theory* 4 (1968) 176 - 180.

GALIL, Z.: A new algorithm for the maximal flow problem, in *Proceedings of the 19th IEEE Symposium on Foundation of Computer Science*, Ann Arbor University, Michigan (1978).

GALIL, Z.; NAAMAD, A.: Network flow and generalized path compression, Report, Department of Computer Sciences, University of Tel Aviv (1979).

GAREY, M.R.; JOHNSON, D.S.: *Computers and Intractability. A Guide to the Theory of NP-Completeness*, Freeman, San Francisco (1979).

GARFINKEL, R.: An improved algorithm for the bottleneck assignment problem, *Operations Res.* 19 (1971) 1747 - 1751.

GARFINKEL, R.S.; RAO, M.R.: The bottleneck transportation problem, *Naval Res. Logist. Quart.* 18 (1971) 465 - 472.

GARFINKEL, R.S.; RAO, M.: Bottleneck linear programming, *Math. Programming* 11 (1976) 291 - 298.

GIFFLER, B.: Scheduling general production systems using schedule algebra, *Naval Res. Logist. Quart.* 10 (1963) 237 - 255.

GIFFLER, B.: Schedule algebra: a progress report, *Naval Res. Logist. Quart.* 15 (1968) 255 - 280.

GLOVER, F.; KARNEY, D.; KLINGMAN, D.: The augmented predecessor index method for locating stepping stone paths and assigning dual prices in distribution problems, *Transportation Sci.* 6 (1972) 171 - 179.

GOLDEN, B.L.; MAGNANTI, T.L.: Deterministic network optimization: a bibliography, *Networks* 7 (1977) 149 - 183.

GONDRAN, M.: Algèbre linéaire et cheminement dans un graphe, *RAIRO Recherche Opér.* 9 (1975) 77 - 99.

GONDRAN, M.: Path algebra and algorithms, in *Combinatorial Programming: Methods and Applications* (Roy, B.; Ed.), Reichel, Dordrecht (1975a) 137 - 148.

GONDRAN, M.: Algèbre des chemins et algorithmes, *Bull. Direction Etudes Recherches Sér. C Math. Informat.* (1975b) 57 - 64 (French version of 1975a).

GONDRAN, M.: L'algorithme glouton dans les algèbres de chemins, *Bull. Direction Etudes Recherches Sér. C Math. Informat.* (1975c) 25 - 32.

GONDRAN, M.: Les éléments p-régulières dans les semi-anneaux, Note Electricié de France HI 1778/02 (1975d).

GONDRAN, M.: Eigenvectors and eigenvalues in hierarchical classification, in *Recent Developments in Statistics* (Barra, J.R.; Ed.), North Holland (1977) 775 - 781.

GONDRAN, M.: Les elements p-reguliers dans les dioïdes, *Discrete Mathematics* 25 (1979) 33 - 39.

GONDRAN, M.: The dioïd theory and its applications, private communication (1980).

GONDRAN, M.; MINOUX, M.: Valeurs propres et vecteurs propres dans les dioïdes et leur interprétation en théorie des graphes, *Bull. Direction Etudes Recherches Sér. C Math. Informat.* (1977) 25 - 41.

GONDRAN, M.; MINOUX, M.: L'indépendance linéaire dans les dioïdes, *Bull. Direction Etudes Recherches Sér. C. Math. Informat.* (1978) 67 - 90.

GONDRAN, M.; MINOUX, M.: Eigenvalues and eigenvectors in semimodules and their interpretation in graph theory, in *Survey of Mathematical Programming* (Prêkopa, A.; Ed.), North Holland, Amsterdam, 2 (1979a) 333 - 348 (English version of 1977).

GONDRAN, M.; MINOUX, M.: *Graphes et Algorithmes*, Eyrolles, Paris (1979b).

GRABOWSKI, W.: Problem of transportation in minimum time, *Bull. Acad. Polon. Sci. Sér. Sci. Math. Astronom. Phys.* 12 (1964) 107 - 108.

GRABOWSKI, W.: Lexicographic and time minimization in the transportation problem, *Zastos. Mat.* 15 (1976) 191 - 213.

GRÄTZER, G.: *General Lattice Theory*, Birkhäuser, Basel (1978).

GROSS, O.: The bottleneck assignment problem, Report P-1630, RAND Corporation (1959).

GRÖTSCHEL, M.: *Polyhedrische Charakterisierungen kombinatorischer Optimierungsprobleme*, Mathematical systems in economics 36, Hain, Meisenheim am Glan (1977).

GUPTA, R.; ARORA, S.R.: Programming problems with maximin objective functions, *Z. Operations Res.* 22 (1978) 69 - 72.

HAAG, W.: Verfahren zur Bestimmung optimaler Branchings, Diploma thesis, Mathematisches Institut der Universität zu Köln, Köln (1978).

HAHN, H.: Über die nichtarchimedischen Größensysteme, *Sitzungsberichte der Akademie der Wissenschaften Wien, IIa*, 116 (1907) 601 - 655.

HAMACHER, H.: Numerical investigations on the maximal flow algorithm of Karzanov, *Computing* 22 (1979) 17 - 29.

HAMACHER, H.: Algebraic flows in regular matroids, *Discrete Appl. Math.* 2 (1980a) 27 - 38.

HAMACHER, H.: Algebraische Flußprobleme in regulären Matroiden, Doctoral thesis, Mathematisches Institut, Universität zu Köln, Köln (1980b).

HAMACHER, H.: Maximal algebraic flows: algorithms and examples, in *Discrete Structures and Algorithms* (Pape, U.; Ed.); Hauser, München (1980c) 153 - 166.

HAMMER, P.L.: Time minimizing transportation problems, *Naval Res. Logist. Quart.* 16 (1969) 345 - 367; 18 (1971) 487 - 490.

HAMMER, P.L.; RUDEANU, S.: *Boolean Methods in Operations Research*, Springer, Berlin (1968).

HANSEN, P.: An O(m log D) algorithm for shortest paths, *Discrete Appl. Math.* 2 (1980) 151 - 153.

HARARY, F.; NORMAN, R.Z.; CARTWRIGHT, D.: *Structural Models, An Introduction to the Theory of Directed Graphs*, Wiley, New York (1965).

HAUSMANN, D. (Ed.): *Integer Programming and Related Areas, a Classified Bibliography 1976 - 1978*, Lecture notes in economics and mathematical systems 160, Springer, Berlin (1978).

HOFFMAN, A.J.: On abstract dual linear programs, *Naval Res. Logist. Quart.* 10 (1963) 369 - 373.

HOFFMAN, A.J.: On lattice polyhedra II: Construction and examples, Report RC6268, IBM Research Center, Yorktown Heights (1976).

HOFFMAN, A.J.: On lattice polyhedra III: blockers and anti-blockers of lattice clutters. *Math. Programming Stud.* 8 (1978b) 197 - 207.

HOFFMAN, A.J.; SCHWARTZ, D.E.: On lattice polyhedra, *Colloquia Mathematica Societatis János Bolyai*, 18 (1976) 593 - 598.

HOFFMAN, N.; PAVLEY, R.: A method for the solutions of the n-th best path problem, *J. Assoc. Comput. Mach.* 6 (1959) 506 - 514.

HOHN, F.E.; SESHU, S.; AUFENKAMP, D.D.: The theory of nets, *Trans. I.R.E.* EC-6 (1957) 154 - 161.

HÖLDER, O.: Die Axiome der Quantität und die Lehre vom Maß, *Berichte über die Verhandlungen der Königlich Sächsischen Gesellschaft der Wissenschaften zu Leipzig, Mathematisch-Physikalische Klasse* 53 (1901) 1 - 64.

HU, T.C.: The maximum capacity route problem, *Operations Res.* 9 (1961) 898 - 900.

IRI, M.; TOMIZAWA, N.: An algorithm for finding an optimal 'independent' assignment, *J. Operations Res. Soc. Japan* 19 (1976) 32 - 57.

ISHII, H.: A new method finding the k-th best path in a graph, *J. Operations Res. Soc. Japan* 21 (1978) 469 - 475.

JANTOSCIAK, J.; PRENOWITZ, W.: *Join Geometries*, Springer, New York, Berlin, (1979).

JOHNSON, D.B.: Efficient algorithms for shortest paths in sparse networks, *J. Assoc. Comput. Mach.* 24 (1977) 1 - 13.

KALABA, R.: On some communication network problems, in *Combinatorial Analysis*, American Mathematical Society, Providence (1960) 261 - 280.

KARP, R.M.: A simple derivation of Edmonds' algorithm for optimum branchings, *Networks* 1 (1972) 265 - 272.

KARZANOV, A.V.: Determining the maximal flow in a network by the method of preflows, *Soviet Math. Dokl.* 15 (1974) 434 - 437.

KARZANOV, A.V.: An efficient algorithm for the determination of a maximal flow, *Ekonom. i. Mat. Metody* 11 (1975) 721 - 729 (Russian).

KASTNING, D. (Ed.): *Integer Programming and Related Areas, a Classified Bibliography*, Lecture notes in economics and mathematical systems 128, Springer, Berlin (1976).

KAUFMANN, A.; MALGRANGE, Y.: Recherche des chemins et circuits hamiltoniens d'une graphe, *Revue Française Recherche Opérationelle* 7 (1963) 61 - 73.

KLEENE, S.C.: Representation of events in nerve nets and finite automata, in *Automata Studies* (McCarthy, J.; Shannon, C.; Eds.), Princeton University Press, Princeton (1956).

KLEIN, M.: A primal method for minimal cost flows with applications to the assignment and transportation problem, *Management Sci.* 14 (1967) 205 - 220.

KLEIN-BARMEN, F.: Über gewisse Halbverbände und kommutative Semigruppen I, *Math. Z.* 48 (1942) 275 - 288.

KLEIN-BARMEN, F.: Über gewisse Halbverbände und kommutative Semigruppen II, *Math. Z.* 48 (1943) 715 - 734.

KOKORIN, A.I.; KOPYTOV, V.M.: *Fully Ordered Groups*, Wiley, New York (1974).

KORBUT, A.A.: Extremal spaces, *Soviet. Math. Dokl.* 6 (1965) 1358 - 1361.

KRUSKAL, J.B.: On the shortest spanning subtree of a graph and the travelling salesman problem, *Proc. Amer. Math. Soc.* 7 (1956) 48 - 50.

KUHN, H.W.: The Hungarian method for the assignment problem, *Navel Res. Logist. Quart.* 2 (1955) 83 - 97.

LAWLER, E.L.: Optimal cycles in a doubly weighted directed linear graph, in *Theory of Graphs* (Rosenstiehl, P.; Ed.), Gordon and Breach, New York (1967) 209 - 213.

LAWLER, E.L.: Matroid intersection algorithms, *Math. Programming* 9 (1975) 31 - 56.

LAWLER, E.L.: *Combinatorial Optimization: Networks and Matroids*, Holt, Rinehart and Winston, New York (1976).

LEHMANN, D.J.: Algebraic structures for transitive closure, in *Theor. Comput. Sci.* 4 (1977) 59 - 76.

LUGOWSKI, H.: Über gewisse geordnete Halbmoduln mit negativen Elementen, *Publ. Math. Debrecen* 11 (1969) 23 - 41.

LUNTS, A.G.: The application of Boolean matrix algebra to the analysis and synthesis of relay contact networks, *Dokl. Akad. Nauk SSSR* 70 (1950) 421 - 423 (Russian).

LUNTS, A.G.: Algebraic methods of analysis and synthesis of relay-contact networks, *Izv. Akad. Nauk SSSR Ser. Mat.* 16 (1952) 405 - 426 (Russian).

MAHR, B.: Algebraische Komplexität des allgemeinen Wegeproblems in Graphen, Report 79-14, Institut für Software und Theoretische Informatik, Technische Universität, Berlin, (1979).

MAHR, B.: A birds eye view to path problems, Report 80-19, Institut für Software und Theoretische Informatik, Technische Universität, Berlin, (1980a).

MAHR, B.: Semirings and transitive closure, in preparation (1980b).

MALHOTRA, V.M.; PRADMODH KUMAR, M.; MAHESHWARI, S.W.: An $O(|V|^3)$ algorithm for finding the maximal flow in networks, *Information Processing Lett.* 7 (1978) 277 - 278.

MARTELLI, A.: An application of regular algebra to the enumeration of the cut sets of a graph, *Information Processing* 74 (1974) 511 - 515.

MARTELLI, A.: A Gaussian elimination algorithm for the enumeration of cut sets in a graph, *J. Assoc. Comput. Mach.* 19 (1976) 58 - 73.

McNAUGHTON, R.; YAMADA, H.: Regular expressions and state graphs for automata, *IRE Trans. on Electronic Computers* 9 (1960) 39 - 47.

MINIEKA, E.: On computing sets of shortest paths in a graph, *Comm. ACM* 17 (1974) 351 - 353.

MINIEKA, E.; SHIER, D.R.: A note on an algebra for the k-th best routes in a network, *J. Inst. Math. Appl.* 11 (1973) 145 - 149.

MINOUX, M.: Structures algébraiques généralisées des problèmes de cheminement dans les graphes: Théorèmes, algorithmes et applications, *RAIRO Recherche Opér.* 10 (1976) 33 - 62.

MINOUX, M.: Generalized path algebras, in *Survey of Mathematical Programming* (Prêkopa, E.; Ed.), North Holland, Amsterdam, 2 (1979) 359 - 364.

MOISIL, G.C.: Assupra unor representări ale grafurilor ce intervin în probleme de economia transporturilor, *Com. Acad. Rep. Pop. Romine* 10 (1960) 647 - 652.

MONGE, G.: Déblai et remblai, *Mémoires de l'Académie de Sciences*, Paris (1781).

MÜLLER-MERBACH, H.: *Operatione Research Methoden*, Van Vahlen, Berlin (1970).

MUNRO, I.: Efficient determination of the transitive closure of a directed graph, *Information Processing Lett.* 1 (1971) 56 - 58.

MURA, R.B.; RHEMTULLA, A.: *Orderable Groups*, Lecture notes in pure and applied mathematics 27, Dekker, New York (1977).

MURCHLAND, J.D.: A new method for finding all elementary paths in a complete directed graph, Report SE - TNT - 22, Transport Network Theory Unit, London Graduate School of Economics (1965).

MURTY, K.G.: *Linear and Combinatorial Programming*, Wiley, New York (1976).

ORE, O.: Theory of equivalence relations, *Duke Math. J.* 9 (1942) 573 - 627.

PAGE, E.S.: A note on assignment problems, *Comput. J.* 6 (1963) 241 - 243.

PAIR, C.: Sur les algorithmes pour les problèmes de cheminement dans les graphes finis, in *Theory of Graphs* (Rosenstiehl, P.; Ed.), Gordon and Breach, New York, (1967) 271 - 300.

PANDIT, S.N.N.: A new matrix calculus, *SIAM J. Appl.* 9 (1961) 632 - 639.

PANDIT, S.N.N.: Minaddition and an algorithm to find the most reliable paths in a network, *IRE Transactions on Circuit Theory CT-9* (1962) 190 - 191.

PAPE, U.: Eine Bibliographie zu "Kürzeste Weglängen und Wege in Graphen und Netzwerken", *Elektronische Datenverarbeitung* 6 (1969) 271 - 274.

PAPE, U.; SCHÖN, B.: Verfahren zur Lösung von Summen- und Engpaß-Zuordnungsproblemen, *Elektronische Datenverarbeitung* 7 (1970) 149 - 163.

PETEANU, V.: Sur la compatibilité du problème central de l'ordonnancement, *Mathematica (Cluj)* 9 (1967a) 141 - 146.

PETEANU, V.: An algebra of the optimal path in networks, *Mathematica (Cluj)* 9 (1967b) 335 - 342.

PETEANU, V.: Optimal paths in networks and generalizations, part I, *Mathematica (Cluj)* 11 (1968) 311 - 327.

PETEANU, V.: Optimal paths in networks and generalizations, part II, *Mathematica (Cluj)* 12 (1970) 159 - 186.

PETEANU, V.: O generalizare a notiunii de centru al unui graf, *Rev. Anal. Numér. Teoria Aproximatiei (Cluj)* 1 (1972) 83 - 86.

PICHAT, E.: Algorithms for finding the maximal elements of a finite universal algebra, *Proc. IFIP Congress Edinburgh*, Booklet A (1968) 96 - 101.

PIERCE, A.R.: Bibliography on algorithms for shortest path, shortest spanning tree, and related circuit routing problems (1956 - 1974), *Networks* 5 (1975) 129 - 149.

PONSTEIN, J.: On the maximal flow problem with real arc capacities, *Math. Programming* 3 (1972) 254 - 256.

PUDLÁK, P.; TŮMA, J.: Every finite lattice can be embedded in the lattice of all equivalences over a finite set, (Preliminary communication), *Commentationes Mathematicae Universitatis Carolinae* 18 (1977) 409 - 414.

PUDLÁK, P.; TŮMA, J.: Every finite lattice can be embedded in the lattice of all equivalences over a finite set, to appear in *Algebra Universalis* (1980).

RADO, R.: Note on independence functions, *Proc. London Math. Soc.* 7 (1957) 300 - 320.

ROBERT, P.; FERLAND, J.: Généralisation de l'algorithme de Warshall, *Revue Française d'Informatique et de Recherche Opérationelle* 2 (1968) 71 - 85.

ROY, B.: Transitivité et connexité, *C.R.Acad. Sci. Paris* 249 (1959) 216 - 218.

ROY, B.: Chemins et circuits: énumération et optimisation, in *Combinatorial Programming: Methods and Applications* (Roy, B.; Ed.) Reidel, Dordrecht (1975) 105 - 136.

RUDEANU, S.: On Boolean matrix equations, *Rev. Roumaine Math. Pures Appl.* 17 (1972) 1075 - 1090.

SALOMAA, A.: *Theory of Automata*, Pergamon Press, Oxford (1969).

SATYANARAYANA, M.: *Positively Ordered Semigroups*, Lecture notes in pure and applied mathematics 42, Dekker, New York (1979).

SCHNORR, C.P.: An algorithm for transitive closure with linear expected time, *SIAM J. Comput.* 7 (1978) 127 - 133.

SCHOLZ, K.: On scheduling multiprocessor systems with algebraic objectives, *Computing* 20 (1978) 189 - 205.

SHIER, D.R.: A decomposition algorithm for optimality problems in tree-structured networks, *Discrete Math.* 6 (1973) 175 - 189.

SHIER, D.R.: Iterative methods for determining the k-shortest paths in a network, *Networks* 6 (1976) 205 - 229.

SHILOACH, Y.: An $O(|V|(|V|+|E|)\log_2(|V|+|E|)$ maximum flow algorithm, Report, Department of Computer Science, Stanford University (1978).

SHIMBEL, A.: Applications of matrix algebra to communication nets, *Bulletin of Mathematical Biophysics* 13 (1951) 165 - 178.

SHIMBEL, A.: Structural parameters of communication networks, *Bulletin of Mathematical Biophysics* 15 (1953) 501 - 507.

SHIMBEL, A.: Structure in communication nets, *Proceedings of the Symposium on Information Networks* (1954), Polytechnic Institute of Brooklyn, New York (1955) 199 - 203.

SHOR, N.Z.: Utilization of the operation of space dilatations in the minimization of convex functions, *Cybernetics* 6 (1970) 7 - 15.

SHOR, N.Z.: Cut-off method with space extension in convex programming problems, *Cybernetics* 13 (1977) 94 - 96.

SLOMINSKI, L.: Bottleneck assignment problem: an efficient algorithm, Report, System Research Institute, Polish Academy of Sciences, Warschau (1976).

SLOMINSKI, L.: An efficient approach to the bottleneck assignment problem, *Polon. Sci. Sér. Sci. Math. Astronom. Phys. Bull. Acad.* 25 (1977) 17 - 23.

SLOMINSKI, L.: Bottleneck discrete problems, in *Proceedings of the Polish-Danish Mathematical Programming Seminar* (Krarup, J.; Walukiewicz, S.; Eds.), Instytut Badan Systemowych, Polska Akademia Nauk, Warsaw (1978) 107 - 123.

SLOMINSKI, L.: Bottleneck assignment problems: an efficient algorithm, *Arch. Automat. i Telemech.* 24 (1979) 469 - 482.

SRINIVASAN, V.; THOMPSON, G.L.: Accelerated algorithms for labeling and relabeling of trees, with applications to distribution problems, *J. Assoc. Comput. Mach.* 19 (1972) 712 - 726.

SRINIVASAN, V.; THOMPSON, G.L.: Algorithms for minimizing total cost, bottleneck time and bottleneck shipment in transportation problems, *Naval Res. Logist. Quart.* 23 (1976) 567 - 595.

STENZEL, L.: Über eine Klasse geordneter Halbgruppen mit negativen Elementen, *Wissenschaftliche Zeitschrift, Mathematisch-Naturwissenschaftliche Reihe, Hochschule Potsdam* 13 (1969) 411 - 415.

SZWARC, W.: The time transportation problem, *Zastos. Mat.* 8 (1966) 231 - 242.

SZWARC, W.: Some remarks on the time transportation problem, *Naval Res. Logist. Quart.* 18 (1971) 473 - 486.

TARJAN, R.E.: Solving path problems on directed graphs, Report, Computer Science Department, Stanford University, Stanford (1975).

TARJAN, R.E.: Graph theory and Gaussian elimination, in *Sparse Matrix Computations* (Bunch, J.R.; Rose, D.J.; Eds.), Academic Press Inc., New York (1976) 3 - 37.

TESCHKE, L.; HEIDEKRÜGER, G.: Zweifach assoziative Räume, Verbindungsräume, Konvexitätsräume und verallgemeinerte Konvexitätsräume, *Wissenschaftliche Zeitschrift der PH "N.K.Krupskaja" Halle* 14 (1976a) 21 - 24.

TESCHKE, L.; HEIDEKRÜGER, G.: Trennung in verallgemeinerten Konvexitätsräumen, *21. Internationales wissenschaftliches Kolloquium*, TH Ilmenau (1976b) 73 - 76.

TESCHKE, L.; ZIMMERMANN, K.: Extremale Algebren und verallgemeinerte Konvexitätsräume, *Ekonom. Mat. Obzor* 15 (1979) 74 - 84.

TOMESCU, I.: Sur les méthodes matricielles dans la théorie des réseaux, *C.R. Acad. Sci. Paris* 263 (1966) 826 - 829.

TOMESCU, I.: Un algorithme pour la dêtermination des plus petites distances entre les sommets d'un rêseau, *Revue Française d'Information et de Recherche Opêrationelle* 1 (1967) 133 - 139.

TOMESCU, I.: Sur l'algorithme matriciel de B.Roy, *Revue Française d'Information et de Recherche Opêrationelle* 2 (1968) 87 - 91

TOMIZAWA, N.: On some techniques useful for the solution of transportation network problems, *Networks* 1 (1972) 179 - 194.

VOROBJEV, N.N.: The extremal matrix algebra, *Soviet Math. Dokl.* 4 (1963) 1220 - 1223.

VOROBJEV, N.N.: Ekstremalnaja algebra polozitelnich matriz, *Elektronische Informationsverarbeitung und Kybernetik* 3 (1967) 39 - 71 (Russian).

VOROBJEV, N.N.: Ekstremalnaja algebra neotrizatelnich matriz, *Elektronische Informationsverarbeitung und Kybernetik* 6 (1970) 303 - 312 (Russian).

WAGNER, R.A.: A shortest path algorithm for edge sparse graphs, *J. Assoc. Comput. Mach.* 23 (1976) 50 - 57.

WARSHALL, S.: A theorem on Boolean matrices, *SIAM J. Appl. Math.* 9 (1962) 11 - 12.

WELSH, D.J.A.: Kruskal's theorem for matroids, *Proceedings of the Cambridge Philosophical Society* 64 (1968) 3 - 4.

WELSH, D.J.A.: *Matroid Theory*, Academic Press, London (1976).

WHITE, L.J.: Parametric study of matchings and coverings in weighted graphs, Doctoral thesis, University of Michigan (1967).

WHITNEY, H.: On the abstract properties of linear independence, *Amer. J. Math.* 57 (1935) 509 - 535.

WONGSEELASHOTE, A.: An algebra for determining all path-values in a network with application to k-shortest-paths problems, *Networks* 6 (1976) 307 - 334.

WONGSEELASHOTE, A.: Semirings and path spaces, *Discrete Math.* 26 (1979) 55 - 78.

WÜSTEFELD, A.; ZIMMERMANN, U.: Nonlinear one-parametric linear programming and t-norm transportation problems, *Naval Res. Logist. Quart.* 27 (1980) 187 - 197.

YEN, J.Y.: Finding the k shortest loopless paths in a network, *Management Sci.* 17 (1971) 712 - 716.

YEN, J.Y.: Finding the lengths of all shortest paths in an N-node non-negative distance complete network using 1/2 N^3 additions and N^3 comparisons, *J. Assoc. Comput. Mach.* 19 (1972) 423 - 424.

YEN, J.Y.: *Shortest Path Network Problems*, Mathematical systems in economics 18, Hain, Meisenheim am Glan (1975).

YOELI, M.: A note on generalization of Boolean matrix theory, *Amer. Math. Monthly* 68 (1961) 552 - 557.

ZADEH, N.: Theoretical efficiency of the Edmonds-Karp algorithm for computing maximal flows, *J. Assoc. Comput. Mach.* 19 (1972) 184 - 192.

ZIMMERMANN, K.: Some properties of the systems of extremal linear inequalities, *Ekonom. Mat. Obzor* 9 (1973a) 212 - 227 (Russian).

ZIMMERMANN, K.: Solution of some optimization problems in extremal vector space, *Ekonom. Mat. Obzor* 9 (1973b) 336 - 351 (Russian).

ZIMMERMANN, K.: Solution of an optimization problem in extremal vector space, *Ekonom. Mat. Obzor* 10 (1974a) 298 - 321 (Russian).

ZIMMERMANN, K.: Conjugate optimization problems and algorithms in the extremal vector space, *Ekonom. Mat. Obzor* 10 (1974b) 428 - 439.

ZIMMERMANN, K.: Einige Aufgaben auf dem extremalen Vektorraum, *Unternehmensforschung und angewandte Statistik* 55 (1976a) T284 - T286.

ZIMMERMANN, K.: Extremalni Algebra, Výzkumná publicace Ekonomicko-matematické Laboratoře při Ekonomickém ústava ČSAV, 46, Praha (1976b).

ZIMMERMANN, K.: A general separation theorem in extremal algebras *Ekonom. Mat. Obzor* 13 (1977) 179 - 200.

ZIMMERMANN, K.: Linearität und Verallgemeinerungen der Konvexität, *Math. Operationsforsch. Statist. Ser. Optimization* 10 (1979a) 17 - 25.

ZIMMERMANN, K.: A generalization of convex functions, *Ekonom. Mat. Obzor* 15 (1979b) 147 - 158.

ZIMMERMANN, K.: Convexity in semimodules (extended abstract), to appear in *Methods of Operations Research*, Athenäum et al. (1981).

ZIMMERMANN, K.; JUHNKE, F.: Lineare Gleichungs- und Ungleichungssysteme auf extremalen Algebren, *Wissenschaftliche Zeitschrift der Technischen Hochschule Otto von Guericke, Magdeburg* 23 (1979) 103 - 106.

ZIMMERMANN, U.: Boole'sche Optimierungsprobleme mit separabler Zielfunktion und matroidalen Restriktionen, Doctoral thesis, Mathematisches Institut der Universität zu Köln, Köln (1976).

ZIMMERMANN, U.: Some partial orders related to Boolean optimization and the greedy algorithm, *Ann. Discrete Math.* 1 (1977) 539 - 550.

ZIMMERMANN, U.: A primal method for solving algebraic transportation problems applied to the bottleneck transportation problem, in *Proceedings of the Polish-Danish Mathematical Programming Seminar* (Krarup, J.; Walukiewicz, S.; Eds.), Instytut Badan Systemowych, Polska Akademia Nauk, Warsaw (1978a) 139 - 153.

ZIMMERMANN, U.: Duality and the algebraic matroid intersection problem, *Operations Research Verfahren* 28 (1978b) 285 - 296.

ZIMMERMANN, U.: Threshold methods for Boolean optimization problems with separable objectives, in *Optimization Techniques* (Stoer, J.; Ed.), Lecture notes in control and information sciences, Springer, Berlin, 7 (1978c) 289 - 298.

ZIMMERMANN, U.: Matroid intersection problems with generalized objectives, in *Survey of Mathematical Programming* (Prêkopa, A.; Ed.), North Holland, Amsterdam, 2 (1979a) 383 - 392.

ZIMMERMANN, U.: A primal-dual method for algebraic linear programming, Report 79-16, Mathematisches Institut der Universität zu Köln (1979b).

ZIMMERMANN, U.: On some extremal optimization problems, *Ekonom. Mat. Obzor* 15 (1979c) 438 - 442.

ZIMMERMANN, U.: Duality principles and the algebraic transportation problem, in *Numerische Methoden bei graphentheoretischen Problemen* (Collatz, L.; Meinardus, G.; Wetterling, W.; Eds.), Birkhäuser, Basel 2 (1979d) 234 - 255.

ZIMMERMANN, U.: Duality for algebraic linear programming, *Linear Algebra and its Applications* 32 (1980a) 9 - 31.

ZIMMERMANN, U.: Linear optimization for linear and bottleneck objectives with one nonlinear parameter, in *Optimization Techniques* (Iracki, K.; Malanowski, K.; Walukiewicz, S.; Eds.), Lecture notes in control and information sciences, Springer, Berlin, 23 (1980b) 211 - 222.

ZIMMERMANN, U.: An algebraic approach to extremal optimization problems (extended abstract), *Methods of Operations Research* 37 (1980c) 295 - 298.

ZIMMERMANN, U.: Some remarks on algebraic path problems (extended abstract), to appear in *Methods of Operations Research*, Athenäum et al. (1981).

AUTHOR INDEX

SUBJECT INDEX